THE
REAL FUTURE

Other books by T. A. Heppenheimer:

COLONIES IN SPACE

TOWARD DISTANT SUNS

THE MAN-MADE SUN (to be published)

THE
REAL FUTURE

T. A. HEPPENHEIMER

1983
Doubleday & Company, Inc.
Garden City, New York

Library of Congress Cataloging in Publication Data

Heppenheimer, T. A., 1947–
The real future.

Bibliography: p. 255
1. Science.   2. Technology.   3. Forecasting.
I. Title.
Q158.5.H46   1983   303.4'9
ISBN: 0-385-17688-0
Library of Congress Catalog Card Number 82-45291

To my children,
Laurie, Alex, Connie.

ACKNOWLEDGMENTS

It is a pleasure to acknowledge the help of the many people who have contributed to this book. Chapter 3 owes much to Robert Bacastow, Cesare Emiliani, Andrew McIntyre, J. H. Mercer, and U. Siegenthaler. In addition, other chapters have benefited from helpful correspondence with Jon W. Gordon, Charles Hall, Nils J. Nilsson, and Harry Weese.

People who have helpfully sent me photos and illustrations include Marjorie Barritt, Roger Beall, Dale Blumenthal, Elizabeth P. Buckley, Don Cabrera, Rose Marie Cline, Sue Cometa, Tom Crouch, Richard E. Drew, Gene Emme, Vickie Grant, Jim Hansen, Uta Hoffman, Roope S. Jussila, Claire Lagarde, Ralene Levy, C. O. Liles, Jackie Lowe, Catherine Markwiese, James D. Maxwell, N. S. Maxwell, Michael J. Monsler, A. Nagao, Richard E. Niswander, Kate O'Hagan, J. R. Owens, Richard R. Peabody, Lease A. Plimpton, Susan P. Plunkett, Winifred M. Reuning, Franklin Riehlman, Mike Ross, Myra Scofield, Ruth Smith, Sandra Tibbets, William Wight, Sylvia I. Willemsz, and Alan Wood.

Special thanks go to the people who particularly came through with what I needed, often in the face of tight deadlines. These included Aryeh Brodsky of New York; Grayce A. Finger of the American Association for the Advancement of Science; and Gayle Westrate of Caltech. In a class by himself is Don Dixon, who as with my previous two books has consistently been there when I needed him.

A number of illustrations are protected by copyright and require credit lines too long to fit conveniently in their captions. "Flatiron Building," figure 49, is a gift of Alfred Stieglitz, 1933. "Gin Lane," figure 58, is available through the Harris Brisbane Dick Fund, 1932. These works of art are reproduced by permission.

Figure 11 is reproduced from figure 17, p. 21 of M. King Hubbert, "Nuclear Energy and the Fossil Fuels," *Drilling and Production Practice*, 1956, American Petroleum Institute. Figure 14 is by Madeleine Leroy-Ladurie and appeared in "Histoire du Climat depuis l'An Mil," Emmanuel Leroy-Ladurie, copyright 1967, Flammarion Press, Paris. These items are protected by copyright and are used by permission. Their copyright holders retain all rights.

Figure 13 appeared as Plate XXI of *Times of Feast, Times of Famine* by Emmanuel Leroy-Ladurie, translated by Barbara Bray. Copyright 1971, Doubleday and Company, Garden City, N.Y. It also appeared as the cover art for *Science*, December 27, 1974. Other material provided by *Science* appears as figure 46 from the cover of the issue of June 4, 1976. *Science* retains all rights and these items are used by permission.

Two figures are used by permission of Scientific American, Inc., and originally appeared as follows. As figure 45, from Joel S. Wit, "Advances in Antisubmarine Warfare," February 1981. As figure 63, from Arthur F. Pillsbury, "The Salinity of Rivers," July 1981. *Scientific American* reserves all rights.

Finally, there are several other people who deserve special thanks. My literary agent, Neil McAleer, was really the one who initiated this book as a project for me; without him I would not have written it. My editor, Jim Menick, was helpful and full of encouragement at every step. Two other editors at Doubleday, Al Sarrantonio and Pat LoBrutto, helped guide this work to completion.

And in all of this I have had the love and companionship of Angela Johnson, whose encouragement and support has been a continual source of good cheer.

Fountain Valley, California
December 14, 1981

CONTENTS

It was the best of times, it was the worst of times, it was the age of wisdom, it was the age of foolishness, it was the epoch of belief, it was the epoch of incredulity, it was the season of Light, it was the season of Darkness, it was the spring of hope, it was the winter of despair, we had everything before us, we had nothing before us, we were all going direct to Heaven, we were all going direct the other way—in short, the period was so far like the present period, that some of its noisiest authorities insisted on its being received, for good or for evil, in the superlative degree of comparison only.

CHARLES DICKENS

THE
REAL FUTURE

1

SOME ROOTS OF THE FUTURE

Not in disaster nor in despair will our futures unfold. There has not been recently, nor should we soon anticipate, any sudden or dramatic change in the world we have known. Our future decades, like our recent ones, will see changes, many of them significant and pervasive. Such changes will leave us puzzled, hopeful, surprised, angry, and moved to vote pro or con, by turns. What they will not leave us is overwhelmed or unable to cope.

The events and trends we may expect offer nothing so world-shattering as the end of our times and the beginning of a noticeably different age. Rather, they will be the kind of developments which together define an era, a continuation of the one we have known. The most significant and lasting themes of the future will have clear relationships to those long ongoing; and when we have lived the next few decades, we will look back and say they were remarkably like those we have already lived.

Does this mean we must echo the gloomy prophet of Ecclesiastes and say that what has been, will be, and there is nothing new under the sun? Hardly. This century has been one of almost ceaseless changes, many of them surprising and startling; enough new things are in history's pipeline to guarantee further changes aplenty. Yet the changes have taken place amid continuity. We have not lost our essential character as peoples, and nations; we have not suddenly invented new civilizations or cultures having no ties to the old. Change amid continuity has defined our times—continuity of ideas, of institutions, of individual people, often of the influences producing change itself. Change amid continuity will define our future.

Our future is a continuation of times past; there is no sharp division, except in our minds. When historians write of our time, they will not mark 1980 or 1982 as the dawn of a new era. Rather, that palm can be awarded to no year more recent than 1945 or more likely 1914. Our time represents a well-defined historical era; but its beginning predates the memories and experiences of all but a very few of us. Its ending, also, most likely will come only after a long while, in some year which few if any of us may see.

Since this is so, this book must treat the future as such a continuation, strongly linked to and deriving much from times past. It then is appropriate to devote a chapter to the history of an appropriately chosen selection of those past times; but this will be history with a difference. This is history presented as a record of trends and developments that are influential today and likely will be in the future; it is history examined to discern roots of the future. So we will begin, with the France of 1760, and with Jean Jacques Rousseau.

Born poor, a wanderer for much of his life, Rousseau was one of those geniuses who blaze like comets across their centuries. He was preeminently a sensitive man; amid the Age of Reason, he wrote of feelings, passions, natural behavior, sentiment, youth, and emotion. His *The New Heloise* (1761) set the theme for a century of romantic fiction and of novels of romantic love. His *Social Contract* (1762) pleaded the cause of the people brilliantly and did more than most to bring on the French Revolution. But he did more, much more. He introduced much of the idealism, many of the ideals of humanistic thought, that even today strongly influence many who write of the future.

What the historians Will and Ariel Durant have written of his 1760s is not so different from what might be said of America in the 1960s:

> Europe was ready for a gospel that would exalt feeling above thought . . . It had heard enough of reason, argument, and philosophy; all this riot of unmoored minds seemed to have left the world devoid of meaning . . . Paris was weary of Paris, of the turmoil and the hurry, the confinement and mad competition of city life; now it idealized the slower pace of the countryside . . . And this proud "progress," this vaunted "emancipation of the mind"—had they put anything in place of what they had destroyed? Had they improved the lot of the poor, or brought consolation to bereavement or pain?

Rousseau was preeminently a lover of nature. Born in Geneva, he had walked and climbed his fill amid Alpine scenery; he described these idylls so vividly that mountain climbing became a major sport. In *Émile* (1762) he anticipated by two hundred years the views of those who today advocate natural, organic foods: "The indifference of children toward meat is one proof that the taste for meat is unnatural. Their preference is for vegetable foods, milk, pastry, fruit, etc. Beware of changing this natural taste and making your children flesh-eaters. Do this, if not for their health, then for the sake of their character."

Again and again, he railed at the corruption of city life and praised the life of the country. In the city, he wrote, "Sincere friendship, real esteem, and perfect confidence are banished from among men. Jealousy, suspicion, fear, coldness, reserve, hate, and fraud lie constantly concealed under that uniform and deceitful veil of politeness, that boasted candor and urbanity, for which we are indebted to the light and leading of this age." Moreover, "The greater part of our ills are of our own making, and we might have avoided them, nearly all, by adhering to that simple, uniform, and solitary manner of life which nature prescribed . . . We think of the good constitution of the savages . . . and reflect that they are troubled with hardly any disorders save wounds and old age . . ."

He evidently knew nothing of yaws, malaria, or infestation with parasitic worms. Nevertheless, his feelings for country life retain an enduring appeal: "Men are not made to be crowded together in anthills, but scattered over the earth to till it. The more they are massed together, the more corrupt they become . . . [Mankind] needs renewal, and is always renewed from the country."

So far, so familiar. But there were other, ultimately less benign, elements to his thought and writings. Again and again he asserted that "man is naturally good"; natural man, free of restraints and fully liberated, could never be evil. In *Émile,* he wrote: "Let us lay it down as an incontrovertible rule that the first impulses of nature are always right. There is no original sin in the human heart . . . There is, at the bottom of our hearts, an inborn principle of justice and virtue . . . and it is this principle that we call conscience . . . Conscience! Divine instinct, immortal voice from heaven, sure guide . . . making man like to God!"

Our times have seen all too many hideous acts wrought by men untrammeled and free to follow those urgings which they have been pleased to call their "conscience." We also have seen the young people raised by Rousseau's prescription:

> Never punish your pupil, for he does not know what it means to do wrong. Never make him say, "Forgive me" . . . Wholly unmoral in his actions, he can do nothing morally wrong, and he deserves neither punishment nor reproof . . . First leave the germ of his character free to show itself; do not constrain him in anything . . . Keep the child's mind [that is, his intellect] idle as long as you can.

Certainly Rousseau was no friend of the sciences, or of reasoned thought; they tended to interfere with the passions that flow from the conscience of natural man. In his *Discourses,* he wrote: "Let men learn for once that nature would have preserved them from science as a mother snatches a dangerous weapon from the hands of her child . . . I venture to declare that a state of reflection is a state contrary to nature, and that the man who thinks is a depraved animal."

With his love of nature, his boundless faith in man set free from society's restrictions, his worship of the voice of youth, and his dislike for the sciences, Rousseau would feel entirely at home today in a mass meeting of environmentalists or of some of our other modern enthusiasts. In addition, not content with midwifing the reaction against thought which has so scarred our age, his Second *Discourse* set the stage for two centuries of attacks on the institution of property:

> The first man who, having enclosed a piece of ground, bethought himself of saying, *This is mine,* and found people simple enough to believe him, was the real founder of civil society. From how many crimes, wars, and murders, from how many horrors and misfortunes, might not have anyone saved mankind, by . . . crying to his fellows: "Beware of listening to this imposter; you are undone if you once forget that the fruits of the earth belong to us all, and the earth itself to nobody."

What he did not know, what would not be appreciated for two centuries, is that "first man" most likely was actually a lungfish, or Devonian amphibian, or perhaps one of the Therapsida—primitive mammal-like reptiles resembling the duckbill platypus. The human institution of private property, we now know, is our version of the instinct of territoriality. This drive, we know today, the drive to gain, maintain, and defend the exclusive right to a piece of property, is an animal instinct approximately as ancient and as powerful as sex.

Nevertheless, this is Rousseau's legacy. Another part of his legacy is the opening sentence of *Émile:* "Nature has made man happy and good, but society has rendered him depraved and miserable." All who are unwilling to take responsibility for their actions, who seek to blame society for their misfortunes, who foster attitudes of guilt and apology in the land for the mess they themselves have made of their lives—all these take their inspiration from Rousseau.

In our time, Rousseau's doctrines and attitudes remain pervasive and influential, as seductive and appealing as ever, particularly in our universities. That they should retain their influence and fresh appeal after two centuries, their hold on our ideals and on liberal thought, testifies to their character as enduring elements of Western culture. In recent years we have seen such ideals and attitudes wax strong, in the antinuclear movement, among the followers of Ralph Nader and Amory Lovins, the advocates of "the greening of America" and of "small is beautiful," and, pervasively, among many environmentalists. Theodore Roszak, Robert Heilbroner, and other influential individuals have attacked the methods and attitudes of science. We have lately seen many books exulting in revivals of the doctrines of Rousseau, proclaiming them the wave of the future. How shall we respond?

We may respond as we would to a revival of religion. We might individually support or reject such a revival; we would join or refuse to join, as we would choose. The one thing we would all know would be that there have been such revivals before. However influential might be the latest one, it would not be something new and heretofore undiscovered, destined to sweep all before it. Similarly, the revived doctrines of Rousseau, however devotedly proclaimed, amount to rediscovering the past and calling it the future. We may be pro or con in our attitudes, the next time we see them reappear; but they will not be new, and however stirring to the emotions, they will not go on to reshape radically our world.

What has reshaped our world in the past two centuries, and what will continue to do so, is of course science, technology, and industry. The general outlines of the Industrial Revolution, which was well under way in Rousseau's time, are quite familiar. What is not so generally known is that the Industrial Revolution grew out of a burst of innovation spurred by an early energy crisis. This crisis arose from the failure of England's supply of solar energy. To be specific, it was the failure of her supply of what we today would call biomass, that is, of fuel wood.

From medieval times onward England, like the rest of Europe, had depended almost entirely on wood or charcoal for fuel. This was true not only in home heating, but in such industries as the manufacture of iron and of iron products such as cannon. Wood also fueled such manufactures as brick, glass, copper, brass, and lead. Moreover, by the mid-1500s these industries were beginning to expand. In his *Pirotechnia* (1540), the Italian Biringuccio had written words to delight a modern strip-mining executive. He advised "whoever mines ores . . . to bore into the center of the mountains . . . as if by the work of necromancy or giants. They should not only crack the mountains asunder but also turn their very marrow upside down in order that what is inside may be seen and the sweetness of the fruit seized as soon as possible."

In England and Wales the population grew from 3 to 6 million between 1530 and 1700; that of London grew from 60,000 to 530,000, making London the largest city in the world. By 1700, 1 Englishman in 4 lived in towns or cities. These years of 1550–1700 thus saw enormous pressure on England's standing crop of fuel wood, a pressure increased by the clearing of timberland for farms and by the demands for timber in building towns and ships. Of 69 great forests in medieval England, 65 were to disappear by the end of the eighteenth century.

The wood crisis years, 1550–1700, saw high inflation in its cost, and England's response was to increase greatly its use of coal. By the 1640s, Londoners were becoming dependent on coal from Newcastle, shipped down the coast; as early as 1599, the Crown had levied a tax on these shipments. By 1700 the production of coal in England was close to 3 million tons per year. By 1750 this increasing use of coal was beginning to darken the sky of London. The English had invented air pollution.

This use of coal in industry ushered in the Industrial Revolution. Coal was a dirty, messy fuel; many products had to be protected from its fumes. It also produced higher temperatures than charcoal. Hence the manufacturers of the seventeenth century could not merely substitute coal for charcoal; they had to invent new furnaces. They devoted much ingenuity to the reverberatory furnace, which "reverberated" or reflected its heat onto the product. The need was for new firebrick, better crucibles, new tools of many types. The demand was for continual innovation and inventiveness. The successes these British craftsmen achieved affected powerfully the attitudes of industry toward technological change.

By 1700 British industry had built up a sizable body of hard-won knowledge concerning coals of different kinds and their uses. Now glass of all types could be manufactured in coal-fired furnaces, as could brick and mortar. The transition from the timbered Tudor architectural style to the Georgian, with its brickwork and leaded windows, reflected this new technology. Coal was also being used in the growing textile industry, and in such industries as gunpowder, candlemaking, sugar refining, the brewing of beer, and the making of scotch whisky. Brass was being made in the reverberatory furnace.

Shortly afterward, industry's continuing thirst for innovation gained a powerful boost from the physicist Robert Hooke. About 1702 Hooke exchanged correspondence with a Dartmouth ironmonger, Thomas Newcomen. Newcomen wanted to use recent discoveries in physics to build a steam engine. Hooke advised him on the principles of the air pump, writing that Newcomen must seek to create "a speedy vacuum" in order to succeed. In 1705 Newcomen succeeded with his steam engine, which worked by condensing steam in a cylinder, allowing air pressure to drive a piston into the resulting partial vacuum. It was the first practical source of power independent of wind, water, or the strength of animals; as noted by the historian John R. Harris, "it might be compared in importance to the invention of the wheel and axle or the discovery of iron." With its rocking beam and piston, the Newcomen engine resembled a modern oil field pumping rig and was soon set to work for a similar purpose: pumping water out of coal mines. By 1733 over a hundred such engines were in use; by the end of that century, some fifteen hundred.

The pace of innovation quickened. At Coalbrookdale, where Newcomen's iron cylinders were cast, Abraham Darby developed a blast furnace to produce iron using coke from coal and laid the groundwork for a vast upsurge in iron manufacture. In Glasgow in 1765, the chemist Joseph Black had been studying properties of steam and formed a friendship with a young instrument maker named James Watt. Studying the Newcomen engine, Watt was surprised at how much coal it wasted. He went on to apply Black's theories to create the steam engine that would change the nineteenth century. In 1773 he entered into a partnership with Matthew Boulton, founding the successful firm of Boulton and Watt, builders of steam engines. In 1776 Boulton would declare to the author James Boswell, "I sell here, Sir, what all the world desires to have—power!" The union of science, technology, and industry was complete.

And the railroad was just around the corner. The advent of cheap, abundant iron meant a hundred new uses for it. In 1763 Richard Reynolds built the first railway with iron tracks, featuring horse-drawn wagons. By 1779 an iron bridge spanned the Severn River; shortly thereafter Thomas Paine, author of *Common Sense*, would invent a better one. In 1784, while Watt was working to improve his steam engine, his foreman William Murdock built a model locomotive to run on roads at eight miles per hour; but Watt did not encourage Murdock to pursue this work. It was Richard Trevithick who built a true road locomotive in 1801. However, Trevithick lacked the leaf-spring suspension needed to assure a suitable ride on rough roads and thus failed to invent the steam-driven automobile. He turned to iron tracks, as an expedient means of smoothing the ride, and three years later invented the railroad locomotive.

During the following twenty years, George Stephenson greatly improved Trevithick's invention, and in 1825 he opened the first commercial line, the Stockton and Darlington Railroad. It offered service over a grand distance of ten miles. Soon afterward the Duke of Wellington, hero of Waterloo and now Prime Minister, secured a government appropriation of £100,000 to build a railroad between Liverpool and Manchester. On September 15, 1830, the Duke and a glittering array of political and social luminaries boarded the eight special trains making the initial run to Manchester. Nearly a million cheering onlookers lined the roadbed as the lead locomotive *Northumbrian* sped by at twenty-four miles per hour.

Thereafter, railroads pushed forward in all directions, their builders knowing they were here to stay. By 1870 there were 15,000 miles of track in Great Britain. In the United States the first railroads opened in 1830; by 1840 there were 3,000 miles of track, by 1860, 30,000. So rapid was the pace of railroad building that despite the destruction of the Civil War, in 1870 the South had 2,500 more miles of railroad lines than in 1860. Between 1880 and 1890 alone, U.S. railroads added over 70,000 miles of new track.

Growth in other industries was equally rapid. As early as 1740 the Sheffield ironmaster Benjamin Huntsman had produced cast steel in crucibles. Not till 1830 would Alfred Krupp in Germany match this achievement; that year the demand for cast steel in all Germany amounted to only five hundred tons. Yet by 1871 the annual production of steel in Germany was half a million tons, and by 1900 had increased further to twelve million. The capital in the Deutsche Bank increased from sixty million marks to seven billion in those same years of 1871–1900. As the century advanced, new industries were rising. In America the production of crude oil went from 2,000 barrels in 1859 to 4,800,000 in 1869 and to nearly 10 million in 1873. Those were the years when petroleum was valued chiefly for kerosene used in lamps and for lubricating oil. Gasoline was often regarded as a waste product to be dumped in the river. Yet even with the automobile still far in the future, there was more than enough demand for oil to promote the growth of that widely feared monopoly, Standard Oil.

Thus, looking backward three centuries from the year 1900, one might have seen all these things flowing from England's early energy crisis. When wood became increasingly expensive, when its supply became subject to shortages, the English did not despair or abandon their way of living. Instead they turned to their new fuel, coal. In seeking to use it most effectively, they were led to prize inventiveness and ingenuity, which brought rich rewards. There developed a wave of industrial creativity that gained them new manufactures, new industries; then, the wave accelerating, new sources of power, new methods of transportation, and an unparalleled command of nature's energies. In time these trends culminated in a great upsurge of production of new wealth. One can only hope that historians of the year 2300 will say that we and our successors will have done as well in meeting our own energy challenges.

While the nineteenth century indeed saw a surge of new wealth, it was distributed most unevenly. Those who worked in the factories and mines were often little more than slaves. The six-day week was standard, the fourteen-hour day far from uncommon. Whole families, including the young children, toiled year after year to eke out the pittance that would keep them alive. In the cold dark hours of a winter morning, one would see large numbers of small children on their way to dim, dank factories located amid slums. Employers prized child labor, since the children operating machines could do nearly as much work as a man, yet could be paid much less. Barbara Tuchman has described these workers:

> They came from the warrens of the poor, where hunger and dirt were king, where consumptives coughed and the air was thick with the smell of latrines, boiling cabbage and stale beer, . . . where roofs leaked and un-

tution and popularly elected parliament. All these developments encouraged hope that the wealth created by industry might reach the people not through revolution, but through law.

Meanwhile, this new wealth was trickling down, albeit slowly. In America between 1860 and 1891, wages in 22 industries increased an average of 68 percent, while prices actually were declining. In Germany, per capita income soared from 417 marks a year in 1890 to 716 in 1911, in a period of very little inflation. Also in Germany the Chancellor, Otto von Bismarck, surprised the world by adopting planks from the platform of the socialists. In 1883, 1884, and 1887 he got the Reichstag to pass the Social Insurance Acts, which provided for payments to workers in case of illness, accident, or old age.* By 1903 these programs were paying out over a hundred million dollars per year in benefits. While minuscule by today's standards (the comparable American programs today pay over a thousand times more), Bismarcks laws buttressed the key socialist principle of the worker being entitled to a share of the wealth he created.

By the late 1890s these developments had only begun to improve the lot of the workers. As the young Winston Churchill remarked to a friend while canvassing amid long monotonous row houses in Manchester, "Fancy living in one of those streets, never seeing anything beautiful, never eating anything savory—never saying anything clever!" Yet for all that their effects were limited, they were sufficient to strike at the heart of the programs of the Marxists. It was Eduard Bernstein, in Germany, who dared to be so bold as to challenge the doctrines of Marx. He argued that Marx was wrong: that the working class was making gains, that wealth was being spread among the people, that people were consuming more sugar, meat, and beer. He wrote of a capitalist society whose economy might expand indefinitely and whose institutions would be flexible enough to prevent Marx's supposedly inevitable collapse. And if the existing order was here to stay, then socialists, abandoning the goal of revolution, might work for liberal reform within the governments then existing.

This doctrine of "Revisionism" brought angry opposition from hardline revolutionaries, but there was no denying its appeal. It meant that rather than be an angry and isolated opposition, socialists might join in their countries' governments. In the succeeding years, in nation after nation, whether or not they officially adopted the principles of Revisionism, the socialists increasingly acted as if they were in fact Revisionists. In 1899 Alexandre Millerand of the Socialist Party became a minister in the cabinet of René Waldeck-Rousseau, who pledged the French govern-

---

* Bismarck was hardly a liberal reformer for all this. He set the retirement age at sixty-five in his social security program, which in 1887 relatively few workers lived to reach.

mended windows let in the cold blasts of winter, where pri
unimaginable, where men, women, grandparents and children
gether, eating, sleeping, fornicating, defecating, sickening and
one room, . . . [where] heaps of foul straw served as beds, . .
sometimes not all the children in a family could go out at one
cause there were not enough clothes to go round, . . . where a cig
and his wife earning thirteen cents an hour worked seventeen hoi
seven days a week to support themselves and three children, whe
was the only exit . . .

Karl Marx was not the only person to look upon such misery
the flames of revolution. In his *Germinal*, Émile Zola wrote of "the
sion of revolution that on some sombre evening at the end of the
would carry everything away. Yes, on that evening the people, ur
at last, would make the blood of the middle class flow, . . . in a t
of boots that same terrible troop, with their dirty skins and
breath, would sweep away the old world." The socialists of the nine
century became almost entirely Marxists. They predicted the
would become increasingly impoverished, driven to the status of p
They anticipated this would bring a social collapse, followed by
tion. It was their intention, then, to lead that revolution and to ush
new society—one based on Rousseau's doctrine that private proper
the root of all evil—a society that would share broadly the wealt
duced by the new industries.

At the root, then, the demand of socialism was that this surge c
wealth reach the people. The anticipated violent means to accor
this demand gained a mystique among many who sympathized wit
poor. Still, year after year, decade after decade, the revolution failed
rive; or rather, to recur. In 1848 and 1849 there had been widesprea
risings and revolts throughout Europe, as people rose against their
cratic regimes. Yet those upheavals, as close to general revolutio
Europe would see in that century, ultimately failed. Partly they coll
in disorganization; partly they were suppressed. Those failures damp
much of the revolutionary enthusiasm in succeeding decades.

While Marxists continued to call for revolution, the spirit o
times was against them. After 1848, despite the failure of that y
revolts, in nation after nation autocratic rule gave way to at least lir
forms of democracy. In England millions of men gained the right to
through the reform bills of 1867 and 1884. (Women would not get
vote for some time yet.) After 1871 France was a republic and Gern
had its Reichstag. The Reichstag, a legislative chamber, was housed ii
impressive-looking building in Berlin to make up for its unimpres
powers, but at least it was elected by vote of all men over twenty-
Even the reactionary Austrian Empire was reorganized under a co

ment "not to content itself to mere political reform, but to embark upon the new path of social reform."

By 1914 the Socialist Party was dominant in French government. The Social-Democrats were the most powerful party in the Reichstag. In Britain, Parliament was under control of a Labor-Liberal coalition, which had forced through a number of new laws and had curbed the power of the conservative House of Lords. In America the reform-minded Democratic Party controlled both Congress and the White House. In nearly all the Western nations change was in the making, reform was in the air. The dread vision of revolution was fading. Marxism was nearly everywhere in retreat; Revisionism was ascendant; the institutions of republics or limited monarchies were proving more than equal to the challenges of the burgeoning Industrial Revolution. In 1914 tens of millions of people could look ahead to the unfolding new century and see prosperity. In no country was there anything resembling today's mass consumer society, yet the trend was clearly in that direction. Indeed, in January of that year Henry Ford announced what was then the astonishingly high minimum wage of five dollars per eight-hour day in his plant. The rise of mass automobile ownership would soon be under way.

Yet already there were national leaders who looked upon the growth of industry and saw, not prosperity for consumers, but weapons for war. In 1888 Kaiser Wilhelm II ascended the throne as German Emperor. His first proclamation, addressed "To My Army," stated, "We belong to each other, I and the Army; we were born for each other." As he later explained to a company of recruits, "If your Emperor commands you to do so you must fire on your father and mother." In 1900, dispatching troops to the expeditionary force that would crush the Boxer Rebellion, he unwittingly gave his soldiers the name by which they would be known: "Even as a thousand years ago the Huns under their King Attila made their name . . . so may the name of Germany stand in China by your deeds for a thousand years!" And, a few years later: "There is no balance of power in Europe but me—me and my twenty-five army corps." During the prewar years he was a frequent visitor to the Krupp artillery proving ground, where to his delight he could feel the ground trembling and see actual shells bursting, targets being destroyed. With a little imagination (and he had a lot), he could envision the real thing.

He wasn't the only one. The Pan-German League, founded in 1891, distributed display signs for merchants' shopwindows reading, *Dem Deutschen gehört die Welt* (The world belongs to the Germans). Philosophers of militarism such as Heinrich von Treitschke and Friedrich von Bernhardi had wide followings. As Bernhardi wrote, "We must strenuously combat the peace propaganda . . . We must become convinced

that war is a political necessity, and that it is fought in the interest of biological, social, and moral progress . . . If it were not for war, we should
probably find that inferior races would overcome healthy, youthful ones
by their wealth and numbers."

They sound like monsters. Yet similar views, expressed less tactlessly
to be sure, were common in England and even in America. Actually, the
Germans were living amid a historic transition. Throughout the history of
Europe, from the late Middle Ages through the nineteenth century,
power politics had been governed by the principle of the balance of
power. No nation could grow too powerful, for any nation that became
too strong would find other states combining in enmity as an unbeatable
coalition. Such coalitions, led by England, had checked the power of
France repeatedly, most recently at Waterloo. But the rise of Germany
marked a transition from an era of merely powerful nations to an era of
superpowers. As the Germans believed, and as they would prove in
1914–18, when war came they would stand against all the powers of
Europe. They would defeat all their European enemies, and in the end
they would yield only to the forces of the one power they consistently ignored and underestimated—the United States.

From the Kaiser on down, the Germans believed they deserved and
could achieve the status of a superpower, predominant in Europe. As the
munitions manufacturer Gustav Krupp put it, "If these aims are achieved,
German culture and civilization will direct the progress of humanity; to
fight and conquer for such a goal is worth the price of noble blood."
Today we are accustomed to seeing the industrialized world, and many
other nations as well, divided into power blocs led by the United States
and the Soviets. It is arresting to think Germany nearly achieved that status and was kept from doing so only by the massed resources of much of
the rest of the world.

With Germany reaching for superpower status, with Russia and the
rest of Europe determined to keep her from achieving it, the resulting war
could only be long, bloody, and utterly destructive of Europe's society. A
half-century later, the Vietnam war would shake America's foundations.
That war cost 57,000 American lives in eight years of fighting. Yet the
British lost 60,000 young men in a single *day*, during the Battle of the
Somme.

Amid the endless mud and stench of the Western front, where shells
whined and explosions mangled the bodies of a generation, the life of the
nineteenth century shattered and met its end. Somewhere in the trenches
of Verdun or Ypres, the spirit of liberal reform died. In its place would
arise the grim spirit of revenge, of retribution; the hobgoblins of fascism,
nazism, bolshevism all arose in that consuming struggle. Terror was born

amid the greatest carnage the world had known, and in much of Europe and in Russia, for decades to come it would rarely be further away than a knock on the door.

Amid the prewar socialist meetings, there had been what the socialist leader Émile Vandervelde described as "a little man with the narrow eyes, rusty beard and monotone voice, forever explaining with exact and glacial politeness the traditional Marxist formulas." This was Nikolai Lenin. In late 1917, following the outbreak of revolution in Russia, the German high command saw an opportunity to use him. They shipped him along with other leading bolsheviks in a sealed railway car from exile in Switzerland to the Finland Station in Moscow. As they were departing, he remarked to his fellows, "Six months from now we shall all be ministers or we shall all be hanged!" The Germans expected Lenin to seize control of the government, withdraw Russia from the war, and make peace on German terms. They were not disappointed. In March 1918, at Brest Litovsk, the new Soviet government surrendered the western regions of Russia, including the Ukraine,† and freed a million German troops for an offensive in France against Russia's recent allies. Amid such treachery Marxism won a nation.

The upheavals Germany launched thus echo down to our own times and into the future; for we have seen how Marxism had been fading in the face of Revisionism. With Marx's doctrines seemingly headed for the museum of antiquated political curiosities of the nineteenth century, the war's destructiveness gave them a new lease on life. To Marx's call, "Workers of the world, unite!," Lenin would add a new and powerful cry, which echoes today: "Oppressed peoples of all lands, rise up!" The ensuing decades would see the dissolution of empires in Asia and Africa, and the nations of the West would often find themselves on the defensive in dealing with the rising power of the Soviets.

The Second World War was largely a continuation of the First. It had the important new element of Japan's drive for power, which led her —at Pearl Harbor and the Philippines—to attack the United States. But the main feature of the war was the struggle of the totalitarians for predominance in Eurasia. Once again it was Germany seeking superpower status, reaching for control of Europe, battling a strong Soviet Russia, with one of these two nations destined to win and to establish preeminence over the other. Once again Germany was too strong to be contained within Europe; it was America that wielded what by now was a global and not merely a European balance of power. Yet the outcome had

† They got it all back, of course, after American forces rolled back the German offensive and Germany surrendered to the Allies.

long been foreseen. In 1835, in his *Democracy in America*, Alexis de Tocqueville had written:

> There are at the present time two great nations in the world, which started from different points, but seem to tend towards the same end. I allude to the Russians and the Americans . . . All other nations seem to have nearly reached their natural limits, and they have only to maintain their power; but these are still in the act of growth . . . ; these alone are proceeding with ease and celerity along a path to which no limit can be perceived. The American struggles against the obstacles which nature opposes to him; the adversaries of the Russian are men . . . The conquests of the American are therefore gained by the ploughshare; those of the Russian by the sword. The Anglo-American relies upon personal interest to accomplish his ends, and gives free scope to the unguided strength and common sense of the people; the Russian centers all the authority of society in a single arm. The principal instrument of the former is freedom; of the latter, servitude. Their starting point is different, and their courses are not the same; yet each of them seems marked out by the will of Heaven to sway the destinies of half the globe.

The two World Wars had begun in 1914, in a virtual recreation of the 1870 Franco-Prussian War, with columns of foot soldiers marching into France accompanied by horse cavalry and artillery. They ended, three decades later, with that lone B-29 in the sky of Hiroshima on a blue August morning. The shadow of those first mushroom clouds hung dark over the world in succeeding decades; the experiences of the twentieth century had made men wary of predicting peace. Yet with the perspective of nearly forty years, we now can say that almost without realizing it, we had passed out of the darkness and into the sunlight. Without quite knowing it, we had made it through the night.

What was upon us, in the decades following 1945, was the brilliant promise of the twentieth century, tragically deferred through decades of war and economic depression, at last bursting forth in full strength. In America prosperity had begun even during the war years. In Europe and Japan, the ravaged cockpits of the destruction, prosperity would wait a decade still, as the nations rebuilt and cast off at least the most visible of their scars. In the Soviet Union it would not arrive at all; the bitter legacy of Marx and Lenin, self-proclaimed friends of the workers, would be to leave those workers falling farther and farther behind. Yet even there was a quickening. If Soviet life could not match that of the West, still it could improve over that of the czarist years, and of Stalin's 1930s.

Through the last decades of the nineteenth century and the early years of the twentieth, prosperity had worked to raise living standards and improve the quality of life. At first there had been limitations on the labor of women and children, reductions in the hours of work, improvements in wages and in the conditions of city life, and a beginning on protection of workers by means of pensions and insurance. These had all predated 1914, at least in Europe. During the years of war and depression the improvements had continued, though often slowly and with reverses; one recalls Will Rogers' quip, that America in the 1930s was the first nation to go to the poorhouse in an automobile. Still, nothing in past history could compare with the wave of prosperity and growth that permeated the United States after 1945, and Europe and Japan after about 1960. In those years, the world's advanced nations embraced the future.

For generations extending well back into the previous century, thinkers and men of ideas had been grappling with the challenges of the abundance and prosperity they could foresee. It was Graham Wallas, in 1914, whose book *The Great Society* sought to define the promise of the new century. Half a century later, Richard Goodwin, one of his intellectual heirs and an aide to John F. Kennedy, stated the promise anew in a speech delivered by President Lyndon Johnson:

> The Great Society rests on abundance and liberty for all. It demands an end to poverty and racial injustice—to which we are totally committed in our time, but that is just the beginning. The Great Society is a place where every child can find knowledge to enrich his mind and enlarge his talents. It is a place where leisure is a welcome chance to build and reflect, not a feared cause of boredom and restlessness. It is a place where the city of man serves not only the needs of the body and the demands of commerce, but the desire for beauty and the hunger for community. It is a place where man can renew contact with nature . . . But most of all, the Great Society is not a safe harbor, a resting place, a final objective, a finished work. It is a challenge constantly renewed, beckoning us toward a destiny where the meaning of our lives matches the marvelous products of our labor.

Coined by thinkers, expanded upon by visionaries, the phrase "The Great Society" passed into political parlance and was inevitably debased. Still, by 1980 the resulting government programs, for all their controversy and waste, had brought America to a milestone. In 1962 a Federal index of poverty showed 22 percent of the American people living below that poverty line. In succeeding years the index was revised continually to reflect inflation. By 1980, taking account of transfer payments provided to persons of low income, only 7 percent of the American people were living below the poverty line. America had gone far toward abolishing poverty.

The future, then, like the past will be keyed to the issues of peace and prosperity. Regarding peace, one may be more than cautiously hopeful. Between 1905 and 1914, a succession of crises and confrontations stretched taut the thread of peace in Europe, till at Sarajevo, it snapped. From 1936 to 1939 a similar succession of crises led inexorably to the Second World War. From the Iran crisis of 1946 to the crushing of Poland in 1981, there has been a continual flow of serious issues involving America and the Soviet Union, which in Korea and Vietnam escalated to war. At times, as in Berlin in 1948 and Cuba in 1962, these crises have brought a direct confrontation between the superpowers. Yet never has there been the sort of ominous concatenation of tensions that preceded 1914 and 1939. Always the superpowers have stepped back from the brink or, more often, declined to approach it. They will likely continue to do so in the future.

The matter of prosperity is more problematical and carries with it a host of questions:

• Much of the world's prosperity has derived from an abundant flow of inexpensive oil. Yet the energy crisis has dramatized physical limits on oil discovery and production, from which have followed political decisions rendering petroleum less abundant and far more expensive. How will the industrialized nations grapple with this problem, which strikes at the root of their prosperity?

• And how will the world respond when fossil fuels prove a faithless friend? There are enough exploitable fossil fuels to increase tenfold the amount of carbon dioxide in the atmosphere. It is becoming increasingly clear that even a much lesser increase will raise global temperatures and particularly the temperatures of the Arctic and Antarctic. This will likely bring about important worldwide climatic changes, including shifts in patterns of rainfall and drought, and melting of parts of the great polar ice caps. Sea levels will rise, islands and lowlands will flood or turn to swampland. How will the world respond, when continued increases in fossil-fuel use become fraught with such climatic peril?

• What uses will the world make of the upsurge and rapid advance in such new technologies as electronics and the manipulation of genes? With 1984 only a year away, and looking more to be just another year than to be Orwell's horror, it is possible to dismiss many of the wildest fears about the control and manipulation of human beings and to see in these technologies great promise. How important will they become, considering that they offer immense new vistas, yet are sparing of energy and resources?

• What will be the importance in the world of the rising economic power of Japan? How will the United States respond to her intensive challenge, not only in such traditional industrial fields as automobiles, but in new areas such as robots? Since the end of the war, Japan has sought no large role in the world's diplomatic councils, nor has she sought deliberately to carve out a sphere of political influence. Yet by her successful example, she stands today to wield more influence than her boisterous neighbor the Soviet Union, which has loudly and repeatedly proclaimed its economic failures to be the way of the future. What will it mean for the world at large to see that Japan, a nation of nonwhites, has risen not only to compete with America and Europe, but to challenge them on their own terms? And with the energetic and capable people of China just a short sea voyage away, will eastern Asia emerge as a center of political and economic power to rival Europe and North America?

• Beyond that, what future will be offered to the billions of poor people of the Third World? To what degree will they make real gains; to what degree will increased wealth in their nations be siphoned off through large and continuing increases in the number of mouths to feed? Will they develop the stable institutions which may direct their progress? Will their leaders retain their frequent infatuations with what they are pleased to call socialism, their guilt-mongering accusations of colonialism, and that sterile agitation for an international redistribution of wealth and income some call the New International Economic Order? Or will more and more of them face up to the internal questions of birth control and of policies favorable to development, learning from the successes of such once-poor nations as South Korea, Singapore, Brazil, and Taiwan, as well as Japan?

These questions and others lie before us, their answers only dimly understood, their exploration the subject of this book. Still whatever the future brings, it will be a future built by ordinary human beings living their lives. Holding to the bright vision of Richard Goodwin's Great Society, at times we may fail and fall back, or find disappointment. Yet the hope will remain, and we will hold to its inspiration. For we know the promising future that is there for us to build, and we know it is real, it is possible, it exists—it can be ours.

2

THE OIL QUESTION

Great crises do not usually produce really fundamental changes. We are sometimes inclined to think of such crises as stemming from the obsolete nature of old forms and practices, so the solution can only lie in bold, new approaches very much at variance with what has gone before. But if a crisis comes upon us with any sort of suddenness, it probably will be met with methods remarkably like those which we have known. There is a reason for this. When the need is for urgent action, people will rarely hazard their fortune to approaches that are bold, or innovative—but untested and unfamiliar. Instead, they will try by all possible means to go with what they have. They will turn to solutions which not only show promise, but which can be accommodated within familiar practices and established institutions. Then, when the crisis is resolved, one finds change, yes; but one often sees a most remarkable continuation of long-established structures and institutions. Crises test the old ways, often lead to their reform, but rarely sweep them away in revolution. Even when there is a revolution, or what some are pleased to call one, the old forms and ways usually can be seen, still strong, still being practiced, but called by new names.

So it will be with the present energy crisis. Recent years have seen any number of projected fundamental changes, which we have been assured must flow from our oil shortages. Books have been written about the "postindustrial society," the "no-growth society," the "zero-sum soci-

ety"; we have seen seers and forecasters writing of societies dedicated to
solar energy, to conservation, or to a rejection of long-established values of
economic growth and progress. Yet we can safely predict none of these
things will happen; at least not within our lifetimes. It is one thing to assert that in theory there are physical limits to growth, that over some centuries the world may evolve toward some sort of steady state. It is quite a
different matter to say that these limits are known today and must
strongly influence public policy. Energy is what we need, but we will meet
this need neither by an overturning of society nor by exchanging our
values for more bucolic ones. What we will do is establish new industries
to draw on the enormous reserves of fossil fuels still existing after the oil
runs out. We will do these things, and life will go on much as before. We
will find these reserves in coal and in synfuels.

The current revival of coal brings us full circle. As in England in the
seventeenth century, we are expanding rapidly our production and use of
it, in the face of rapid depletion of more convenient but less abundant
fuels. As in the England of that time, we find a premium put on ingenuity, inventiveness, resourcefulness. In four centuries coal has never
changed its character as a difficult and demanding fuel. Neither, however,
has it ceased to offer rich rewards to those industries and nations who
apply their brains to face its problems.

The litany of these problems was summed up in the early 1970s by
S. David Freeman, now chairman of the TVA: "There are two things
wrong with coal today. We can't mine it and we can't burn it." More
specifically, the growth of coal use has been hampered by such difficulties
as the restoration of strip-mined land and the removal of coal's sulfur,
which produces corrosively acid pollution. Yet today its growth is proceeding apace.

Coal exists in profusion in many parts of the United States, but the
greatest and most accessible reserves lie in the West. Today's coal frontier
is the Powder River Basin of northeastern Wyoming. The country there is
rolling rangeland, arid high plains where ranchers run herds of cattle, or
buff-colored upland thinly overgrown with grasses. As in much of the
West, this is often a vast and starkly empty land, little changed from the
days of the Conestoga wagon trains, where snowcapped mountains loom
in the distant blue haze and faraway railroad trains outline themselves
against the prairie stretching away into the distance. To this open
rangeland have come the coal miners, with their hard hats and faded
denim jeans; as some have put it, the real cowboys wear tennis shoes so
you can tell them apart from the coal miners who wear cowboy boots.
Gillette, Wyoming is the capital of this area. Nearby are immense strip
mines with names like Caballo, Rawhide, Wyodak near the Dakota state
line, and Black Thunder.

Here and elsewhere in the northern Great Plains, conditions for mining approach the ideal. The coal seams are fifty feet thick, sometimes much more; near Gillette they run in places to two hundred feet thick. Often these seams lie beneath only thirty to forty feet of overburden. Strip miners can remove over two hundred feet of overburden when they have to; at one Oklahoma mine, ninety-five feet of material has been removed to reach a seam eighteen inches thick. Fifty billion tons of coal lie near Gillette alone, with energy equivalent to all that America will use for the next fifteen years, and in the West there are a number of places like Gillette.

A Western strip mine is like something out of Jules Verne, with equipment of heroic size. It is a place where the very rock beneath one's feet is taken apart, stacked in neat piles, and put back together again, over an extent of many square miles. At first, large bulldozers scrape off the topsoil and push it into piles, laying bare the underlying rock. Drillers bore holes down to the level of the coal seam, packing the holes with charges of dynamite; when the explosives are set off, an expanse of cliff is at once reduced to a pile of rubble. Then the dragline moves in. It may stand three hundred feet tall and weigh thirteen thousand tons, and this virtual land battleship does not merely stand in one spot; it moves on massive supports. It uses a huge bucket to scoop up the rubble, suspending the bucket from its long boom with steel cables, then pulling the bucket upward and toward the base of the boom as it scoops from below. It dumps this blasted overburden into the canyonlike pit opened by previous mining, amid plumes of dust blowing in the wind and the sunlight.

What is at the bottom of this canyon or trench, of course, is coal—a broad and deep black seam. Following more drilling and dynamiting, the coal is blasted into chunks, and power shovels go to work, loading twenty tons and more at a time with shovels the size of a bus. Each shovel load is dumped into a haulage truck several times larger than a railway locomotive. These trucks have rubber tires ten feet across; their operators sit in enclosed cabs above the engine radiator and off to one side, fifteen feet above the ground. As each truck is loaded it trundles off with a hundred and fifty tons of coal.

When the seam has been dug, the land must be carefully restored. Again it is the bulldozers which do the work, grading the filled-in trenches and scooping the piled topsoil back into place. The topsoil will also be shaped and graded, then spread with fertilizer and planted. As one executive at the Black Thunder Mine has said, "We'll just widen the valley some and drop the hills, and the land will just be forty feet lower."

Across these sunbaked plains black trains a mile long snake their way, nearly lost in the immensity of the land. These unit trains meet the coal-laden trucks from the mine. A conveyor, angled starkly upward and feeding a hopper, loads the train's hundred or more hopper cars with over ten thousand tons of coal. Then, its five or six diesel locomotives straining, the train begins its long haul; perhaps to Houston, perhaps to the Mississippi Valley. Along the main line of the Southern Pacific or Burlington Northern, through prairie railroad towns like Alliance, Nebraska, such trains pass every half hour with their rattle and roar.

Many such trains will unload at coal-fired generating plants. In previous decades such plants, with their tall smokestacks, would have blackened the sky with their soot. Nowadays it often is hard to tell whether or not a particular plant is operating. For starters, it has fly-ash precipitators, which use static electricity to collect the coal ash. Also it very likely has stack-gas scrubbers to clean and scrub the sulfur compounds from the coal gases. These remove most of the noxious components of the coal smoke, so almost the only time a casual observer can see anything is on a cold day when a long plume of vapor extends downwind from the stack.

Some of the coal, particularly from Kentucky or West Virginia, goes to the export market. In Norfolk, Virginia, are two of the world's largest railroad yards; in each of them tracks laid side by side span over a mile of waterfront. Together the yards can accommodate over twenty thousand coal-hopper cars. A pier a third of a mile long features the world's largest and fastest coal-loading facility. Its two traveling loaders, each as high as a seventeen-story building, handle eight thousand tons per hour. With car dumpers and conveyors, these loaders combine coal of different grades, blending shipments to order for the customer. The shipments include high-grade metallurgical coal bound for Japan and steam coal en route to Europe.

For all this, this coal harbor must be expanded and improved. In the Newport News ship channel the colliers wait, coal-carrying ships looking very much like oil tankers, forty and fifty of them at a time. They incur demurrage or waiting charges of up to fifteen thousand dollars a day; and they wait three weeks and longer. Half a world away are similar scenes at Ras Tanura, the Saudi Arabian oil terminal where the tankers wait, but this is Virginia. Our nation is the Persian Gulf of coal.

American coal reserves are immensely large, preposterously large. The Bureau of Mines has estimated 437 billion tons can be mined with 1975 technology at 1975 prices. The energy equivalent is close to ten thousand quads, the equivalent of 1.6 trillion barrels of oil and enough to supply all our energy needs at current levels into the twenty-second century. At the risk of a bad pun, this may just be scratching the surface; no doubt there are huge reserves that have not been found, or the mining of which will prove economical in the future but not today. In the world at large, the *Survey of Energy Resources 1976* of the World Energy Conference has estimated the total resources of all types of coal to be 11.5 trillion tons. M. King Hubbert, whose studies correctly predicted our oil production would peak in 1970, has prepared similar forecasts for world use of coal. His conclusion: the production peak may not come till 2140. It is no wonder that even Carroll L. Wilson, a member of the Club of Rome and a leading advocate of limits to growth, has called coal the "bridge to the future" which will tide the world over till it develops replacements for fossil fuels.

We have been stepping up its use. In 1974 our production was 603 million tons and it supplied only 17.8 percent of the nation's total energy. In 1980 production stood at 840 million tons, of which 710 million was consumed domestically. This sufficed to furnish 20.6 percent of the nation's total energy, with the *Oil and Gas Journal* projecting a rise to 22.3 percent in 1982. In 1980 total production rose at a very fast rate of 8 percent, with nearly half the increase being exported. Further increases lie ahead. The National Coal Association, in its May 1980 report to the President, forecast 1985 U.S. production as high as 1,080 million tons. In the eight years from 1980 to 1988, some 350 electric-generating units are scheduled to start up, about half of them coal-fired. Based on that alone, the annual demand for coal by U.S. utilities should jump from 525 million tons in 1979 to over 850 million in 1988.

This burgeoning growth will demand huge investments. The growth of coal will be paced by improvements in its transportation, as railroads invest large sums to rebuild tracks and facilities. In addition, a network of slurry pipelines may grow to supplement the railroads. These will carry pulverized coal in a slurry with water. We will see continuing progress in cleaning up coal and controlling its combustion to prevent pollution, or removing its sulfur by such techniques as solvent refining. In this process coal is dissolved in a solvent, then treated to remove the sulfur and filtered to remove the ash. Most importantly, we will see the birth of significant new industries, dedicated to transforming coal into gas and into convenient liquid products which can run our automobiles and diesel engines.

This is the matter of synthetic fuels, of fuel products chemically synthesized from coal. Coal is not pure carbon, but resembles petroleum in that both are complex mixtures of hydrocarbon compounds. The difference is that coal hydrocarbons contain less hydrogen than do those of petroleum. Hence producing synfuels from coal amounts to chemically reforming its compounds by adding more hydrogen. For example, when coal is solvent-refined, the solvent is a liquid rich in hydrogen. In Gulf Oil's SRC II process, the heated solution is pressurized with still more hydrogen; the solvent acts to transfer it to the coal, turning the coal to oil. Other processes, which add even more hydrogen with the aid of catalysts, transform coal into methane, or natural gas.

The technology of synfuels has been known for decades, but in an era of cheap oil this technology has been used only by nations rich in coal and fearing cutoff of their oil supplies. Prior to and during World War II, Germany built some fifteen coal-hydrogenation plants, partly with American help. As late as 1939 Standard Oil of Indiana built a synfuels plant for aviation gasoline for the Luftwaffe. These plants, relying heavily on slave labor, reached peak production of some 110,000 barrels per day in March 1944 before being destroyed by Allied bombing.

After the war U.S. scientists combed through the German documents describing this effort, and the Bureau of Mines built an experimental synfuels plant at Louisiana, Missouri in 1949. It soon shut down in the face of low prices for fuels from petroleum. The South African government was not so prodigal. In 1950 it launched that country's first synfuels plant, Sasol I, using adaptations of the German technology. The plant opened in 1955 and at once encountered very great problems. All system operations had been successfully tested in pilot plants, but the full-scale plant's gasifiers had to be shut down every two months or so for lengthy servicing, and some control valves wore out in days.

For five years engineers wrestled with such difficulties before gaining success. The plant then went on to serve as a nucleus of a versatile chemical complex supplying much of South Africa's need for nitrogen fertilizers, plastics, and pipeline gas. Situated amid a huge coalfield, the plant today performs well above its design standards. It converts an inferior coal containing 35 percent ash into a broad spectrum of products, including alcohols, ethylene, propylene, butylene, and gasoline; in 1979 this last item cost fifty cents per gallon at the refinery gate. The plant now produces some 10,000 barrels a day of product, about 4 percent of South Africa's total requirements.

Though production has been modest, the experience gained has been priceless. In 1973 the first oil embargo drove South Africans to a stepped-up effort. They went to work on a much larger plant, Sasol II, designed to produce 55,000 barrels a day, mostly of gasoline. It went into operation in 1980. Significantly, it not only is much more efficient—1.78 barrels of oil per ton of coal vs. 1.26 with Sasol I—but is much cleaner. It recycles waste water rather than dumping it in rivers and thus is designed to attain zero liquid waste. Covering more than a square mile of treeless high veld near Johannesburg, with huge gasifier vessels stretching for a quarter-mile, it resembles nothing so much as an oil refinery complex the size of a city.

It was completed just in time. With the 1979 fall of the Shah of Iran, South Africa lost its long-term oil contracts and was driven onto the spot market with its high prices. Very quickly they contracted with the Fluor Corporation, builder of Sasol II, to begin work on the very similar Sasol III. When completed in 1984 the three Sasol plants together will produce 112,000 barrels a day, nearly half of South Africa's needs. The complete effort is costing her $7 billion, equivalent to a $300 billion synfuels program in the United States.

We have not pursued such a program yet; we lack South Africa's situation of necessity. Nor do we have her cheap coal ($6 per ton vs. $29 in the United States), her black labor force working for low wages, or the low capital investment costs of a plant like Sasol I, built decades ago. But as in South Africa, our goal is to build the commercial plants which will give us valuable experience along with the gasoline. Then we will be able to build on this experience and launch a rapid expansion, when necessity compels.

Today the largest operating U.S. synfuels plant is a $200 million facility on the banks of the Big Sandy River near Catlettsburg, Kentucky. It is run by Ashland Oil and its partners and demonstrates the "H-coal" process, which uses a catalyst to speed a hydrogenation process similar to solvent refining. It went into operation in June 1980, converting 600 tons of coal per day into 1,800 barrels a day of synthetic crude oil. This indicates the magnitude of any serious U.S. synfuels effort, for 600 tons a day was the capacity of the largest German wartime production units. In this country, however, such a capacity qualifies it only as a pilot plant.

The U.S. synfuels industry to date has largely been a government-subsidized experimental program and will continue in this character for some time. In June 1980 Congress established the Synthetic Fuels Corporation, with the goal of supporting growth of this new industry to a level of half a million barrels of synfuels per day by 1987, 2 million a day by 1992. To get there, Congress authorized it to commit $20 billion up through about 1984, with another chunk of money, as much as $68 billion, being appropriated then to finish the job. It takes six years to build a plant, and to meet these goals, a new project twenty-five times larger than the H-coal plant at Catlettsburg would have to get under way every two or three months, from 1981 through 1986.

Nothing of the sort has been happening, and as the consultant Erich Reichl predicts, "Federally backed projects will take three times as long to construct and they won't run." Other critics have pointed to heavy reliance by the industry on contracts from the Department of Energy, as well as high costs and a general reluctance by industry to commit its own funds on a large scale. As Richard Corrigan has noted in the *National Journal*, "If these proposed plants are still regarded as lemons in the eyes of the private sector, it does not follow that they merit subsidies from the government." No one denies that building a synfuels industry will be costly and risky; but at least we are pursuing a diverse variety of approaches. A review of some of them will indicate the frontiers being explored, as well as their somewhat mixed results:

• Coal gasification. In 1972 four natural-gas companies announced plans to build four synthetic-gas plants, each with a capacity of 250 million cubic feet per day. Over the next eight years they fought Indians over coal leases and bankers over financing terms, watched their construction-cost estimates double, filed thousands of pages of documents with regulatory agencies, and spent $120 million without turning a spadeful of dirt. Three of the companies dropped out; the survivor, American Natural Resources, cut its planned capacity to 125 million but soldiered on. It drew in four other firms as partners, and after Congress set up the Synthetic Fuels Corporation, President Carter announced their first $250 million loan guarantee. Subsequently President Reagan approved a loan-guarantee package totaling $2.02 billion, with $500 million being risked by the companies themselves. Construction finally began in 1980 in North Dakota, and the plant will be operating in 1984.

• Solvent refining. In 1973 Gulf built a $32 million pilot plant at Fort Lewis, Washington, to test its SRC II process. The Department of Energy nurtured it through the years, and until 1981 it was liquefying thirty tons a day, producing seventy barrels. Gulf then went on to assemble support from Japan and West Germany, in addition to its federal support, to launch a 20,000 barrel-per-day production plant, intended as the first of twenty to fifty coal liquefaction plants. Its costs mushroomed from $700 million in 1979 to $1.6 billion in 1981, and Gulf by then was estimating a cost of $3.4 billion to build it and operate it for five years. With this the three governments pulled out, and the project collapsed. The Fort Lewis plant was shut down, too. It was nice while it lasted.

• Sasol technology. In 1980 Texas Eastern Synfuels got $24 million in start-up money from the Department of Energy to build a duplicate of the Sasol II plant. Total cost was estimated at $3.5 billion, with a target date of 1986. If built, it will produce gas, liquid fuels, and petrochemicals, in the equivalent of 50,000 barrels per day.

• Synthetic chemicals. Early in 1980 a subsidiary of Eastman Kodak announced plans to build a plant to synthesize acetic anhydride from coal. Acetic anhydride is used in producing rayon as well as photographic films. The plant is to go into operation in 1983, with production the equivalent of 2,700 barrels of oil per day. Despite this modest level, the plant is of great importance. It marks a first step in shifting the chemical industry from oil and gas to coal.

• Synthetic gasoline. In a landmark discovery, Mobil Oil's "Mobil-M" process uses a catalyst to transform methanol into high-octane gasoline costing $1.20 per gallon at the refinery gate. Methanol, wood alcohol, can be made inexpensively from coal; the Sasol and Eastman Kodak plants do so, and the Department of Energy is backing W. R. Grace in studies of a plant to produce methanol from coal at 35,000 barrels per day. Mobil has run a pilot plant in New Jersey since 1976 and now is planning a 45,000 barrel-a-day plant costing $2.5 billion. West Germany is also building a Mobil M plant, and New Zealand has selected the process as well. New Zealand has large reserves of natural gas, which can be converted into methanol costing only sixty-five cents per gallon. In fact, the Mobil M process will convert any alcohol to gasoline, including, as their chemists have discovered, a cup of brandy. The phrase "one for the road" may have new meaning.

Synfuels embrace an even broader scope than this. They include the very large field of fuels made from living plants or other life forms, a topic we will defer to Chapter 8. They also include tarry or rubbery substances found in abundance, but requiring a stage of industrial processing before they can be run through a refinery as synthetic crude. The most important of these are heavy oils or tar sands, and oil shale. Canada particularly has pioneered the production of tar sands, described as the poorest quality petroleum reserve being developed anywhere in the world. Yet the resource is huge. Venezuela, for one, may have 2 trillion barrels. Canada's confirmed deposits in Alberta alone total at least 1.35 trillion barrels, a hundred times more than the oil of Prudhoe Bay and over four times the oil used by the entire world in its history to date. By the 1990s, Canadian production could approach a million barrels per day, a third of her requirements, making Canada the next major nation after South Africa to switch to synfuels.

The huge Athabasca and Wabasca tar sands deposits in northeast Alberta can be strip-mined. They are covered by up to twenty feet of muskeg, a morass of water and decaying vegetation. In the summer it swallows up land vehicles; in the winter it freezes hard as rock. To reach the tar sands one begins by draining the water two years before digging, then removing the vegetation while it is frozen. For all this, muskeg removal is one of the simpler problems. The tar sands themselves are a mixture of sand, water, and a sticky black material called bitumen; on a hot summer day it oozes from fresh tar sands and trickles in viscous glistening rivulets. Tar sands resemble road asphalt and will stick to anything they touch, frequently clogging machinery. The bitumen can also dissolve the rubber tires of the operation's huge trucks, which resemble those of a Western strip mine.

The sands are mined with bucket-wheel excavators nearly the size of a strip-miner's dragline, but these sands are highly abrasive. During the first summer of mining, the hundred-pound bucket-wheel teeth of hardened steel, all 120 of them, wore out in four to eight hours of digging. Heavy equipment sank into the softening, sticky deposits underfoot. In winter, at forty below, undisturbed deposits of frozen tar sands resembled concrete but were tougher. Bucket-wheel teeth glowed red hot; the teeth were torn from their sockets, and the buckets' quarter-inch-thick steel plates were ripped apart. This problem was solved by "fluffing" the tar sands: dynamiting them in the summer so they would not freeze so solidly.

These problems were faced by the pathbreaking tar sands company, Great Canadian Oil Sands, Ltd., a subsidiary of Sun Oil. Construction of the first plant began in 1964, at what today would count as a low capital investment, $300 million. It began producing 50,000 barrels a day of synthetic crude oil in 1968, but with its operating problems it operated at a loss until 1976. It was followed by a larger plant, Syncrude Ltd., which opened in 1978. Syncrude cost $2.2 billion, seven times as much as the first plant, yet was designed to produce 129,000 barrels a day, less than three times as much. Syncrude today is the model for other plants currently planned.

The Syncrude project is a dramatic sight. There are 4 draglines, each with a boom 360 feet long and a bucket which scoops 80 cubic yards at a bite. The excavated pit stretches for three miles. A power station, large enough for a city, cogenerates electricity and process steam, dominating the scene with its 600-foot stack. Two conveyor belts, 6 feet wide, carry the tar sands three miles to a processing plant as large as a refinery. There the bitumen is washed out in what has been described as "the world's largest washing machines," distilled, and blended into a synthetic crude oil and shipped by pipeline to the refineries in Edmonton to the south.

The Athabasca experience is not directly applicable in the United States, but we too have heavy oil sands; 30 billion barrels worth, largely in Utah. They thus are larger than our proved reserves of petroleum. They are mostly too deep for strip mining, but can be gotten out when heated with steam. Here too we are learning from the Canadians. Similar deep reserves of bitumen lie in the Cold Lake region of Alberta, where Imperial Oil, a subsidiary of Exxon, has been working on a pilot-plant scale since 1964. In our own country, the first heavy oil-sands production plant may be built near McKittrick, California, forty miles west of Bakersfield. Getty Oil has been working there and has proposed to produce 20,000 barrels a day, perhaps as soon as 1986.

It is oil shale, however, which may define our future in synfuels; and as with coal, much of this future will be written in the West. The borders of Utah, Wyoming, and Colorado meet on a windblown high plateau, sweeping from the Rockies westward toward snow-peaked mountains in Utah. Spring comes late and fall comes early; the summer sun heats the arid buttes and plains above a hundred degrees, killing all but the hardiest desert plants. The winter's icy winds sweep the land with bitter blasts at twenty below. Few people live in this inhospitable region. Yet the land is wealthy. It contains some 2 trillion barrels of shale oil, locked away in immense formations. This is in the form of kerogen, a rubbery solid that turns into oil when heated to nine hundred degrees. But to tap this resource has not been easy.

As Thomas Maugh wrote in *Science* in 1977, "For most of this century, this oil shale has been considered a major potential source of energy. But exploitation of that treasure has proven elusive. The potential oil shale industry has been riding a dizzying roller coaster . . . Every two or three years, it seems, this embryonic industry reaches a peak where the need for oil from shale seems imperative, the economics viable, and the problems minimal. And then the cost of petroleum comes down. Or the cost of construction skyrockets. Or water resources appear too meager. Or air pollution standards appear too strict. Or Congress backs off from incentives that the industry thinks necessary. And every time the roller coaster comes downhill, one or two more companies bail out, and the ever fewer hardy survivors . . . look forward to that next glorious high, hoping that this time . . . they will reach that seductively beckoning goal of commercial production."

One hardy survivor is California's Union Oil Company, which has owned Colorado shale land since 1922. In 1957 Union tested a pilot plant processing up to 1,000 tons of shale per day. It featured a "rock pump" forcing a stream of shale through a cylindrical retort, where hot gases released the oil. In 1958 low crude prices forced the plant to shut down. Following the 1973 embargo, Union offered to build a 10,000 barrel-per-day production retort if Congress would lower the cost of shale oil by enacting a tax credit of $3 per barrel. This measure failed in the House of Representatives by one vote in 1976. Two years later, at the instigation of Senator Herman Talmadge, the Senate passed the tax credit but it failed in conference committee with the House. Finally, in 1980, Union got its tax credit. They announced that not only would they build this Colorado plant, to be completed in 1983, but they would then expand the plant by adding four more retorts for a total of 50,000 barrels per day, at a total cost of $1.5 billion. However, they soon not only had their tax credit, they also had the windfall-profits tax, and they had to seek additional backing from the Synthetic Fuels Corporation. In 1981 President Reagan approved a contract providing up to $400 million for government purchases of jet and diesel fuel from their shale oil, at a guaranteed price, and at last Union could go ahead.

Overshadowing Union's aboveground process is a subterranean one developed by Occidental Petroleum, featuring an underground oil mine in the prime deposits of Colorado's Piceance Basin. Three mine shaft headframes, holding ventilators and elevator hoists, tower above the terrain. The largest one is taller than the Statue of Liberty and will lift 50,000 tons of rock a day out of its mine shaft. The elevators also will lower equipment, including trucks to be reassembled in underground machine shops. A thousand feet down and more, miners will drill and blast an array of tunnels extending for miles, crisscrossing the formation at several deep levels to separate it into a collection of immense blocks, each half an acre square and three hundred feet high. These blocks will be made into the chambers where oil will be extracted.

To prepare each chamber, crews will hollow out each block, working from the tunnels to excavate large cavities. The shale they remove, some 20 percent of each block, will be hoisted and processed in aboveground retorts resembling those of Union Oil. Then half a million pounds of explosives will be planted in boreholes drilled through the blocks. Their explosion turns the shale into rubble, which takes up more space and expands to fill the cavities. This blasting is the key to the operation for the chunks of rubble must be of the right size. It is the development of its blasting techniques that has cost Occidental its years of experimentation.

With this, the chamber is ready for production. Pipes are drilled into place from the surface above, bringing in diesel fuel, air, and steam. The top of the chamber is set on fire and its burning controlled. The fire heats the shale rubble, driving out the oil, which percolates to the bottom and is pumped to the surface from a collecting tunnel at the bottom of the chamber. Like a slow cigar burning downward, the chamber will stay afire for a year. Sixty such chambers burning simultaneously, along with processing in retorts of shale taken from the cavities, are expected to yield 95,000 barrels per day. The cost may be less than that of petroleum. As Occidental chairman Armand Hammer has said, "Oil's getting expensive to find. With shale you never hit a dry hole."

Yet even these plans pale before a $500 billion effort proposed by Exxon in 1980. It called for digging six enormous open-pit shale mines in the Piceance Basin and in Utah's Uinta Basin. As the columnist Neal Peirce has described this work, suggested for completion by the year 2010:

> Each would rank among the greatest man-made declivities of history—a half-mile deep, 3.5 miles long, 1.75 miles wide. Total excavation in the pits would equal a Panama Canal a day. Each pit would produce 1 million barrels of fuel daily from the crushing and high-temperature "cooking" . . . of 3.7 million tons of oil shale. Each pit operation would require 22,000 miners and 8,000 refinery workers. Another 2 million barrels a day, Exxon suggests, could be produced by additional operations heating the oil shale . . . in vast underground pillars . . . Where in the arid West is there water to process such quantities of shale . . . ? Again, Exxon believes it has an answer: building three 10-foot pipes to transfer

(pumping uphill) 1.7 million acre-feet of water a year from the . . . Missouri River in South Dakota, more than 600 miles as the crow flies.

Such breathtaking projects may appear virtually as science fiction, but they are what America will need, and what we will attain. Yet if the future is to bring no sudden epiphany, no dramatic revelation setting things to right for all time, if we are to be left instead with the continued difficult efforts of imperfect human beings trying to do the best they can with what they have, then in a couple of centuries, maybe sooner, we will find our energy situation showing some interesting resemblances to what it is today. For coal and other fossil fuels, no less than oil, can be mined out and depleted. Their very large abundance means that their history will be measured in generations or even centuries, not in mere decades, but they are just as susceptible to Hubbert's curves. The day will come when world coal production, or production of bitumen or shale oil, will peak out and thereafter decline.

This is hardly a new observation. In 1865 William Stanley Jevons published a book, *The Coal Question*, which described how England's coal reserves were subject to eventual exhaustion. This sparked considerable debate and discussion and led to predictions of an energy crisis. For example, in 1885 the American journal *Science* offered the following observations:

In 1861 . . . Mr. Hull estimated that the available coal in Great Britain represented a total of 79,843,000,000 tons . . . In 1881, 154,000,000 tons were extracted . . . The output shows considerable fluctuation from year to year . . . but on the whole, a very rapid increase; the output for 1875 being double that of 1854, and that for 1883 double that for 1862; and if the amount increases at this rate (3,000,000 tons annually), the supply must be exhausted in the year 2145 A.D. . . . One of four things must then happen,—either some new source of energy must be supplied, or a larger percent of the coal must be utilized, or coal must be imported, or England must give up her manufactories. It is doubtful if any new source of energy . . . will be discovered, unless some explosive be used for the purpose . . . While it is hardly possible to use less coal, we may get more out of it . . . But instead of a decrease in the waste, there is likely . . . to be an increase; for each year faster speed is demanded by rail, and steamships are rapidly replacing sailing vessels . . . The idea of importation is hardly practicable, for the nearest coal-mines of any extent are in Canada and the United States . . . To supply England with the necessary coal, 2,100 ships . . . , each carrying 6,000 tons and making thirteen trips a year, would be required. The cost would be necessarily greatly increased . . . If coal becomes scarce, there will be no way of paying for food, emigration will begin, the death-rate will increase, the birth-rate decrease, and England will change once more to an open, cultivated country, devoid of all other industries.

A forecast such as this offers more than simply a sense of *déjà vu*. It illustrates the point that when we seek to write about developments a century or more ahead, we cannot avoid being caught up in the parochialism of our own time. We will often overlook or at best dimly perceive insights those future times will take for granted. It also illustrates that a forecast so far ahead can be correct only in very broad, general terms. Any attempt to predict specifics will be seen eventually as telling far more about the predictor and his times than about the future under discussion.

As we look ahead to the year 2100 or thereabouts, and to the second energy crisis, it is tempting to imagine a repeat of the drama of an international cartel, an embargo or revolution, and a succession of manipulated shortages and price hikes. There are several reasons why this is unlikely and why oil is a poor model for the economics of other fossil fuels.

Far more than coal or any other commodity, oil has always been subject to either glut or shortage and to wild price swings. In 1859 in Pennsylvania, crude oil cost $20 a barrel, but in 1861 the price fell to ten cents. Then in 1864 again it was at $12. In 1932, during the Depression, the discovery of the great East Texas field set off an oil rush which drove prices again to ten cents a barrel; the governors of Texas and Oklahoma had to call out the National Guard to shut the fields down. The influential economist Paul Frankel for many years has argued that beginning with John D. Rockefeller and his Standard Oil Trust, the oil industry has repeatedly tended to fall under monopoly control as the only way to control prices. As Frankel wrote in 1945, "What began in Pennsylvania of the seventies developed according to a pattern which we can detect in the oil industry all over the world down to this very day: the ascent of one concern or a group of concerns which, by centralization of control . . . , attains in due course a paramount position."

The sheer size of the reserves of coal, shale oil, and bitumen also will stabilize their future markets. Unlike oil, these fuels do not exist in a few immense fields in some area resembling the Persian Gulf. To the contrary, very great reserves of one or more of these fuels are found on every continent and in many of the world's nations. A country seeking to rely on such an energy source will usually find what it needs either directly within its borders or in one or more of the nations with which it maintains close and friendly commercial ties. There will be little prospect of politics and geology brewing a witches' cauldron wherein some of the world's most desirable fuels are in the hands of some of the world's most unstable nations.

The choice of alternative fuels also brings in some interesting comparisons. In much of the world, oil has been and still remains almost incredibly cheap to produce. In Saudi Arabia the cost of drilling, pumping, and moving it from the desert oil fields to the Persian Gulf ports is about thirty cents per barrel. In Texas and elsewhere in the United States, prior to 1973, domestic crude cost $3.50 a barrel and this still was well above the production cost. The huge price increases after 1973 have largely reflected oil's uniqueness in this respect. There simply has been no readily available and comparably convenient fuel capable of being produced in quantities and at a cost which would put a cap on OPEC's prices.

The alternatives to oil, whether fossil or nonfossil, by contrast all have in common that they cost a lot. This is why, until recently, they have not been developed. Yet by the same token, with a number of competing energy sources, at higher but more or less comparable prices, we will not again see so many of our energy eggs in one basket. Our economies will not be linked to a single fuel of exceptional cheapness and convenience. Rather, they will be linked to a most diverse mix of replacements for oil, including not only fossil fuels but various forms of solar energy as well as of nuclear power. A world of several competing fuels will be much more secure than one dominated by that petroleum which, in turn, is dominated by a monopoly.

With world oil production having apparently peaked in 1978, the message of OPEC's rude price hikes is that we must turn to these alternatives, all of which are more costly. When Arabian oil was hiked above $3.00, we did not have the luxury of finding large new supplies of replacement fuel costing $6.00. But with other fossil fuels we will indeed have this luxury; there will be no such clear-cut distinction on the basis of cost. Once we have dug up all the rich oil shale assaying at twenty-five gallons per ton, we still will have plenty of shale assaying at fifteen gallons. When we have mined the coal lying within two hundred feet of the surface, there will be lots more down to five hundred feet.

The upper regions of the earth's crust contain a quantity of burnable carbon amounting to some twenty quadrillion tons, the equivalent of over a hundred quadrillion barrels of oil; in other words, more than a thousand times the world reserves of coal, or a hundred thousand times the size of the Athabasca tar sands. To be sure, only a small part of this represents exploitable reserves of fossil fuels, but still this offers the pleasant possibility that we will successively exploit poorer and poorer grades of shale and other fossil fuels. This is exactly what we have done with reserves of iron and copper ore and with other minerals. As with these minerals, then, the future may see fossil-fuel prices that stay reasonably steady for long periods or that increase slowly and moderately. The wild OPEC price hikes then will appear as an aberration in a long-term pattern of steady, predictable energy prices.

From this perspective, a phrase like "the second energy crisis" represents a misnomer, an example of the parochialism of our times at which future generations will smile indulgently. We may envision a world turning to fossil fuels other than oil, building great industries for their supply and use, and, with appropriate pollution controls, going its merry way for at least a century, perhaps even for several centuries. The subsequent slow, gradual shifts in industrial patterns in no way would qualify as a crisis. Yet there is another and very real way such fossil-fuel use could bring on a genuine crisis, or at least an extended and difficult era of painful adjustments. There could be a second energy crisis arising not out of problems with its supply, but from consequences of its use.

3

ANTARCTICA AND MIAMI

When fuels are burned, they release carbon dioxide, CO_2. This gas then builds up in the atmosphere and acts to trap heat in the lower portions, raising global temperatures. Often called the greenhouse effect, this actually is part of the normal work of the atmosphere's naturally present water vapor and CO_2, both of which act to trap heat. This is fortunate; if it were not the case, the sun would heat the earth's surface to a global average of only —4.2 degrees Fahrenheit, and the world would be shrouded in ice. Because of the heat-trapping action of these atmospheric gases, the earth is considerably warmer; on average, 58.8 degrees.

As early as the turn of the century, and sporadically for several decades thereafter, various scientists suggested man's industrial activities might be affecting the climate. However, to prove this turned out to be one of the difficult problems of modern science. Despite many attempts, the first really good solution to this problem came only in 1975. That was when Syukuro Manabe and Richard T. Wetherald published the results of studies with their three-dimensional general-circulation model. What this meant was that they had to solve equations for the behavior of the entire atmosphere, at all levels and over the entire surface of the earth, both land and sea. Only through a lengthy computer solution of these equations, treating a large variety of subtle and interrelated effects influencing atmosphere, ocean, land, and polar ice, could they make headway.

Their specific problem was to consider the effects of doubling the atmospheric concentration of CO_2, from 300 to 600 parts per million. They concluded this would raise the mean surface temperature by 5.3 degrees. They were far from the only scientists to study this problem, but theirs was the work which set the standard. In May 1979 Frank Press, science adviser to President Carter, commissioned a review of this question by the National Academy of Sciences. They concluded that Manabe and Wetherald were right. As their report stated, "We have tried but have been unable to find any overlooked or underestimated physical effects that could reduce the currently estimated global warmings."

The next question was how long would it take the atmospheric CO_2 to double. As early as 1958 Charles D. Keeling, of Scripps Institution of Oceanography, had initiated regular month-by-month monitorings of its atmospheric concentration. Keeling and his associates made their measurements atop Hawaii's Mauna Loa and at the South Pole, both places being far from any center of industry. They could compare their results with the world's production records for oil, coal, and natural gas, and they found a strong, marked rise in the year-by-year CO_2 concentration.

In 1958 this concentration stood at 314 parts per million. In 1981 it was 340. Further, this rise had been following closely the world's industrial use of fossil fuels. These results meant that of the CO_2 released by industry, 52 percent was winding up in the atmosphere. Oceanographers could account for another 37 percent; it has been dissolving in the oceans. The rest, about 10 percent, apparently has been taken up by forests whose growth (or regrowth) has been speeded by the extra CO_2 in the atmosphere. With worldwide fossil-fuel use showing a historical growth rate of 4 percent per year, the National Academy group could conclude that this CO_2 doubling would take place by the year 2030. Even a halving of this growth rate would delay the doubling only till 2050. As early as the year 2000, the increase in concentration will probably be 20 percent over the preindustrial value, with an attendant mean temperature rise of 1.4 degrees.

To put this into perspective, it is necessary to appreciate the really unique climate the world has been enjoying lately and to understand what even a few tenths of a degree can mean when we look at global averages. There is no way to understand the earth's long-term climate other than in the light of a highly unpleasant but inescapable fact: we are still living amid the Ice Ages.

Everyone has heard of the last great advance of the glaciers, when Chicago lay beneath a massive slab of ice a mile thick. We are all familiar with geologists' reconstructions of the advance of such glaciers, as standing walls of ice thousands of feet high advancing relentlessly, grinding down pine forests, razing the hills, spreading with awesome implacability to entomb half of Europe and North America in a dark and frozen fastness. But we associate such scenes with the extinct woolly mammoth and saber-toothed tiger, with the primitive Neanderthal men and their rough stone tools. We flatter ourselves with the fancy that such fates are not for us, that we have advanced beyond them.

In recent years we have gained for the first time a record of ancient temperatures and their changes. Two leaders in this effort have been the glaciologist Willi Dansgaard and the marine geologist Cesare Emiliani. They have been able to take advantage of a "fossil thermometer": isotopes of oxygen. Depending on the temperature, marine plankton take up these isotopes at different rates, and Emiliani has traced in these shells a record of ancient temperatures in the Caribbean. These have not been ordinary seashells found on a beach, of course; Emiliani has found them in long cores of sediment drilled from the seabed. Similarly, Dansgaard has worked with long cores of ice drilled from the Greenland ice cap. Depending on temperature, snowflakes form with different isotope ratios, and layers of ice then can also serve to record these ancient temperatures.

These temperature records show we are living at the end of a warm period without parallel in the last 100,000 years, with a return of the glaciers close at hand. The last time temperatures were warmer was 124,000 years ago. Not only do we have the temperature evidence; we have the fossil remains of the warmth-loving hippopotamus found in, of all places, Great Britain. The times of the past 10,000 or so years, the times since the glaciers retreated, have not marked their final defeat or their abandonment of the now-sunny lands these glaciers once submerged. Instead, these 10 millennia have been an interglacial, a time between successive major advances of the glaciers. We have the geologic and temperature records of past interglacials, and we know several important things about them.

They have been few and far between. In Emiliani's temperature curve, there have been only nine warm periods in the past 700,000 years, each an average of some 80,000 years apart. Nor have they lasted long. None of them, with a climate comparable to our own, has lasted more than 10,000 years. Ours has lasted just over 10,000. Moreover, the time of peak temperatures, the climatic optimum, occurred some 6,000 years ago and the climate has been slowly deteriorating since. Nor is there anything special about our interglacial, except that we happen to be living in it. Geologically recent records of temperature changes show no great difference from those past interglacials. And the end of an interglacial is not a gentle, pleasant affair. Hippopotamuses may have grazed in England during the last major one, but studies at Barbados show that within 5,000 years the sea level was dropping, dropping, dropping to reach a level over 230 feet below the present one. As the Barbados sea level was falling, in other places ice was piling up.

In January 1972 at Brown University, a group of internationally prominent researchers gathered in a conference with the interesting title, "The Present Interglacial, How and When Will It End?" Reporting on the conference proceedings, climate specialists George Kukla and R. K. Matthews stated,

> When comparing the present with previous interglacials, several investigators showed that the present interglacial is in its final phase (Emiliani, Imbrie, Lozek, Mörner, Wright) and that if nature were allowed to run its course unaltered by man, events similar to those which ended the last interglacial should be expected to occur perhaps as soon as the next few centuries . . . At the end of the working conference, the majority of the participants agreed to the following points: The global environment of the last several millennia is in sharp contrast with climates that existed during most of the past million years. Warm intervals like the present one have been short-lived and the natural end of our warm epoch is undoubtedly near . . . Global cooling and related rapid changes of environment, substantially exceeding the fluctuations experienced by man in historical times, must be expected within the next few millennia or even centuries.

One may say, "So what"; even a few centuries is quite far off. But geologists are a cautious lot, reluctant to go out on a limb. The conclusions of the Brown University conference echo, perhaps with unconscious irony, a 1975 report by three geologists who had been studying evidence of ancient eruptions of an inactive volcano in the Pacific Northwest. They offered a forecast: "The repetitive nature of the eruptive activity . . . with dormant intervals typically of a few centuries or less, suggest that the current quiet period will not last a thousand years. Instead, an eruption is likely within the next hundred years, possibly before the end of this century." Their volcano was Mount St. Helens. It blew up in 1980.

But how much cooling does it take to bring on an ice age; alternately, how much warming serves to end or prevent one? Here too we have the evidence. Not only do we have the temperature curves of Emiliani, Dansgaard and others, we have the remarkable calculations of a project known as CLIMAP. This project undertook to assemble enough geologic evidence to map the surface of the earth at the height of the last Ice Age, 18,000 years ago. They then used those data to study the global climate in July of that era, by means of a general-circulation atmospheric model resembling that of Manabe and Wetherald. Their conclusion matched that of other geologists using very different methods: the ice-age climate on average was only 8.8 degrees cooler than the present one.

Even much smaller temperature changes can dramatically affect a glacier. In the French Alps the village of Argentière stands at the base of a glacial valley. It is a picture-postcard village, with a tall church steeple and heavy-timbered houses with stout roofs; pine forests run up the nearby mountainsides. This Alpine village has changed little in its outward appearances in over a century. The same cannot be said of the Argentière Glacier. An engraving of the town made about 1860 shows the glacier as massive and threatening, extending the full length of its valley and poised just beyond the village limits. During the next century the climate warmed by just about one degree. A mid-1960s photo of the same scene shows the glacier had retreated, its terminus then barely one third of the way down from the top of the valley.

From this standpoint then, a CO_2-induced climate warming appears not only beneficial but necessary. A new ice age would be as destructive as a nuclear war, and would be every bit as severe a catastrophe. If we are warned of one on good evidence, and we have been so warned, we are justified in doing whatever is necessary to prevent or delay it. And how fortunate we are, if to do this we require no surge of global effort, no massive endeavors marshaling the world's concerted energies; nothing more than to continue with the growth of industry, the rise in living standards, and their attendant use of fossil fuel.

It is worthwhile keeping in mind what a new ice age would be like. If it were like the last one, it would completely destroy Canada, Scandinavia, the United States as far south as the Ohio River, much of Britain and of northern and central Europe, the Soviet Union north of the Ukraine, southeast Australia, New Zealand, and southern Chile and Argentina. All these lands would be trod under by glaciers and snow. Adjacent to these buried lands would be wide belts of land ruined by being turned to tundra with its permafrost. Seaports and coastlines would be high and dry, hundreds of feet above the diminished oceans; the continental shelves, no longer submerged, would stretch in places for hundreds of miles toward the sea. The climate would be much drier, with less rainfall. What was left of humanity would have to fall back largely on the tropics, and much less of the world's land would be available for agriculture.

Even without such disheartening prospects, a CO_2-induced warming will have its uses. Inevitably, the winds of climatic change will blow some people good fortune indeed. As Dansgaard and others have shown, the climate of recent decades has on the whole been exceptionally favorable; now we may hope to keep it so. Although the climate has been deteriorating for some thousands of years, it has not been all downhill; there have been occasional episodes of warmth. The years from 1930 to 1960 were unusual in this respect. As the climate expert Reid Bryson has stated of this period of supposedly "normal" climate, "The normal period is normal only by definition. There appears to be nothing like it in the past thousand years." Indeed, Dansgaard has shown that global temperatures have been oscillating with periods of 78 and 181 years. Such oscillations would have made a worldwide temperature drop since 1960 much more severe than it has actually been, except that the CO_2 buildup has been arresting its decline.

In the Soviet Union and other cold lands, as well as in India, it can only be a good thing if CO_2 works to preserve the warm temperatures of recent decades, even to make them warmer still. In northwestern India before 1920, in a time of cooler temperatures, the monsoon would fail from time to time and bring less than half of normal rainfall. This happened on average every 8.6 years. Between 1920 and 1960, however, amid warmer temperatures, the rains failed only every 14 years. During the unusually warm decade of the 1930s the monsoon was even more reliable, failing the equivalent of only once every 18 years when averaging over an array of stations. In recent years India has become self-sufficient in grain. So long as the rainy seasons come and the climate is prevented from deteriorating, India is likely to remain so.

By the same token, in cold lands with a short growing season, even a slight temperature drop can be disastrous. Bryson has shown how such slight changes affect the growth of crops in Akureyri, Iceland, which like much of the Soviet Union stands near the northern limit of agriculture. If the mean temperature drops 1.8 degrees, the growing season shrinks by two weeks. The cooler days inhibit plant growth, moreover, and yields will fall 27 percent. Let the temperature fall further, by another 2.5 degrees. The growing season will be less than four months long, and yields will be down a total of 54 percent. In a country close to the limit of success in agriculture, this can mean starvation.

In his history of Iceland, Thoroddsen has chronicled only 12 famine years in the five centuries between 975 and 1500 A.D. Yet there were 34 famine years between 1600 and 1804. In these famines, people and livestock died in winter and the meager crops and pastures failed in the brief summer. These tragedies befell Iceland during periods of cold climate. Yet how cold were they; what was the difference between the famine centuries and the good ones? The answer is about one and a half degrees.

Since 1940, world temperatures have declined by no more than half a degree. Yet this has been sufficient to play havoc with the agriculture of the Soviet Union. In 1954 that nation's premier, Nikita Khrushchev, launched a daring scheme to make farmland of the "virgin lands" in central Asia. These lands were marginal, largely underlain by permafrost. But based on records of previous decades, the Soviet leaders had reason to predict, in any decade, six bumper crops and only two crop failures. Their virgin-lands program soaked up much of their meager agricultural resources, finally maturing about 1960. In 1962 and 1963 there were crop failures, and succeeding years saw this bold program fail quite thoroughly.

So a CO_2-induced warming offers many a pleasing prospect. It offers even more. The Manabe-Wetherald calculations show that along with the global temperature rises, there would be a worldwide increase in rainfall. From a global average of some 39 inches of rain per year, the temperature rises would produce an increase of 3 additional inches. Yet such averages are of scant comfort during a drought, and in some areas, this is what the CO_2 might bring. Bryson has estimated that between 1957 and 1970, the CO_2 buildup produced a warming of 0.2 degrees. Using climatic theory, he has shown that in the Sahel region of West Africa, this would suffice to reduce the annual rainfall by 3.4 inches. In fact, from meteorological records the actual falloff in those years was 3.8 inches. During the early 1970s the Sahel was wracked by drought, and reports were widespread of severe misery and starvation.

Most definitely such climatic changes will affect the United States. The agronomist Louis M. Thompson has shown what may happen to yields of corn, soybeans, and wheat. In the Corn Belt, the highest yields of corn and soybeans have come in summers with temperatures lower than normal. When the daytime temperature rises above 90 degrees, the weather is unfavorable for these crops. Corn yields, averaging 110 bushels per acre in normal weather, would drop by up to 8 bushels in case of a 5-degree temperature rise. Wheat yields are even more sensitive. With no change in rainfall, a temperature rise of 3.6 degrees would cut yields of Oklahoma wheat from 25.2 to 21.4 bushels per acre. In most wheat-growing states, less rainfall would mean an even larger falloff in the yields.

And the rainfall would be expected to decrease. During Bryson's "normal" period of 1930–1960, temperatures were higher than in 1850–1870. The Great Plains and Rockies were then getting nearly 20 percent less rainfall for the entire year than in 1850–1870 and even less during the summers. In those earlier years the high plains swarmed with extensive herds of buffalo. Buffalo hunters killed them off with their rifles, but Bryson has suggested that even without these hunters, the decline in rainfall would have reduced the size of the herds by at least half and perhaps by as much as 75 percent.

What of the rest of the country? All over North America, normal patterns of rain are governed by the seasonal presence of zones of high and low pressure. Any shift in these patterns can steer storms far to the north or south, leaving a drought along their normal track. Reporting on an East Coast drought in 1980, according to *Newsweek*, "A low-pressure region on the East Coast . . . sent Gulf storms spinning far out over the Atlantic, bypassing thirsty Easterners like New York taxicabs at rush hour." Any global warming will produce many such shifts of high- and low-pressure regions. Manabe and Wetherald may have predicted a 7 percent increase in global rainfall, but this is hardly a 7-percent solution. Most regions and localities have built their reservoirs and water-supply systems in the expectation of continued normal rainfall. Any shift in their rainfall patterns, any falloff in their expected precipitation, can leave them in the position of New York City in the mid-1960s. A series of dry years nearly emptied the reservoirs there, and during the worst of that drought, these reservoirs were down to 25 percent of normal capacity.

The coming decades of CO_2 rise will be the very decades when America stands to experience water shortages, even in the best of conditions. Much of the water America uses comes from a network of underground aquifers, water-bearing formations of porous rock. These are similar to the formations where oil is found, and as with oil, we have been depleting this bounty. Every day the nation takes 82 billion gallons from its reserves of ground water; every day nature replenishes only 61 billion. The world's largest underground reserve of fresh water may well be the great Ogallala Aquifer, which underlies much of the Great Plains. Yet every year farmers withdraw more water from this formation than the entire flow of the Colorado River. At today's rates of usage, the Ogallala on average has no more than 40 years of useful life remaining.

Even though CO_2-induced climatic changes may exacerbate such shortages, they will come in any case, and America will cope. One important solution, indeed, droppeth literally as the gentle rain from heaven. Every day some forty times more rain falls than the 106 billion gallons of water we use for all purposes. Even a large falloff still would leave us with much more than we use. Moreover, America wastes much of what water it has. As with energy prior to 1974, we have been guzzling water most profligately. It takes 15,000 gallons to grow a bushel of wheat, 60,000 gallons to produce a ton of steel, 120 gallons to irrigate the chicken feed for a single egg at breakfast. A steak at dinner represents about 3,500 gallons, while the entire steer probably has fattened on enough irrigated corn to represent water sufficient to float a ship.

Invariably, we will adjust. As with South Africa's Sasol II, our industries will recycle their water, reflecting its value. It may be possible to recharge aquifers with surplus water, available during the spring runoff. The most significant savings will come in agriculture, which accounts for 83 percent of our total water use. Conventional irrigation methods, employing trenches or ditches, slosh water all over the landscape and waste a great deal. Fly over the Midwest; you will see green circles on the land below, which are a newer type of irrigated field. These employ center-pivot irrigation systems, in effect enormous rotating lawn sprinklers a quarter-mile long. Yet even these waste water on the land between the individual crop plants. The ultimate irrigation system is just now coming into widespread use. Called drip irrigation, it is especially suited for arid lands. It uses plastic tubes to deliver carefully measured flows of water to each individual plant. Not only does it conserve water; it prevents salt buildup, a serious problem in many irrigated fields. Moreover, the steady, even supply of drip-irrigation water raises these yields up to 70 percent. For even greater water savings, farmers can survey their fields with infrared instruments to determine crop conditions. They then can see if the plants are getting enough water and hold back on their irrigation until they find it is actually needed.

Yet in so complex a matter as the world's climate, there can be no final victories. We may prevent the climate from deteriorating, even stave off an ice age. But in doing so we may nevertheless bring about other problems, both serious and destructive. Indeed, these problems may be so great that only a new ice age would exceed them; but from avoidance of this ultimate horror, we will not have it as a common standard for comparison. If we avoid an ice age only to subject ourselves to other problems, less severe but still very serious, then we still will face a crisis, not as destructive a one as nature left alone would bring to us, but a crisis nonetheless.

What would bring on this crisis is a rise in sea level, brought on by a melting of part of the Antarctic ice cap. The Manabe-Wetherald calculations show an average global warming of 5.3 degrees, but this is just an average. Near the poles this warming is amplified, and it increases. Above the eightieth parallel of latitude this increase amounts to over 18 degrees, which is quite sufficient to bring vast changes to the Arctic and Antarctic.

It is not hard to understand this. In the Canadian Arctic, the limit of the northern snowcap is near the northern reaches of Hudson Bay. In the Keewatin District of the Northwest Territories, north of Baker Lake, snowdrifts stay on the ground until late summer. August is cool and frequently cloudy; the snows begin in September. With even a slight warming trend, the snows would come later and melt sooner; the limits of permanent snow cover would retreat northward. This retreat would expose barren land and empty sea, both of which are much darker than snow or ice. Being darker, they would absorb more sunlight, grow warmer still, and send the snow limit even farther northward. Not only does this happen in the Manabe-Wetherald calculations, there is also a reduction in atmospheric turbulence. This reduction keeps the warmth close to the polar surface rather than mixing it through the atmosphere. The result of this warming can only be a rise in sea levels.

Nor is this simply theory. As noted, the last time temperatures were significantly warmer than today was 124,000 years ago. We have the geological evidence for the sea level of those days: fossil coral-reef terraces in Hawaii and Barbados. They show that at that ancient time, the sea stood at least twenty feet higher than it does today. That was over 100,000 years ago, true. But today glaciologists know of a body of polar ice which could send sea levels surging to nearly that height, an ice mass that is unstable and at serious risk of melting in any time of global warming. This is the West Antarctic Ice Sheet.

The West Antarctic sheet has the peculiar feature that, unlike the ice sheets of Greenland and of East Antarctica, it rests upon bedrock up to a mile and a half below sea level. If the ice were to melt, that part of Antarctica would simply disappear, leaving no rocky land mass exposed above the ocean. This ice sheet is protected, however, by fringing shelves of ice over a thousand feet thick, which float upon the sea. The largest such shelves are the Ross Ice Shelf as well as the Filchner and Ronne ice shelves. The Ross shelf, for one, is a floating slab of ice the size of Texas. These, plus smaller ice shelves fringing the Antarctic Peninsula, act as an icy barrier or buttress preventing the main ice sheet from surging or collapsing.

Any environmental change which destroys these ice shelves will also destroy the marine ice sheet extending below sea level. The ice shelves are vulnerable, however, and can melt if sea or air temperatures should rise. Indeed, ice shelves are absent along the coasts of the Antarctic Peninsula where summer sea temperatures rise above 29.3 degrees. (Seawater freezes at 28.5 degrees.) They also are absent where summer air temperatures rise above the freezing point, 32 degrees. Along the fronts of the Ross and Filchner-Ronne ice shelves, in this land of the midnight sun, midsummer air temperatures average 23 to 25 degrees. Thus, Manabe and Wetherald's 18-degree predicted temperature rise would be quite sufficient to destroy these ice shelves, not only by melting but by enhancing lines of weakness such as rifts and bottom crevasses.

Today, along the ice shelves' outer edges, immense icebergs the size of islands, flat-topped and steep-sided, continually are breaking off to float northward. This breaking off of icebergs is called calving, and J. H. Mercer has described what would be expected to happen:

> Once this level of comparative warmth had been reached, deglaciation would probably be rapid, perhaps even catastrophically so, because even a small additional rise in temperature would affect a large expanse of gently sloping ice shelf. As Hughes graphically describes it, "a relatively minor climatic fluctuation along the ice shelf calving barrier can unleash dynamic processes independent of climate that cause calving bays to remorselessly carve out the living heart of a marine ice sheet."

With the ice shelves disintegrating, it would be the turn of the main ice sheet. Its deep glaciers would surge forward, collapsing into the surrounding ocean in a matter of decades. Geologists know of just such a surge and collapse at the end of the last Ice Age. At that time much of North America was still covered by the retreating Laurentide Ice Sheet, part of which overlay Hudson Bay. This Hudson Bay dome of ice was vulnerable to destruction just as is the West Antarctic Ice Sheet today and is believed to have collapsed completely in less than 200 years.

It may have collapsed much more quickly, for geologists find it difficult to resolve events which have taken place on shorter time scales. Emiliani has found that when the breakup of the Laurentide Ice Sheet was at its height, a flood of meltwater flowed down and diluted the Gulf of Mexico, where he has studied the fossil record. As he has written, "The concomitant, accelerated rise in sea level, of the order of decimeters per year, must have caused widespread flooding of low-lying areas, many of which were inhabited by man. We submit that this event, in spite of its great antiquity in cultural terms, could be an explanation for the deluge stories common to many Eurasian, Australasian, and American traditions." In other words, the melting Laurentide glaciers sent sea levels rising at rates close to a foot per year and spawned the tale of Noah's Ark.

The melting of the West Antarctic sheet would raise sea levels some sixteen feet and would permanently change the world's map. West Antarctica would no longer show broad ice shelves and an extensive ice sheet the size of a subcontinent; it would be reduced to a scattering of islands. What must concern us, though, is that if the last glacial retreat was impressive enough to lead to legends of a deluge among primitive peoples the world over, what may a similar rise do to the not-so-primitive peoples of the next century?

The greatest hardships, of course, would hit such locales as the Netherlands and New Orleans. Both of these have large portions of land already below sea level and are protected only by extensive dikes and levees. The inhabitants would have no recourse but to build these works higher, and hope. Wendell Rawls of the New York *Times* has reminded us of what we would stand to lose:

> The flags of a half-dozen nations have flown over the city in its nearly three hundred years of life . . . New Orleans was always a more international city than others in the new nation, with a diversity of languages and cultures, pigments and life-styles, that remains unique in America. Ships came up the river with the gold of Ferdinand and the silver of Maximilian, the wool of King George III, the furniture of Louis XVI, the law of Napoleon and the languor of the Caribbean. And the ships took away cotton, indigo and sugar, and fond memories of fast cards and carefree music. Spanish and French and British, white and black, mulatto and octoroon, Caribbean and Creole and Cajun and Confederate and carpetbagger, gambler and harlot, planter and pirate. This was—and is—New Orleans.

Along the Gulf Coast, nearly a third of Louisiana would be submerged. In Texas the cities of Galveston, Corpus Christi, Beaumont, and Port Arthur are located on flat terrain and are already subject to flooding by hurricanes. They would be inundated permanently. Florida would virtually have to be written off. All its southern part, from Key West to Lake Okeechobee, would drown. So would all but four of its cities with 1970 population above 25,000. We would lose Savannah, Georgia; Charleston, South Carolina; four of Virginia's eight largest cities and a sixth of the District of Columbia. Washington National Airport and the Lincoln Memorial would be swamped; the Mall would be flooded, and it would be possible to launch a boat from the White House lawn and row to below the West Front of the Capitol building. The Washington Monument would resemble nothing so much as a lighthouse.

Altogether, the United States would lose some 50,000 square miles of coastal lowlands, with a value in today's money of several trillion dollars. This would include much of New York City and Boston, as well as the Sacramento River flood plain which includes the city of Sacramento. All this is merely the situation for the continental United States; it will be the same around every other continent around the world. One must especially think of the hundreds of millions of people who live around the heavily populated deltas and estuaries of such great rivers as the Nile, the Niger, the Ganges, the Yangtze. In China, in Egypt, in Nigeria, and most certainly in the lowlands of Bangladesh, throughout vast areas all will be swept away.

Can there be protection in seawalls or dikes? These will surely be valuable in limited areas such as New York and Boston. Yet in such Northeast cities, foresight and bold vision are scarcely the order of the day. A common practice is deferred maintenance, which means no maintenance. Their water mains are leaky, their roads potholed, their subways barely surviving from one patch-up job to the next. Such cities are too busy coping with today to think of tomorrow. They will probably begin to think of levees or other works only when the Antarctic ice sheets begin their surge and the water is already rising. By then it may be too late.

The favored lands of Florida would have a different problem. Any seawall built to protect Miami would be a massive affair, stretching ninety miles from West Palm Beach to south of Coral Gables. However, there is no high ground in southern Florida to which such a seawall could be anchored. Instead, it would have to loop around to the west of the Gold Coast, surrounding its cities entirely like a castle wall on the inside of a moat. One may think of the monastery of Mont-St.-Michel on the French coast, which at high tide turns into an island and is connected to the mainland by a causeway; but a Miami causeway would be over a hundred miles long. Nor will Miami be alone; New Orleans and other Gulf cities will be in the same boat. And in any of these places, the first good hurricane to come along might drive storm tides that would breach the seawall and set the whole effort largely at nought.

Still it is an ill wind of climatic change that blows nobody good. At least one powerful nation would benefit most handsomely from these developments, namely, the Soviet Union. The history of Russia has been characterized as a search for a warmwater port, and in this search, both czars and commissars have been less than brilliantly successful. This quirk of history may yet stand the Soviets in good stead. A rise in sea level would not threaten a huge extent of invaluable coastline, dotted with great cities and ports. Instead, the inundations would be quite localized. Leningrad would suffer, as would parts of her Baltic and Black Sea coasts, but the warming trend would benefit mightily the country as a whole. Her agriculture would expand enormously, perhaps even outstripping the power of her ham-handed bureaucracy to hold it back. Moscow would no longer see snow begin to fall before the end of September. Most enticing of all, the Arctic pack ice would diminish greatly. Much of her Siberian coastline might become navigable, with incalculable advantage to the economic development of that vast land. It would be one of history's little ironies if Soviet industry, seeking to aid that nation in gaining new warmwater ports, were to do more toward this goal simply by belching CO_2 into the air than by forging weapons for wars of conquest.

Faced with such scenarios for the future, one is reasonably entitled to ask two questions. How seriously should we take such predictions; what can we do about them?

It is easy to dramatize the effects of a CO_2 buildup, and it is worthwhile asking what is the directly observable evidence that these things indeed will be happening. Perhaps significantly, the answer today is that all we observe is the measured buildup of CO_2 in the atmosphere, as recorded by Keeling's group since 1958. Everything else represents inferences drawn from theory. We have every reason to believe the theories are quite good. For example, the general-circulation models of the atmosphere closely resemble the mathematical models used in weather forecasting and have been tested many times against actual data. Still, an important test of these theories would be the actual detection of a polar climatic warming. At the National Center for Atmospheric Research, Roland A. Madden and V. Ramanathan have searched for such a warming in meteorological records. As they have stated in reporting their results, "The results indicate that the surface warming due to increased carbon dioxide which is predicted by three-dimensional climate models should be detectable now. It is not."

There may be reasons for this. In 1975 Wallace S. Broecker, of Lamont-Doherty Geological Observatory, published a climate prediction:

we are on the brink of a pronounced global warming, but it will not become detectable till about 1985. He obtained this prediction by combining CO_2 warming projections with Dansgaard's climate cycles as recorded in the Greenland ice cap. Thus, between 1940 and 1985 the warming effect of the CO_2 would have been offset by a natural cooling trend. In addition, the 1979 review by the National Academy emphasized an effect that could act as a giant heat sponge, soaking up the warming and delaying cautionary signals until matters are past remedy. This is the ability of the deep ocean waters to act as a heat reservoir, absorbing surface heat through the pumping action of patterns of ocean currents. If the deep ocean is indeed acting as such a heat sponge, it would delay the onset of large-scale warming by a few decades. Once the sponge soaks up its heat, however, there would be nothing to prevent the climate from warming rapidly.

Fortunately, we still have time before Miami slides beneath the waves. If these climatic catastrophes were to take place during a single decade, the results would be disastrous indeed. But in fact they will more likely be spread out over 50 years, possibly much longer, and may not even begin till well into the next century. Certainly we can hope for ample warning. The most dramatic (and destructive) event would be the surge and collapse of the main body of the West Antarctic Ice Sheet itself. But long before this happens, we will have a clear warning in the breakup of the Ross Ice Shelf. Even before that happens, we will be able to detect the polar warming trend in meteorological records. We will have time to prepare, and time to adjust.

Mankind, however, is no great respecter of warnings, and quite probably the nations and their policymakers will go on much as before, paying little heed to even the strongest words of caution. Scientists studying the CO_2 problem often display the attitude that if only they can give an alarm which is sufficiently strong and clear, mankind will take action to avoid such climatic disruptions. Such an attitude takes far too sanguine a view of people overcoming their tendency to ignore problems, to muddle through, and to give little attention to a predicted disaster until it looms up and hits them in the face.

Could we hope, even in principle, to fine-tune the effect? Could we release enough CO_2 to keep global temperatures toasty warm without placing Antarctica (and Miami) at risk? This might mean allowing the CO_2 in the atmosphere to increase up to some level but not beyond; say, to increase by 150 parts per million, or 50 percent over preindustrial values. At the University of Bern in Switzerland, climate specialists U. Siegenthaler and H. Oeschger have calculated what it would take to do this. They would allow world industry, and CO_2 production, to grow by about 50 percent by the year 2000. But after that, world fossil-fuel use must decline sharply, returning to 1970 values perhaps as early as 2030, and dropping to rather less than half of 1970 values well before the year 2200.

The world could do this if the only available fossil fuel was petroleum, but it is absurd to think the industrialized nations will overlook their shale oil, coal, and reserves of heavy oil sands. To suggest this is as much as to say that billions of people will turn away from autos, home heating, and industry, merely for the sake of mathematical calculations published in the *Journal of the Atmospheric Sciences*. There is a tremendous inertia to societies, a deep reluctance to make changes unless they are absolutely forced to; the pleasures of burning fuels next year will always outweigh the pitfalls of sea levels rising some years later. What's more, even long before the seas begin to rise there will be no turning back. Once the CO_2 is in the air, it will not be removed simply by shutting the factories down. Before the seas begin to rise there will be no immediately apparent reason not to use fossil fuels; the predictions of disaster will still be merely predictions. After the seas begin to rise, there will be nothing gained in ceasing to burn fossil fuels; such an act of global self-abnegation will come far too late to make any difference. In the end, rising seas will engulf the still-smoking factories and plants that will have been the cause of it all.

No, the human response to a CO_2 warming will be quite in keeping with our attitudes toward other dangers. We build our cities on flood plains or near fault lines. The flood waters come in time, and the earthquakes; we mourn our dead, rebuild what we have lost, and life goes on. Even in our personal decisions the proven danger of cancer often counts for nothing against the modest social or physical pleasure of the moment when we smoke a cigarette. For some 30 years there has been ample evidence that smoking causes cancer and other diseases, yet per capita consumption among U.S. adults has stayed close to 4,000 cigarettes per year.

Our reserves of fossil fuels will not last longer than a few centuries, and in time we may well be driven to base our societies upon renewable energy sources, including all the various forms of solar and nuclear energy. Even today we appreciate this fact, and there is intensive pursuit of these alternatives. We cannot overlook the chance that they will become available so soon, so abundantly, and at such low cost as to cut short our rising use of fossil fuels. During this century's middle decades, low-cost petroleum drove down the use of coal. Will low-cost solar energy do the same?

This is a pleasing prospect, but not a very realistic one. With all the recent rises in the cost of energy, the 1980 average price for U.S. coal still was $29 a ton. Considering its energy content, this was the equivalent of oil at $7.50 per barrel. Not even the most enthusiastic proponents of solar energy hope to deliver their BTUs at anything remotely like that cost. Further, even to the degree that there are nonfossil alternatives, these are subject to the law of supply and demand. If demand grows in some nations for the renewables, they will tend to increase in price while the cost of fossil fuels will tend to drop. Other nations, less worried perhaps about their coastlines, then will have the opportunity to secure fossil-fuel supplies at lower cost, and the CO_2 rise will continue apace. In any case, while economists concern themselves greatly with energy prices and supplies from year to year, glaciologists will find it difficult to predict the breakup of Antarctica with an accuracy much better than 20 years. And in practical economics, a predicted event 20 years off might as well be a thousand.

Can there be a role for moral suasion? In recent years the United States has taken a leading role in seeking to discourage worldwide use of nuclear power. We have pursued this policy on grounds of nonproliferation and with similar arguments of high moral standing. But in this immoral world, we have not prevented the use of nuclear power from growing. All we have done is see our own leadership in this key technology, which we ourselves pioneered, pass instead to such nations as France, Japan, and the Soviet Union. Suppose that in our councils of government the CO_2 issue becomes predominant, and we launch a crusade against use of fossil fuels, recommending instead such available alternatives as— nuclear power. Will other nations follow us? Or will this be widely seen as one more moralistic spasm of an inconstant government unable to steer a steady course from one administration to the next, let alone from decade to decade?

So in this matter economics will prevail, along with the world's demand for coal and oil to speed its economic growth. There is every reason to believe that in the next few centuries the nations will burn all the fossil fuels they can recover. Siegenthaler and Oeschger have told us what will happen then. The CO_2 in the atmosphere will rise, not merely to twice the preindustrial value, but to ten times or more. The mean world temperature also will rise, not by the 5.3 degrees of Manabe and Wetherald, but by at least 17.5 degrees. As the last dregs of coal or shale vanish into the world's furnaces, the temperatures and the CO_2 will reach this peak. Thereafter they will fall, but slowly, slowly, as CO_2 dissolves in seawater and is cleansed from the air. After several hundred years the oceans and atmosphere will reach a new steady state, which by the year 3000 will see the CO_2 in the atmosphere at a permanent level of about 1,200 parts per million, or four times the preindustrial value. The resulting permanent temperature rise will average some 10 degrees.

Over the succeeding thousands of years, these warmer temperatures would very probably melt the rest of our planet's polar ice. Even a natural cooling, which in the absence of man would bring back the Ice Age, will likely prove insufficient to do more than ameliorate the man-made warming. Any such melting of the ice, of the vast and sweeping glaciers of Greenland and of the rest of Antarctica, would raise sea levels not by a mere sixteen or twenty feet, but by close to two hundred feet. This would send the sea lapping against the foothills of the Appalachians, flooding deep the states of the Gulf Coast. The skyscrapers of Manhattan would be left sticking out of the Atlantic like so many poles in the canals of Venice.

Fortunately, this melting would take place over thousands of years. Because the Greenland and East Antarctic ice sheets rest upon rock above sea level, they would not be subject to the powerful marine gnawing which could destroy West Antarctica in mere decades. The initial sea rise, perhaps during the next century, would indeed be dramatic, forcing many changes within the lifetimes of the people who experience it. The subsequent rises would be at rates of geologic change. Nations and societies could accommodate themselves to the fact that from one generation to the next, for some thousands of years into the future, the sea would be rising by yet another foot.

Twenty-five hundred years ago Aeschylus wrote, "Hear the sum of the whole matter: every art possessed by man comes from Prometheus." Indeed, in his *Prometheus Bound* one finds the exchange,

PROMETHEUS: I caused men to cease foretelling doom.
CHORUS: To cure this malady what did you bring them?
PROMETHEUS: I set and placed within their hearts blind hopes.
CHORUS: That was a great gift that you brought to them.
PROMETHEUS: Besides that, by the way, I gave them fire.
CHORUS: Does man yet wield now that fervent flame?
PROMETHEUS: Yes, and 'twill teach him all technologies.

The name of Prometheus means the Forethinker, and perhaps we can imagine that when he brought us the lesser of his two gifts, he entered into a bargain: "I shall give you fire, and with this gift you shall build civilizations and nations untold. Your arts and comforts, your lives and works, all shall flourish beyond measure. And you will lift the curse of your ancestors, for you shall prevent the return of that Ice Age which was their tyranny. But you in return must suffer the seas to rise, and in time to drown the cities you shall build upon the shore."

That will prove a harsh bargain in the next century or so. But when we have made our adjustments and rebuilt our coastal lowlands, and when the time for the return of the glaciers is upon us, we will appreciate its merit. Perhaps, in the end, we may think of Prince Henry of Navarre,

who in 1593 accepted Catholicism and renounced his Huguenot faith to
become King of France. He said at the time, "Paris is well worth a Mass."
In a similar spirit, we may judge that avoiding another ice age is worth
Miami.

4

NUCLEAR DECISIONS

Any discussion of the prospects for nuclear power must take note of the names of two places, Hiroshima and Three Mile Island. The events those names call to mind, the fears and the issues they raise, certainly will continue to color the nuclear enterprise. So it is appropriate at the outset to deal with these matters.

There is irony in Hiroshima; the tragedy of that city amounts to an accident of history. The first atomic bombs were ready in the early summer of 1945, but not in the early spring. Had this been otherwise, the first city to blaze in nuclear fire would most likely have been Berlin, or perhaps an industrial center like Essen. In view of the thousand-year loathing with which that Reich will be remembered, can anyone believe that those cities, like Hiroshima, would today inspire a depth of emotion which past generations reserved for Jerusalem or Rome? In Berlin today stands the ruin of the Kaiser Wilhelm Memorial Church, its steeple laid open by a British bomb. It is a war memorial fully as gaunt and moving as the Hiroshima cenotaph, with its dome of naked and twisted girders, yet is far less well known. It is not hard to see why. We have largely forgiven the Japanese their wartime trespasses; we have not been so forbearing toward the Germans. The Pacific war, at least for the Allies, was for the most part a naval and colonial war, with many similarities to the Spanish-American conflict of 1898. And naval wars are not the stuff of which lasting enmities are made.

The name of Hiroshima has meaning for the nuclear enterprise, but not only for the reasons usually invoked. It has been said that Hiroshima was unnecessary, that there were other and more humane ways to gain the Japanese surrender. The historian knows the alternative to Hiroshima and Nagasaki was not some beautiful idyll of love and peace. The alternative was Operation Olympic and Operation Coronet. These were the planned invasions of the Japanese home islands, scheduled for November 1945 and February 1946. Our military planners anticipated a million American casualties and two million Japanese; and these would merely have been what in nuclear parlance would be called the "prompt casualties." The ensuing desperate struggles, hand-to-hand and place-to-place, would have cost many more. Hiroshima and Nagasaki mercifully ended the war, sparing the vast majority of those lives. That we used the atom bomb is a thing to regret; we mourn the dead, we remember the fallen of that nation, once an enemy, now a friend. But we did what we had to do. Hiroshima speaks of a world of imperfect choices, where nature offers no guarantee that the best option is invariably a desirable one. It reminds us of hard decisions that at times become unavoidable and cannot be postponed. And in this world, nuclear power, with all its difficulties and risks, still may be the best available choice.

Moreover, on that same August 9 when a plutonium bomb mutilated Nagasaki, the Soviets declared war on Japan. Having won victory in Europe, the Soviets no less than the Americans were prepared to transfer vast forces to the East. Had the war dragged on till 1946, it would have been impossible to deny the Soviets a major role in the final invasions, with incalculable consequences for Japan's postwar political and economic future. In no way could General Douglas MacArthur have been Japan's sole postwar proconsul. In no way then would he have had the opportunity to perform his magnificent act of statesmanship, bringing the laws and values of a liberal Western democracy to a dazed and defeated Japan that in five thousand years had known only emperors and autocracies. Today's strong, prosperous, unified, democratic Japan simply would not exist. At best, Japan today would be a Korea, riven in two wary and mistrustful halves, each heavily armed, each ruled by a latter-day shogunate. At worst, she would have been an Asian Czechoslovakia, ruled for a few years by a coalition government before succumbing to a communist coup. It is not easy to think of the discoverers of nuclear fission as the ultimate authors of Japan's liberty and prosperity. But the world has seen far stranger twists of fate, and that may well be history's ultimate judgment.

The historian remembers more. In July 1941, in response to widening Japanese aggression in China and Indochina, President Roosevelt froze Japanese assets and placed an embargo on sales to Japan of strategic materials, including oil. The British and Dutch followed in kind. At one stroke, Japan was cut off from all supplies of oil and left with reserves sufficient to last only through 1942. Japan responded in December with greatly widened attacks upon Pearl Harbor, the Philippines, and Malaya. The main goal of her aggression, however, was none of these places, but instead was the Dutch East Indies, with its rich oil fields. This was the Pacific war that ended in the shadow of the mushroom cloud. Today, all nations fear the possibility that under another oil embargo or shortage, some country will be driven by desperation to make war for the conquest of oil fields, a war that will escalate into nuclear conflict. Hiroshima reminds us that this nightmare of the 1980s was the reality of the 1940s.

Three Mile Island is different; that Mark Twain-sounding name speaks not to what was, but to what might be. The risks of nuclear power cannot be denied; they can merely be allowed for, designed for, and guarded against with unsleeping vigilance. Only time will assuage the fears these risks elicit; time to increase our experience and familiarity with nuclear power, time to prove whether Three Mile Island will be worse or, alternately, less severe than tomorrow's nuclear accidents. But Three Mile Island must be seen in perspective. To David Howell, Britain's Secretary for Energy, it brought reassurance: "It showed that when some stupid errors were made, and the system was put under great stress, safety was still maintained." To the economist Samuel McCracken, citing such accidents as proof of nuclear peril "is like trying to prove that cars are too dangerous by citing examples of seat belts working." To the columnist George Will,

> After the [accident], one network still referred to the Harrisburg "calamity." What language does that network reserve for events that kill people? . . . The blueprints of industrial society, its dams and bridges and transportation, were drawn in blood. Think of the ships that broke up or blew up before maritime engineering matured . . . Think of railroad and airplane accidents during the infancy of those technologies. Now consider the safety record of commercial reactors, including the fact that defense in depth did not fail at Three Mile Island and will be refined in light of that experience . . . The biggest public-health hazard is the public's behavior when eating, drinking, smoking and not exercising. The second biggest hazard is in your garage, gentle reader. Many of this year's 50,000 traffic deaths and two million disabling injuries would be prevented by prohibiting left turns. All in favor say "aye."

Hiroshima and Three Mile Island, however, do not exhaust the litany of nuclear energy. There is a third place, and with time it may yet become a name most influential of all and most hopeful. That name is Creys-Malville.

The village of Creys-Malville in France, about thirty miles east of

Lyons, is one of those places our century has left largely untouched. Chickens wander through the lanes between stone farmhouses. Cows and other cattle calmly chew their cud, grazing in fields along the Rhone River, which runs fresh and clear. The village is so small (population fifty) one would hardly even go there to find a pink-cheeked country demoiselle. Yet this unlikely place is important. Nearby, in a large fenced-off area along the Rhone, the world's most advanced nuclear power plant is nearing completion. This, Super-Phénix, is being built by the French-led European consortium, Novatome. Here the scenes are very different from Creys-Malville's rustic idyll. A domed concrete cylinder, 275 feet high, dominates the scene like a lone turret of some towering medieval castle. Tall construction cranes, perching on their long slender single legs, hover nearby. Within the tower, a 750-ton steel cauldron is already in place; it will house the reactor core. When Super-Phénix begins operating, an event set for August 1983, it will stand as the first commercial-size fast-breeder reactor.

Super-Phénix is the centerpiece of the French nuclear program, a program extending far beyond Creys-Malville; indeed, extending throughout the entire country. France, the home of Pierre and Marie Curie, has taken the decision to become the world's leader in the use of energy from the atom. Other nations may dither and debate; the French are building. As the 1980s proceed, that nation increasingly will be seen as a showcase for nuclear power.

The French nuclear enterprise begins at the uranium mines. Within that country are easily tapped deposits totaling 100,000 tons. An additional 140,000 tons are in the African nations of Gabon and Niger, where France's nuclear-materials company, Cogema, has a controlling interest. At Tricastin in the Rhone Valley, about midway between Lyons and Marseilles, another Cogema-led European consortium has built Eurodif, one of the world's largest plants for uranium enrichment. It can produce nearly 3,000 tons a year of reactor-grade uranium, enough for a hundred large power plants. The sprawling Eurodif complex itself requires four nuclear plants, each rated at 930 megawatts, to fulfill its electrical needs. This is enough electricity to serve a city of four million. For France, Eurodif is more than a triumph for her ingenuity; it is a guarantee of energy independence. For many years the United States and Soviet Union have monopolized the world's market in enriched uranium. Now France will never again be reliant upon those outside powers.

Then there are the reactors. The national utility Électricité de France has been following a policy of *tout nucléaire*, "all nuclear," a policy of building no more oil- or coal-fired power plants but only nuclear ones. A single large manufacturer, Framatome, builds the plants in what amounts virtually to mass production. Significantly, they are of American

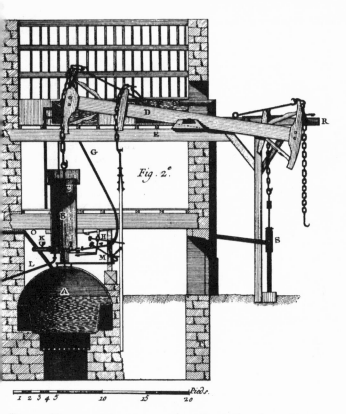

Fig. 1. *Newcomen's steam engine. Engraving from Denis Diderot's* L'Encyclopédie, *1751.* (Courtesy New York Public Library.)

Fig. 2. *William Frith's* "The Railway Station," *London's Paddington Station about 1840.* (Courtesy Royal Holloway College, London.)

Fig. 3. *Warrens of the poor*. (Courtesy National Archives, Washington, D.C.)

Fig. 4. *War for the control of Eurasia: house-to-house fighting in the ruins of Stalingrad, 1942*. (Courtesy National Archives, Washington, D.C.)

Figs. 5, 6. *Europe rises from the ashes. Berlin lies in ruins in May 1945, following 363 Allied bombing raids. In the background is the Kaiser Wilhelm Memorial Church. A year and a half later the streets had been cleaned up and electricity and streetcar service restored, but little else had been done. Today prosperous West Berlin extends in all directions from the Memorial Church, preserved as a war monument.* (Photos courtesy German Information Center, New York.)

Fig. 7. *Dragline in a Montana strip mine.* (Courtesy National Coal Association.)

Fig. 8. *Loading coal in a strip mine. The off-road truck can haul 120 tons.* (Courtesy National Coal Association.)

Fig. 9. *Unit train hauling coal in Wyoming.* (Courtesy National Coal Association.)

Fig. 10. *Bucket-wheel excavator at Canada's Syncrude project in the Athabasca Tar Sands.* (Courtesy American Petroleum Institute.)

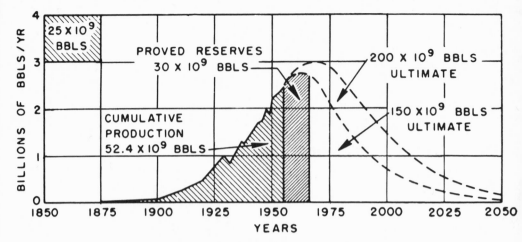

Fig. 11. *Hubbert's 1956 curves, predicting that U.S. domestic production of oil would peak out by 1970.* (Courtesy American Petroleum Institute.)

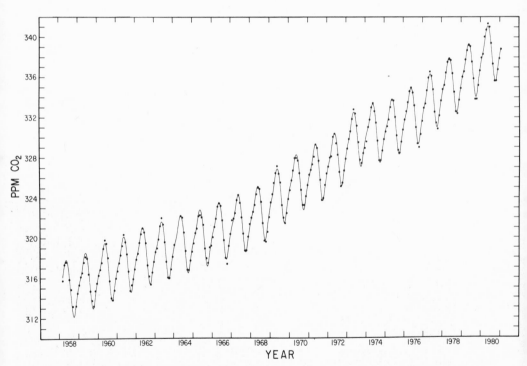

Fig. 12. *Measured changes in the amount of carbon dioxide in the atmosphere, as observed by Keeling and associates at Mauna Loa. Annual oscillations reflect the growth of vegetation.* (Courtesy Charles D. Keeling.)

Figs. 13, 14. *What a difference a degree makes: views of Argentière and its glacier in 1860 and 1966. The retreat of the glacier, from the outskirts of the village nearly to the top of the mountain, was produced by a climate warming of only about one degree.* (Engraving courtesy American Association for the Advancement of Science. Photo courtesy Flammarion Press, Paris.)

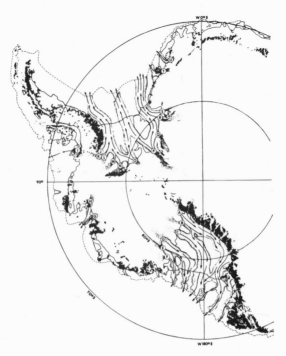

Fig. 15. *West Antarctic ice streams. These streams are immense, slow-moving glaciers which carry ice from the mainland to the ice shelves, from the margins of which it breaks away as icebergs. Should the ice shelves melt, these streams will greatly speed up and drain West Antarctica of ice.* (Courtesy T. Hughes.)

Fig. 16. *Washington, D.C. after the flood.* (Art by Don Dixon.)

design: pressurized-water reactors built under license to Westinghouse. Framatome has for over a decade regarded these Westinghouse reactors as a standard, well-proven design and has been assembling them in extensive nuclear parks, usually in clusters of four. Early models were rated at 900 megawatts, but Électricité de France has switched to ordering 1,300-megawatt units. It will soon be the world's most experienced operator of such reactors, each large enough to serve well over a million people. As of mid-1982, twenty-nine reactors were operating. Every two months a new one comes on stream, which is as fast as Framatome can build them. By 1985 there will be fifty-two such reactors, producing 55 percent of France's electricity and one fifth of all her energy. According to goals set in April 1980 by Minister of Industry André Giraud, in 1990 this will rise to 70 percent of the electricity and 30 percent of the energy. By the same date, oil's share of all French energy is to drop to 30 percent. France will then be the first nation to get more energy from the atom than from oil.

In the United States, spent uranium taken from reactors is regarded as waste, the storage or disposal of which pose difficult problems. In France these same spent fuel rods are regarded as a valuable resource since, although they contain nuclear waste, they also contain plutonium and unreacted uranium, which can be recycled to fuel additional reactors. Accordingly, at La Hague near Cherbourg, Cogema has the world's only commercial-scale reprocessing center for fuel from pressurized-water reactors. The United States has built such centers only to shut them down, but the French plant is designed to reprocess 400 tons a year of spent fuel, and will be expanded to handle 1,600.

The true nuclear waste is in the form of fiercely radioactive fission products from the split uranium products. They include the notorious strontium 90, which in the days of atomic fallout used to turn up in children's milk. In Marcoule near Avignon, about ten miles down the Rhone from Eurodif, is the world's first industrial-scale treatment plant for nuclear waste. The waste arrives in liquid form from the reprocessing plant, ablaze with radioactivity. The first thing done at Marcoule is to let it cool off for five years, while the radioactivity begins to decay. Then the waste is dessicated, dried to a form resembling freeze-dried coffee. Under remote control, this powdery stuff is mixed with borosilicate powder, melted, and poured into steel canisters. It hardens into cylinders of black solid glass resembling obsidian, permanently encasing the radioactive wastes in a form which cannot leach or dissolve in water. Eventually the canisters probably will be buried in a final resting-place in solid granite, but for now they are being stored underneath the floor of this Marcoule plant. By 1980 there had already accumulated some 70 tons of this black glass. Nevertheless, in this vitrified form the waste is rather manageable. A working reactor generates in a year only about as much of this high-level waste as would fit under a dining-room table. The entire French nuclear industry during the balance of this century will produce no more vitrified waste than would fill a five-story building.

Even the best reprocessing, however, will not stretch uranium reserves forever. Standard reactors rely on uranium 235, the fissionable isotope, which exists as less than 1 percent of natural uranium. The available reserves of French uranium would then begin to run low early in the next century. To avoid this, the French for years have been developing the fast-breeder reactor, an effort in which they now lead the world. The breeder works with uranium 238, which makes up the other 99 percent but is nonfissionable, and converts it into fissionable plutonium 239. In fact, it can produce more fuel than it consumes. In this manner, all uranium becomes an energy resource, not merely the rare uranium 235, and the breeder stretches uranium's available energy fiftyfold. The breeder thus is the key to a long-range nuclear program such as that of France, which is building large and building to last. As former president Valéry Giscard d'Estaing has said, "If uranium from French soil is used in fast-breeder reactors, we in France will have potential energy reserves comparable to those of Saudi Arabia."

The program has impressive domestic support. In its present form it was initiated by President Giscard, whose Minister of Industry, André Giraud, had previously headed France's powerful nuclear energy agency. In the Chamber of Deputies, all four main political parties support the program; indeed, they have not even debated the program since 1975. During the 1981 French election campaign, President Giscard attacked his rival, François Mitterand of the Socialists, as one under whom France would say "goodbye to nuclear independence." But on being elected president, Mitterand's administration reviewed the program, elected to defer three of the planned reactors, and in other respects opted to go along with the policy of Giscard. Some Socialist deputies have opposed building reactors, but this party as a whole, now the largest in the Chamber, is pronuclear. The Gaullist, Giscardist, and Communist parties have endorsed the program without reservation. With this support, the program has not been swayed by outside pressures. Indeed, a week after the accident at Three Mile Island, Giscard boldly announced he was stepping up the construction of nuclear plants. Although his Giscardists lacked a majority in the Chamber, he met no opposition.

The legal system in France has offered little cheer to that country's antinuclear activists. In France, as elsewhere, activists have broad legal standing to challenge siting or construction decisions in court. Nuclear plants require a special operating permit, in addition to a variety of other permits required by all industrial plants, all of which may be challenged in court. In addition, Électricité de France must file an environmental impact statement to receive a valid construction permit. Moreover, the regulatory authorities will not allow the introduction of new technology, such as the fast-breeders, with a lower standard of safety than existing technology. However, all such regulatory decisions are in the hands of the nuclear energy agency. There is no independent agency for nuclear safety such as our own Nuclear Regulatory Agency. There is no law resembling our National Environmental Protection Act, empowering the courts to review the adequacy of an impact statement. Nor is there a law explicitly defining safety standards, such as Germany's Atom Law which mandates that safety must take precedence over economic issues and that safety precautions must be taken to the limit of science and technology. Instead, France has regulations and administrative decrees imposing the less stringent standard of "adequate precautions." A court challenge to a permit or license may take two years and until the courts render their decision, the contractors are free to continue work. Indeed, the French courts have not touched the issues of safety or environmental impact, but have acted only on questions regarding the procedures followed in granting a permit. Even if the courts do revoke a permit, Électricité de France need only apply for a new one. In a celebrated 1978 case, environmentalists demanded suspension of a permit on grounds that that utility had filed no impact statement. The court agreed. However, the utility calmly prepared a second application, attached the required statement, and proceeded with the work. The environmentalists had no way to bring a further challenge on the grounds that Électricité de France had been acting with an unseemly haste likely to compromise safety.

The French experience thus points up the argument raised by some nuclear critics that nuclear power must be closely associated with large government organizations and central authorities, whose actions stand remote from the individual citizens they affect. Such an argument is suspect, of course; many of the same activists who quail at large government bureaucracies for energy are enthusiastic advocates of even larger government bureaucracies to levy taxes, redistribute income, and force social change. Nevertheless, the question of energy supply is so important that many nations will heed the counsel of the poet Alexander Pope: "For forms of government let fools contest, Whate'er is best administered is best."

Moreover, France is far from the only country where the atom is admired. The British also have a long and distinguished nuclear tradition, with much of their effort centered at Dounreay, Scotland, on the windswept and rocky coast of the North Sea. On a clear day in Dounreay you can see the Orkney Islands. There the British have built another successful 250-megawatt fast-breeder power station, the Prototype Fast Reactor. It was completed in 1974, shortly after Phénix, but experienced a lengthy series of equipment failures, which at least had the virtue of making Britons the world's experts on steam generator leaks in breeders. Today the British are proceeding with work on a Commercial Fast Reactor the size of Super-Phénix. Under Prime Minister James Callaghan they had plants operating or under construction, sufficient to produce 20 percent of their electricity by the early 1980s. The Tory government of Margaret Thatcher has stepped up this program with plans for twenty more such plants, and the atom may produce half that nation's electricity by 2001.

The Japanese, even more dependent than the French on imported oil, share with France a steely determination to do something about it. Japan has had terrifying experiences with nuclear explosions; they also have endured enormously destructive earthquakes. They know that no part of their land is safe from future earthquakes. Still they have been building power reactors. Some thirty such plants are operating or nearing completion, and as in France, the plants they are building are licensed from American designs. These plants have been unfazed by quakes measuring up to 7.5 on the Richter scale. In November 1980 the Cabinet voted to step up this program to the level of an all-out effort. They now plan to double their previous nuclear commitments to reach a level of 52,000 megawatts of capacity by 1990, nearly as much as the entire United States has operating today. This will then provide a quarter of their electricity.

Japan's nuclear program includes more than reactors. In October 1980 an enrichment plant went into operation, employing centrifuges for uranium separation, in a process more advanced than that of Eurodif and requiring less electricity. It now has four thousand centrifuges whirling. The Tokai Reprocessing Plant is also open for business, and a second reprocessing plant is under way. As for breeder reactors, in 1978 Japanese engineers started up their experimental "Joyo" plant, and they are preparing to build the much larger "Monju." What's more, they are moving ahead on a program to dispose of nuclear wastes at the bottom of the Pacific Ocean. Obviously Japan has come a long way since November 1945, when American occupation authorities seized Japan's five cyclotrons, hacked them to pieces, and dumped the remains in Tokyo Bay.

Japan, like other nations, has had to contend with antinuclear opponents, but these have mainly been active on the local level when they learn of plans to build a plant nearby. Typically, a utility company searches for a site in total secrecy and announces its selection only after having purchased as much of the land as possible. Then, confronted with lawsuits, the utility fragments the opposition by offering large out-of-court settlements, "cooperation money," which is all some opponents had wanted in the first place.

No such opposition exists in the Soviet Union. That country has been described as "a nuclear engineer's paradise," but engineers like to build things that work, and Soviet reactors have not always done so. Their approach has been to build the reactor first, then see if it can be made to work. Thus, at Shevchenko near the Caspian Sea, the Soviets in 1972 completed the world's first full-size fast-breeder reactor, the 150-megawatt BN-350. However, they apparently did no testing of the steam generators, and three of the six installed soon failed dramatically, releasing clouds of gas that were reportedly spotted by a U.S. spy satellite. Three years later they were still working on the problem. They persisted, however, and applied what they were learning to a larger breeder, the 600-megawatt BN-600 at Beloyarsk, near Sverdlovsk in the Ural Mountains. It went into operation in March 1980; until the completion of Super-Phénix, it will be the world's largest breeder. They also have begun building a breeder of 1,600 megawatts, larger than any planned in France.

The Soviets have been pressing forward vigorously. With a policy that is virtually *tout nucléaire*, in the regions west of the Urals where four fifths of the people live, some three fourths of the new generating capacity being installed is nuclear. In Volgodonsk, the Atommash reactor-manufacturing plant is being expanded to turn out 1,000-megawatt reactors at the rate of eight a year. By 1990 they hope to have over 80,000 megawatts of installed nuclear power, a quarter of their electrical capacity. They also are bidding strongly for the export market. By 1990 their domestic and export capacity may total as much as 180 reactors and 120,000 megawatts, about the same as there is in the whole world today. In all this they have sanction from on high, for as Lenin said, "Communism is socialism plus electrification."

Clearly, the Soviets have confidence in their reactors. Not only are they building big nuclear parks with as many as six reactors at one site; they also are using them for cogeneration. They are building reactors in cities, not out in the country. In Gorky and Voronezh, reactors are being built to pipe steam to provide district heating for a quarter-million people each.

And there is more. One may look to Sweden, where in 1976 Prime Minister Olof Palme and his Social Democrats were toppled from office in an election turning largely on the nuclear question. In March 1980 the public voted in a referendum to go ahead with nuclear power, and Sweden will more than double its nuclear capacity by 1988. Or one may look at Canada. That nation has accomplished the remarkable feat of developing its own successful nuclear technology, a reactor system with the upbeat name of CANDU (Canadian Deuterium Uranium). It runs on natural or unenriched uranium and heavy water. Its Pickering Station, on the shore of Lake Ontario just outside Toronto, will comprise eight reactors with a total power of 4,000 megawatts. Even larger is the nuclear park at Bruce,

Ontario, on Lake Huron north of Detroit. Its eight reactors will deliver some 6,000 megawatts, enough to run New York City. Also at Bruce is a large plant extracting heavy water from the adjacent lake.

As of early 1982 the world counted twenty-three nations operating power reactors, 76 being in the United States and 199 in other countries, with 151 more under construction. Their continued safe operation, not always free from accidents but invariably safe from radioactive releases that would endanger the public, is testimony to the care and caution of their builders and operators. This worldwide industry has often been attacked as unacceptably risky, but it needs no defense. It has written its record and continues to write it strong and clear in the day-by-day uneventful work of these hundreds of reactors, with more being added every year.

What of the future? If this record continues, current fears of the atom can only diminish and fade. Fear is born of unfamiliarity, which will increasingly lessen. With growing familiarity, there will be much less credence given to scare stories of nuclear plants exploding and releasing radioactivity that kills twenty thousand people. The likelihood will be seen as on a par with the chance that a fully loaded Boeing 747 will crash with flaming fuel tanks into a stadium during a football game. That too is a possible event which would kill twenty thousand, but its possibility has not hindered the growth of commercial aviation. In coming to accept a new technology, familiarity and demonstrated performance count for everything.

There remains the question of the risk of war from the peaceful atom. This gained new urgency in 1974, when India exploded a bomb made with plutonium from a Canadian-supplied reactor. Plutonium, of course, is first-rate bomb material, but it has to be of the right isotopic composition. The Nagasaki bomb, for one, contained six kilograms of pure plutonium 239. It is easy to "spike" reactor plutonium, adding isotopes to greatly increase its radioactivity while reducing and making erratic its explosive power. Such "hot" plutonium would be an inferior bomb material, difficult for a power-mad dictator to work with, and would be lethal to thieves or terrorists. There already is a force of worldwide nuclear inspectors under the Vienna-based International Atomic Energy Agency. Sale of reactors, and particularly of breeders, may be limited to those nations which accept its inspections and, in addition, which implement the safeguards of the Nuclear Nonproliferation Treaty. Reprocessing plants, which produce plutonium, could be located only in reliable nations, reprocessing fuel for their customers. The Soviets already do this, requiring that spent fuel from exported reactors be returned to their country for reprocessing.

In the end, no safeguards can stop a country with the means and the determination to develop nuclear weapons. But safeguards greatly increase the difficulty, decrease the chances of secrecy, and prevent the thing from being done easily and with little effort. It is worth remembering that to the would-be bomb owner, far more attractive than any commercial nuclear materials are the world's stores of existing weapons. Yet for over three decades military forces have successfully guarded hundreds of tons of bomb-grade material, much of it in the form of bombs. In an era of breeders and reprocessing, that experience will be most helpful.

What then, finally, of the nuclear enterprise in the United States? Having ranged the world, it is appropriate to come home and examine matters closer at hand. On the surface, with its 76 reactors licensed to operate and some 67 on order or under construction, the United States would appear to have a program fully as vigorous as that of the French. In fact, the U.S. effort is dead in the water. The lifeblood of an industry is its new orders, and so far from calling for additional reactors, between 1979 and 1982 utility companies canceled 37 reactors already ordered, while indefinitely postponing further construction on 15 partly completed ones. The last surge of orders was in 1973, when 41 new plants were ordered, and in 1974, with 26. Then new orders fell precipitously: 4 in 1975, 3 in 1976. The last new orders were in 1978, when Chicago's Commonwealth Edison, with perhaps the nation's strongest commitment to nuclear plants, ordered 2 more. Even then, however, they gave themselves an out. Their contracts with Westinghouse specified they could withdraw on payment of a negligible cancellation fee. As the nuclear critic Amory Lovins said even before Three Mile Island, the U.S. nuclear industry was like a dying brontosaurus; its spine was broken but its head had not yet heard the news.

How could an enterprise begun so prosperously turn out so miserably? It is easy to blame the antinuclear movement, and to a degree they deserve it. While claiming to represent the interests of the public, the rancor of their well-publicized attacks has often set at nought an important public interest: knowledgeable, informed, serious debate of complex and controversial issues. They have recklessly sowed fear and mistrust, and their specific charges have often been grossly exaggerated; one thinks of a group of ladies agitatedly discussing the cancer-causing perils of nuclear plants while they puff away on their Virginia Slims.* Nor have antinuclear activists been so pure of heart as they might wish. As noted in

---

* Uranium and thorium are found in coal. Ten large coal-fired power plants, operating normally for a year in large cities, will produce as much (or as little) radiation hazard to the public as did the entire Three Mile Island "catastrophe." As George Will has written, "Nuclear plants, like anything else, can malfunction and become hazardous. Coal plants are hazardous when functioning normally."

Britain's 1976 "Report of the Royal Commission on Environmental Pollution,"

> Nuclear power provides a dramatic focus for opposition in some countries to technological development and we have no doubt that some who attack it are primarily motivated by antipathy to the basic nature of industrial society, and see in nuclear power an opportunity to attack that society where it seems likely to be most vulnerable, in energy supply.

Still, to give them their due, while nuclear critics have exacerbated that industry's problems, they have hardly caused them. These problems run deeper than that. The most significant have been falling demand, rising costs, lengthening delays, and national indecision on the important matter of nuclear waste.

No utility will build a power plant if there is insufficient demand for its electricity. Between 1948 and 1973, the use of electricity grew at 8 percent a year, increasing nearly sevenfold in that quarter-century. In a sense, in 1973 there were seven light bulbs where in 1948 there had been only one. Those were the years when the nuclear industry grew and prospered. Since then the rising cost of electricity has slowed its annual growth to 3.5 percent. At the same time past construction commitments, often left over from those years of rapid growth, have given the nation an average excess generating capacity of 35 percent, compared to the electricity we actually use. In such a situation, few new power plants of any type will be ordered.

Then there are the rising costs. The experience of one Connecticut utility, Northeast Utilities, will tell the tale. Their first nuclear plant, Millstone Point 1, went on-line in December 1970 and cost $198 per kilowatt to build. That cost was typical for both coal and nuclear plants completed about 1970, though to be sure, Millstone 1 was built under a fixed-price contract and its builder, General Electric, lost money on the deal. Millstone 2 was completed in 1975. Its cost was more than twice as high, $523 per kilowatt. The third unit in 1979 had an even higher estimated cost, $1,050 per kilowatt, but it is still under construction and will not be in service at least till 1986. By then, of course, it will cost even more. Its name is appropriate; such plants surely are millstones around the necks of their owners. Moreover, plants begun in 1979 would have cost $1,600 per kilowatt, some being much higher. In April 1979 the New York State Power Authority canceled a proposed plant and cut its losses after the price tag soared from $1,500 to $2,600 a kilowatt.

What has driven up the costs? Inflation has helped, but four fifths of the increase has been due to continuing changes in design requirements, as regulatory agencies have demanded stricter and stricter safety standards. One firm has counted up a total of 145 directives, many of them as thick as telephone books, issued by the Nuclear Regulatory Commission since 1968. If the top of this stack of regulations were to stay put, so would the nuclear costs; engineers are very clever at meeting even very lengthy and complicated requirements, so long as they remain known. But with new regulations issuing almost monthly, a plant under construction must be continually rebuilt with new equipment to meet the changing rules. That costs money, lots of it. Even after completion a plant is far from immune to demands for further costly changes. One of the early successful nuclear plants was Indian Point 1 on the Hudson River, owned by Con Edison. Today it stands idle, its owners unable to bear the costs of new systems required by changing regulations. If you wonder who ultimately winds up paying those costs, go take a look in the mirror.

Then there are the delays. From the handshake sealing an order to the ribbon-cutting which opens a completed plant, the time required is at least six years for a coal plant, more often eight to ten. For a nuclear plant it is at least ten years, sometimes fourteen or fifteen. The reason is not the slow pace of construction, but paperwork and legal battles; the environmental-impact statement alone requires at least two years. Power plants are built with borrowed money and do not begin to earn back this investment till their generators begin turning, and during these long years, the meter is ticking every day on interest charges and construction costs. By imposing these delays the antinuclear people do their greatest damage. Time and again during those years, utilities and their contractors must secure appropriate licenses, permits, and approvals, often from several agencies. Each document requires a public hearing, and few are granted without a challenge in court. That is why plants remain unbuilt as the regulatory process goes on day after day, while attorneys for the plaintiffs drive Mercedes and eat steak, night after night.

But the biggest problem for the nuclear industry is not in the law courts. It is in the storage and disposal of nuclear waste, particularly spent fuel rods. The French, with far fewer reactors and less operating experience, have their reprocessing plant at La Hague and the waste-vitrification plant at Marcoule. We have neither. Instead we have "temporary" storage tanks, some of which are over thirty years old. Every nuclear plant has a large open tank closely resembling a swimming pool, where the spent fuel rods are stored. The water glows a bright, vivid blue recalling the famous Blue Grotto of the Isle of Capri; but this glow is due to gamma rays from the radioactive waste. The problem is that these pools have never been intended as more than a temporary expedient. Many of them are now rapidly filling up.

There is no dearth of carefully studied proposals for permanent waste disposal. As in France, a widely preferred approach is to solidify the waste and vitrify it, then encase the resulting glass blocks in steel canisters. Or the waste could be sealed in blocks of ceramic, which some experts prefer to glass. It could then be permanently buried in a deep, stable geologic formation. As early as 1957 the National Academy of Sciences recommended burial in underground salt beds, and this has remained the most popular choice among the experts. Other proposed formations include basalt, granite, shale, and beds of red clay at the bottom of the deep blue sea. The problem is that nothing has been done. None of these alternatives has even been tested, let alone adopted for routine use; not even the salt beds, a quarter-century after the National Academy's report.

While the waste continues to pile up, the natives are getting restless. In 1976 California adopted a law prohibiting future construction of nuclear plants until the state "finds that there has been developed, and that the United States through its authorized agency has approved, and there exists a demonstrated technology or means for the disposal of high-level nuclear waste." Since then a number of other states have passed similar laws. The science-fiction writer Jerry Pournelle has said that to get rid of the waste, "bury it out by yonder hill!" But states and local governments have repeatedly told would-be waste disposers to dig somewhere else, and the question of waste disposal has become as politically controversial as any other aspect of nuclear power. At least nine states have already prohibited the burial of wastes within their borders. Seven of them have even barred surveyors or geologists from prospecting for disposal sites.

Again, it is worth noting that in calling for waste disposal, no one is proposing a nuclear Love Canal. There is no suggestion that highly radioactive liquids be transported in easily corroded steel drums, to be dumped casually on the open ground near a water supply, or buried under a thin landfill. Instead, all proposals call for concentrating the waste into a solid form, a glass or ceramic, which in turn would be encased as large impermeable blocks in canisters of thick stainless steel. Burial would be thousands of feet below the surface, in unfractured rock formations which geologists find to have been undisturbed for millions of years, and which are far from aquifers or other water supplies. It should hardly be impossible to go ahead and do these things, yet nothing has been done. Meanwhile, the true threat of a nuclear Love Canal lies in those temporary storage tanks, some of which date to just after World War II. Some of them have already leaked.

What will it take to restore the nuclear enterprise? The most important requirement is the widespread public belief that nuclear power is necessary and is something on which the nation must move forward. If a widely perceived necessity should arise, then as in France or Japan, we would rise to the challenge. We would go ahead and make the necessary decisions. We would cease our endlessly prolonged debates. We would stop demanding that all nuclear technologies conform to some never-never-land of perfect freedom from risk, a standard from which any shortcoming is treated as an open road to mass murder. Instead, we would learn from the French. Their attitude is that they will build and operate the best systems they can, but they expect to see mistakes, they expect to learn from them and to make improvements, and they anticipate that these mistakes will not be lethal. We will regard nuclear plants as we do airliners, for in aviation the search for safety is unstinting, but the rare crash of a commercial jet does not lead to demands to shut down the airlines.

We lack the urgency of a France or Japan. Those nations in 1980 imported respectively 73 percent and 90 percent of their energy supplies. For the United States the comparable figure was 17 percent. If the predominant public sentiment is that nuclear power is something we can get along without, then the future may be marked by half-completed power plants from which the owners will have simply walked away, and by operating plants which will have been shut down and mothballed for want of means to deal with their waste.

Between these extremes of general rejection or embrace, however, there are a few developments we may hope to see. As anyone who remembers Senator Joe McCarthy will attest, fashions in hysteria do change. Short of a national nuclear commitment, we must at least hope for an exhaustion and considerable lessening of the excitement over nuclear energy. The several dozen plants under construction should be completed and licensed for operation. The reprocessing plant at Barnwell, South Carolina should be opened, reversing a decision taken early in the Carter Administration. Similarly, the Department of Energy should move off dead center and initiate a program for waste treatment and burial. The Clinch River breeder reactor will be built. Looking to the future, the Nuclear Regulatory Commission should standardize and codify its regulations. Any utility ordering or building a nuclear plant will then have a clear and fixed design to work with, not one which must continually change to hit a moving regulatory target.

This will not bring a nuclear renaissance, but it will bring something that at least is possible: the consolidation of nuclear power. It will preserve the gains we have made, tie up the loose ends of waste disposal and uncompleted plants, stretch our uranium supplies, and hold open the option of a major nuclear commitment in the future, if we should want it. By holding open this option, we would avoid the possibility that the nuclear enterprise will pass completely into the hands of other countries, leaving us able to do little more than to mine and export uranium. Instead, we would stay able to compete on an equal footing in the growing overseas market. Nor would we stand hopelessly outclassed in the realm of breeder reactors. Most importantly, nuclear power here as in so many other advanced countries would be supplying, by 1990, close to 25 percent of the electricity. Even for a nation so lavishly endowed as ours is, that would be valuable security against the day when the oil is gone.

Beyond these practical concerns, nuclear power illustrates some significant features of our national life. It is often said that we live in an age of big organizations, that the individual can count for little, that bureaucratic and technological behemoths will continually roll over the fragile humans who stand in their path. The experience of nuclear power argues the contrary: the big corporations may rage in vain against the determined opposition of even a small group of people who stand and will not be moved. For all their raucousness, all their obstructionist damage to the nation's energy programs, these obdurate oppositionists have done the nation a service. At a time when the nation needed the reminder, they showed once again that you can fight City Hall. They gave new meaning to the vision of the lone individual who stands his ground against powerful forces and says, "No." In this, America has again been shown that when an issue becomes controversial, the controversy must run its course until it either dies away or is resolved in a new consensus. There is no way to force an issue, to cut short a national debate, to reach a decision until the people affected have made up their minds.

Nuclear energy shows how people can change their minds with experience. The odyssey of Tom Hayden, an influential leader of the political left, will illustrate. In 1962, while he was at the University of Michigan and nuclear power was in its infancy, he founded Students for a Democratic Society. Its "Port Huron Statement" supported nuclear power and called for its development, as a means to deflect the atom into peaceful applications. Fifteen years later Hayden had become an effective antinuclear leader in California and had married the actress Jane Fonda, famous for her movie *The China Syndrome*. In this there is a clear lesson for advocates of solar energy. While today it is hailed and praised, come 1995 there may be protestors picketing and demonstrating against the big windmills or power towers that now attract so much favorable comment. A new technology may be welcomed while it is in a stage amounting to laboratory experiment, but there is simply no way to tell how it will fare till it has had the chance to age a bit. Nuclear energy says to solar that until the first few thousand megawatts of solar energy are in routine use, we will have no way to say what it will contribute to the nation, or whether it will justify its high hopes. After all, in 1954 Lewis Strauss, chairman of the Atomic Energy Commission, declared that atomic energy would be too cheap to be worth metering. People would pay a low monthly charge and use as much as they wished.

The nuclear enterprise offers a perspective on another popular cliché, that rampant technology is proliferating so rapidly as to outrun man's capacity to tame it. In fact, one of the most difficult things in the world is to come up with a technology which is really new, as opposed to being a reasonably straightforward development of what is known. Such a new technology often will need billions of dollars and years of time before it reaches the stage where it can begin to do some of the world's work, play some role in the nation's life. At that point, even as its use begins to grow, it will be held up to a searching scrutiny to seek out and lay bare its faults, subjecting them perhaps to sensationalism and exaggeration. Because it is new, it will receive a blare of attention probably far out of proportion to its actual status. Only then, after the attention has died down and the excitement has diminished, will its growth proceed apace, allowing it to rise to gain its place among the nation's activities.

But if there is no technical juggernaut rolling relentlessly onward, neither is there a technical imperative driving us inexorably toward nuclear power, or toward any other particular technology. If small is not always beautiful, big is not necessarily better. In the end, nuclear power comes down to a matter of switching on a light or appliance and receiving a monthly electric bill. In such everyday terms, not in the sweeping rhetoric of juggernauts or imperatives, the atom must find its eventual role. The rise of any new technology results from a creative tension, a continuing challenge and response, as inventive and able people compete for markets or seek to solve problems, or deal with customers and critics. In all this, success is never guaranteed; all that is guaranteed is the chance to try. It is this endless testing and sifting, this winnowing of possibilities, that produces what we see as technical progress.

Can we look ahead, then, to something more than mere consolidation of the nuclear enterprise? The answer surely is that we can. In recent years there has been much talk of the "risk-free society." Such a society never existed and never will, for the world is chancy and uncertain, and this is not likely to change. Foolhardy recklessness or incaution will always carry their penalties, and we will continue to see much effort at removing needless hazards. But there is no way to swath the nation in cotton wool, to assert that nature made the world benevolent and kind, and if it is otherwise it is someone else's fault. The world will never belong to those who temporize, who hold back from commitment or decision, who endlessly agonize for fear of making a mistake, who show evident unease with the times. The world will always belong to the bold, to those who take decisions and act upon them, who seize opportunities willingly and with full knowledge of the significance of their actions, who weigh risks with prudence but will not delay a choice longer than is necessary. In such a world, nuclear power indeed will have a bright future.

In this spirit of boldly pursuing our opportunities, we can already look forward to the advent of fission's successor. This is fusion, the energy source of the sun. The term "fusion" covers any process in which light nuclei fuse into heavier ones, or at least into different ones, with release of energy. The most interesting of these reactions employ deuterium and tritium, D and T, the heavy isotopes of hydrogen. Deuterium, D, is present in seawater at a proportion of about an ounce per ton. For many years it has been extracted in a straightforward way, in quantities measured in tons, for use in such nuclear reactors as Canada's CANDU. In principle it is possible to build a fusion reactor that uses only D. If all U.S. electricity were generated by such reactors, the needed deuterium would amount to about twenty pounds per hour. (The corresponding figure for coal is about 200,000 tons per hour.) This deuterium could be had from a small separation plant, whose intake would simply be a two-inch pipe feeding in ordinary water. However, the all-D reactions are difficult to ignite. The reaction employing D + T, deuterium plus tritium, is also difficult but is notably less so, and has received the most attention. Tritium does not exist in nature, but it can be produced from lithium in a nuclear reactor. Fortunately, there is plenty of lithium, particularly in seawater and in certain heavy brines.

With such an extraordinary treasure available, no farther away than the beach, it has certainly been worthwhile to put forth a large effort to tap it. However, controlled fusion has been the most complex and intractable problem yet faced in physics. To make headway, the fusion fuels must be handled as a plasma, an ionized gas at temperatures of some 100 million degrees. Then, right at the outset, the plasma must be made to stay put, by means of confining magnetic fields. The problem is that the plasma resists confinement. When penned up by such fields, it tries with all the tricks at its disposal—and it has quite a few—to writhe, wiggle, leak, or push its way out. Thus, while serious fusion research has been proceeding since 1950, real progress has been gained only in recent years. For many years the problems of confining and heating the plasma appeared so severe that there was no guarantee fusion could ever be made to work. But the 1970s saw rapid progress, and a key experiment in 1978 removed the last doubts that fusion could be attained.

For thirty years, Princeton University has been a leading fusion research center, and during the 1970s it had the largest and most successful experimental system, the Princeton Large Torus. Its construction was completed in 1976, and it was working well a year later. In the fall of 1977 the Princeton managers brought to it neutral-beam injectors, large boxlike devices heating the plasma by shooting powerful beams of atoms into it. Injecting over 1 megawatt of beam power, these helped the PLT set a new plasma temperature record in December of 46 million degrees Fahrenheit. However, this was not good enough for the plasma contained impurities that held the temperature back. Then the PLT managers arranged for some key parts of the PLT, made of tungsten, to be replaced with new parts made of graphite. Early in 1978 they installed the water-cooled graphite parts, and added two more neutral-beam injectors for good measure. In July they turned on the power, shooting 2 megawatts of beam power into the heart of the plasma.

The results were very exciting. The plasma lost its reluctance and responded, rising to 115 million degrees. This meant the new graphite parts had successfully suppressed the formation of harmful impurities. It was a temperature much hotter than they had hoped for and had a special appeal; for the first time they had exceeded the temperature needed to ignite a fusion reaction. But what was best of all, the plasma remained stable. There was no evidence for a particularly worrisome class of instabilities predicted by theorists, even at the hottest temperatures. Then, just to show these results were no fluke, the Princeton people boosted the beam power to 2.5 megawatts. The temperature reached 135 million degrees and again the plasma stayed stable.

What was more, the reaction could have produced fusion energy. For simplicity's sake the tests were run using deuterium only, but had they used D + T they would have produced 50 kilowatts of power—a good start on the road to break-even, a condition wherein a fusion machine puts out as much power as is needed to run it. The news was dramatic enough to hit the newspapers. The Washington *Post* ran a banner headline, "U.S. Makes Major Advance in Nuclear Fusion." The Knight-Ridder wire service, in a story carried by fifty to a hundred newspapers, wrote that "Scientists at Princeton University have produced a controlled thermonuclear fusion reaction that experts are hailing as a major technical breakthrough."

The PLT was of a class of fusion machines known as tokamaks, the name being a Russian acronym for "toroidal magnetic chamber." The tokamak was invented in the Soviet Union and since 1970 has represented the leading approach in fusion work. Today the successor to the PLT is beginning operation, also at Princeton. This is the Tokamak Fusion Test Reactor. Essentially a scaled-up PLT, and twice as large, it comes equipped with up to 24 megawatts of neutral-beam power. Its mission is to reach break-even with a plasma of D + T. For the first time, therefore, it will have equipment for handling radioactive tritium and will produce plasmas that emit floods of neutrons and helium nuclei from their fusion reactions, just as in a real reactor. As an experimental reactor, the TFTR will produce up to 20 megawatts of fusion power, in short bursts.

The TFTR today exists as a complex of buildings and facilities, sprawling across Princeton's Forrestal Campus. Seeing all this, one is struck by the feeling: they are really doing it. This is not just another physics lab; this is a new industry inchoate. The main reactor building has walls of concrete, several feet thick, to keep in the neutrons. Within those walls are structural beams the size of bridge girders, made entirely of stainless steel. Magnet coils form huge rings, twenty-five feet across. The water mains, carrying cooling flows to these magnets, would be suitable for a city waterworks.

Other large tokamaks are also nearing completion: Japan's JT-60, Moscow's T-15, and the Joint European Torus. This JET is even larger than TFTR and will go beyond TFTR to produce a more intensely burning or reacting plasma. It also will produce more fusion energy and approach more closely to conditions in an operating reactor. The JT-60 will not burn tritium, but will simulate break-even conditions in a nonfusing plasma, and will have divertors, special magnet coils to remove impurities from the plasma. The T-15 also will not burn its plasma but will test an equally important development, large superconducting magnets. Within such magnets, an electric current once started will flow indefinitely and will need no continuing current supply. They also will need no cooling water.

Beyond this will be the first true tokamak reactors. After the TFTR our next big project will be either the Engineering Test Reactor, described by the Department of Energy as "the first fusion power-generating device," or the Fusion Energy Device, defined by act of Congress as "a magnetic fusion facility which achieves at least a burning plasma and serves to test components for engineering purposes." To support this work, there will be a Center for Fusion Engineering. Also there will be an international project, known as INTOR, International Tokamak Reactor. These reactors will sustain a burning plasma for many minutes at a time, injecting frozen pellets of fresh D + T for refueling and using divertors or other techniques to remove the helium "ash" from the spent fuel. In addition they will qualify as the first ignition tokamaks, sustaining an ignited plasma. Such a plasma will keep itself hot by its own reactions, requiring no neutral-beam injection as a continuing energy source; the neutral beams will flash only momentarily at the beginning. These reactors will breed tritium and will generate some 100 megawatts of electricity. Among the nations planning such tokamaks, the Japanese plans are most advanced. Their Fusion Experimental Reactor, to be operating by about 1995, will be the last step before proceeding to a commercial prototype.

Sometime in the 1990s, the U.S. program will reach a similar milestone: "the assessment of fusion's full potential in the light of a broad scientific and technological base following from our basic strategy of parallel physics and technical development." Following this assessment, we too will move forward on the demonstration plant, "a prototype energy system which is of sufficient size to provide safety, environmental reliability, availability, and ready engineering extrapolation of all components to commercial size." This prototype plant will be fusion's counterpart of Super-Phénix. Like Super-Phénix, this prototype fusion plant will have to produce continuous, reliable power without a small army of Ph.Ds in continuous attendance. Once it passes this test, there will then exist a fusion industry, building and operating plants and reactors.

What will fusion be like, when it finally arrives? The European energy analyst Wolf Häfele has written that "everything that the fusion reactor is supposed to be able to do someday, the breeder can already do today." But in fact fusion will be a vast improvement over the breeder. In nuclear energy the three questions arousing the most concern are reactor safety, nuclear waste, and proliferation of nuclear weapons. A fusion reactor will be quite incapable of going out of control and releasing massive quantities of radioactivity. Unlike a nuclear plant, when shut down a fusion reactor will stay shut. It will have no fiercely radioactive fission products generating so much heat as to require special cooling, lest they melt the reactor. Fusion thus may need none of the emergency core cooling or the elaborate plumbing for safety systems that make nuclear plants so complex and costly. Nor will fusion reactors produce the blazing nuclear waste, including radioactive iodine and strontium, that today appears so hazardous. Their interior parts will become radioactive, true; but they will be of metal, which is much more manageable than liquid nuclear waste. If the alloy is vanadium-aluminum, even this radioactivity will soon decay. Finally, a fusion reactor uses no bomb-grade materials like plutonium. Nor can it be made to produce plutonium, except with major and highly visible modifications.

Beyond the tokamak fueled with $D + T$ will be a whole range of approaches which to Harold Furth, director of the Princeton fusion lab, "seem to be ordered in terms of increasingly interesting economic potential and diminishing technical evidence." These have names like Elmo Bumpy Torus, Tokatron, Surmac, Spheromak, Field-Reversed Mirror; in twenty years one of them might be as exciting as tokamaks are today. The field-reversed mirror is designed to produce a dense, high-pressure plasma with a shape resembling a smoke ring. Such a plasma cell, the size of a beach ball, might produce tens of megawatts of fusion power. Its development would stand as the culmination of work that for years has been going forward at Lawrence Livermore Laboratory, east of San Francisco. As an alternative to the tokamaks, the people there have been developing the magnetic mirror machines, which confine plasma in long tubes rather than in the doughnut-shaped torus of a tokamak. Such tubes have their ends plugged with magnetic fields, which like mirrors reflect escaping particles back into the plasma. They promise higher plasma temperatures and higher pressures too and may be much simpler and straightforward to build than the tokamaks, thus giving a more compact and less costly reactor. To plug the ends successfully has been no easy task, but Livermore's Mirror Fusion Test Facility B, now under construction, promises to be virtually the TFTR of mirror devices. It features superconducting end magnets as tall as a two-story building, which were too large to be transported even by railroad flatcar, and which had to be assembled on site. When completed, it too will be expected to reach break-even.

And beyond all this is the prospect of fuels other than D + T. After that reaction is mastered, the next least difficult is D + D, or fusion of nuclei in a pure deuterium plasma. Its successful use would fulfill the prediction of Jules Verne in *The Mysterious Island*, that "water may be the coal of the future." The D + T reaction produces tritium, which could be recycled back to the old-fashioned D + T tokamaks; but it also produces helium 3.† Moreover, if tritium is aged in the wood, it too takes on the pleasing flavor of helium 3. Radioactive tritium has a half-life of only twelve years, and if stored in casks (no, not wooden ones) it decays to that helium isotope. There is a reaction, D + ^{3}He, that is a bit more difficult than D + D but produces no neutrons or other radioactivity. It releases its energy entirely as charged particles. This opens the prospect of directly converting fusion energy to electricity, by tapping off these charged particles on electrodes. The mirror machines would be particularly useful for this, and at Livermore researchers have done experiments to tap off particles leaking through a magnetic mirror, demonstrating direct energy conversion efficiencies up to 86 percent.

Carried to their conclusion, these trends could mean the fusion reactor of the future would be small enough to fit into a household basement, yet produce enough electricity to run the World Trade Center. It would be quiet, simple, and efficient, and might indeed function largely without moving parts. Nor would it have radioactivity, elaborate cooling systems, turbogenerators, massive radiation shields, or any of the other baggage we associate today with nuclear power. Such devices would fulfill the prophecy of H. G. Wells, of a world set free. They would be the universal small, neat, compact energy source. They would be flexible enough to be made to release their energy as neutrons when desired, or as electricity, as microwaves, as X rays or ultraviolet, or as charged particles. They thus could serve to produce hydrogen from seawater, as an inexhaustible fuel resembling natural gas. Or their energy could be used to produce synfuels. They could power airplanes and ships. For thousands of years past, seafarers carried no fuel but relied upon the wind in the air. For thousands of years to come, seafarers may again carry no fuel, but rely upon deuterium taken from the ocean. And at Lawrence Livermore Laboratory and

† Helium 3 is chemically identical to the helium in balloons, which is helium 4. The atom of helium 4 has two protons and two neutrons in its nucleus; that of helium 3 has one less neutron.

elsewhere, it has not gone unnoticed that fusion would be a fine power source for spaceships and for starships.

As Harold Furth has stated, "The prospect of adapting fusion power perfectly to the needs of society and the environment does not appear to be limited by any narrow physical constraints, only by the incompleteness of scientific knowledge and the limits of technological imagination." The central problem in fusion today is plasma physics, and when we have mastered the techniques of plasma production, heating, and confinement, then the plasma will become simple. The design of new fusion devices will become straightforward. In this respect, fusion today is like aviation earlier in this century, and the rise of the fusion reactor is akin to the development of the jet engine. To enter the jet age it was necessary to understand the flow and behavior of air and hot gases within a turbojet and to have metals that would stand up to jet heat. To enter the fusion age it is necessary to understand the flow and behavior of ionized gases within a magnetic field and to have metals that will stand up to fusion neutrons.

In seeking insight into the future of fusion, aviation indeed provides a most useful metaphor. Within the topic of nuclear energy is fission and fusion, which will bear comparing to two approaches to aviation: lighter-than-air flight and heavier-than-air. Both date to before 1800, and like fusion and fission in the earliest days of nuclear energy, for a time they developed almost side by side. Like fission, the balloon gained the dramatic, early successes. The brothers Montgolfier had their hot-air balloon in 1783, quickly followed by hydrogen balloons and cross-Channel flights. There even are those who say those early balloons might have changed the course of a war. Had Napoleon been of a less Boeotian turn of mind, he might have had a balloon corps, and aerial observers might have given early warning of the approach of those Prussian forces of General Blücher that cost him victory at Waterloo.

In heavier-than-air flight Sir George Cayley was the prophet of his day, breaking with four centuries of attempts to invent machines that would fly by flapping their wings. In 1799, to quote the historian Charles H. Gibbs-Smith, "he arrived at a correct and mature conception of the modern aeroplane." It had fixed wings, fuselage, a tail unit of elevator and rudder, and an independent propulsion system. His papers of 1809–1810, "On Aerial Navigation," laid the foundations for the modern sciences of aerodynamics and flight control. But to achieve his vision was not easy, for it called for vast advances not only in engines, but in aircraft stability and control. (Always the difficult technical problems have involved stability and control.) As a youth he had been inspired by the exploits of the balloonists. In later years he would inspire such friends as W. S. Henson, whose 1843 design for an "Aerial Steam Carriage" today looks like a Victorian's prediction of a twin-engine high-wing monoplane out of the

1930s. But true flight, like fusion, was an intrinsically recalcitrant problem and like fusion was long delayed.

Pursuing this metaphor, we might then say that today's proud tokamaks will be tomorrow's version of the biplane. They will serve to establish fusion as a force in being, demonstrating its promise for all to see; but the tokamak may be quickly superseded by such advanced approaches as the magnetic mirrors. Since 1950, the history of fusion has been a story of dedicated experts persevering through the most chilling difficulties, at last winning through to the promise of success, and to their just reward. But in the next century, this history may be largely forgotten. The brief spurts of barely controlled fusion that we anticipate in the 1980s may by then loom no larger than the brief spurts of barely controlled flight that Hiram Maxim, Samuel P. Langley, and other steam-powered aviators achieved in the 1890s. For most people in this century, aviation began with the Wright brothers. For most people in the next century, fusion will perhaps have begun with INTOR and the Engineering Test Reactor.

Pursuing the aviation metaphor even further, we may recall the state of aviation in 1940, following a century and a half of aeronautical progress. Lindbergh had flown the Atlantic. The propeller and piston ruled the skies; this would be literally true in the coming war. Passenger service was spreading in all directions, with the DC-3 in this country, with the Clipper flying boats across the oceans. The leaders of aviation could well be proud. Yet in a few short decades, all this would be swept away. For what would the aviation leaders of 1940 have made of the Boeing 747? Or of nonstop jet service, New York to Tokyo? (Not till 1957 did the DC-7 offer nonstop service, New York to Los Angeles!) Or airports like Dallas-Fort Worth and Narita (Tokyo), sprawling for miles and annually serving more people than all of 1940's ships out of New York Harbor? Or air-traffic control systems with radar and computers? For all that existed in aviation in 1940, its true story did not even begin until the war. Similarly, the fusion advances that may unfold in our lifetimes may be no more than a prelude to fusion's true promise, which may not even begin until new and hardly dreamed-of technologies take shape, in a century or more.

5

SUBARUS AND SUBWAYS

Everybody knows about the Great American Dream Machine. Long, low-slung, wide as a freeway lane, hung with silver-bright chrome, and with a roaring engine suitable for Indianapolis, this was the vision that lured the customers to the showrooms. If you were around in the 1950s, you surely remember the Tail Fin Age, as automakers vied to create auto rear sections which would increasingly resemble a twin-tailed fighter plane. That particular craze was fairly long-lasting, as these things go; Cadillacs first sprouted little nubs of tail fins as early as 1948, and the last new models with such fins came out in 1966. During those same years engines added more vroom, approaching 450 horsepower in some cases, the limit of what two rear wheels and a transmission could comfortably transmit. Even those who wanted something more suited for day-to-day use, a Plymouth, Chevrolet, or DeSoto, almost invariably got what today would qualify as a gas-guzzling behemoth. But gas was cheap, families were often big, and Detroit was inclined to give the customers what they wanted.

We shall not see their like again. Oh, there are still quite a few of those beloved old beasts on the freeways, and it will be a while yet before such big, luxurious cars stick out in traffic like an ocean liner surrounded by tugboats. And people with big, luxurious wallets can continue to buy cars to match. But America's parking lots and highways are well on their way to looking like those of Europe, where the small and simple car has from the start been a way of life. This is certainly not surprising, of course. For fifteen years the cacaphony of TV ads about Detroit's glamourmobiles has been matched by a comparable volume of concern over the auto as a public issue, and nowadays everyone is well aware that the industry and its products are in the midst of profound changes. Just what is really happening, however, tends to get lost in the shuffle of daily headlines.

The U.S. auto industry is in trouble, but that is hardly a new observation. What is happening is that after thirty complacent years, beginning with its conversion back to civilian production in 1946, the world has finally caught up with it. Those three decades saw virtually no significant technological changes in cars. The auto of 1970 was almost entirely the auto of 1940, except for styling changes and larger engines and that options like radios and automatic transmissions, rare in 1940, were commonplace three decades later. Amid that complacency, the industry became inbred, locked into patterns and practices which left it ill equipped to face the future.

That future began to arrive with consumer advocate Ralph Nader and the legislation of the early 1970s, which mandated sweeping controls on auto pollution as well as measures to improve auto safety. These represented the first substantive changes in autos since modern car designs came of age in the 1930s. The industry bitterly resisted these changes. Government agencies responded by dragging them kicking and screaming into the 1970s. The controversy was vociferous, but what the industry was fighting was not unfair deadlines or impossible technical requirements. It was fighting the demand that it break its fossilized habits and accept the need to change with the times; it was fighting for the right to go on building gas-guzzling dinosaurs according to its own lights. In this fight, to its initial advantage but eventual detriment, it had a powerful supporter in President Richard Nixon.

On August 15, 1971, Nixon moved to fight an inflation rate then hitting an apparently intolerable level of 4 percent. He imposed wage and price controls. Among these were controls on the price of domestic petroleum and its products. His wage-price controls brought at least a little temporary relief, but during the election year 1972 he used these controls as a cover, under which Federal Reserve Chairman Arthur Burns sought to stimulate the economy by pumping up the money supply at what would prove to be a highly inflationary rate. The controls in any case were riddled with exceptions and inequities, and when they were removed in 1973, inflation promptly spurted upward. But one set of controls remained in force: the controls on oil.

Then came the oil embargo and price rises. Even before then, the auto industry had detected a shift in consumer preference toward smaller cars and was already moving modestly in that direction. All through the

1970s, however, Nixon's oil-price controls remained in force and prevented the price of gasoline from rising to match the rising world price of crude. This artificial price suppression was justified as a way to protect the consumer, but what it really did was send false signals to consumer and auto industry alike. The public, encouraged by the promise of continued price controls, soon became convinced the oil crisis was a hoax trumped up by the oil companies; by 1975 car buyers were back clamoring for gas-guzzlers. The industry in turn found its move to smaller cars was ill timed. For example, General Motors was having trouble selling its Chevettes, then the most fuel-efficient car on the market.

Meanwhile, Congress had put the industry in a bind by passing the 1975 Energy Policy and Conservation Act, which mandated new fuel economy standards but at the same time stimulated the appetite for big cars by keeping gas prices down. The industry response was to prepare to invest massively in new technology to make the big cars fuel-efficient. They did not make a strong move toward the small cars that were proving unpopular with the public. In any case, such small cars had never been popular in the corporate boardroom, where the byword was, "Mini-cars mean mini-profits." Thus, early in June 1978, Ford Motor's top executives met to give final approval to new designs for a line of small cars. Henry Ford II, the chairman, was not enthusiastic about risking huge sums of money on such a chancy new line. After all, outside the company offices in Dearborn, the local gas stations in Michigan were selling unleaded for only sixty-three cents a gallon; Toyotas and Datsuns were piling up at the docks unsold. So Ford did what any prudent board chairman would do. He cut the small-car program in half, thus saving his company from having to borrow a billion dollars from its banks.

Then came the Iranian revolution and the new oil crisis. Gas prices shot up; the demand for big cars plummeted. Chrysler, which had been emphasizing mid-size cars, was out in the cold. Ford and GM had been supplying a good part of the European small-car market from their overseas manufacturing facilities but were unprepared for another sudden shift in the domestic market. Japan had had years of experience building cars to run on gasoline at an uncontrolled $3.00 a gallon and was poised and ready to flood the market. And the Big Three, rather than continuing with their genteel intramural competition, were suddenly faced with a rampant challenge from what would soon be the world's largest auto producer, with the world's strongest and most aggressive economy.

Japan, like Great Britain, is a smallish collection of islands lying off the coast of Eurasia; and as the nineteenth century belonged to the British, so the twenty-first may belong to the Japanese. Certainly they have long excelled at learning from Western nations. In 1543 three Portuguese seamen carrying matchlock rifles were the first Europeans to reach Japan.

No Japanese had ever before seen Western firearms. The local warlord bought the guns by paying in gold and set his chief swordsmith to copying them. Thirty years later, at the Battle of Nagashino, Japan's Lord Oda led an army with ten thousand matchlocks. They annihilated the charge of the noble samurai warriors, who were armed only with swords.

During the ensuing centuries Japan largely shut herself off from the outside world, but that isolation lifted with Commodore Perry's naval visit in 1853. Japan then set out to build herself into a modern nation, under European guidance. In 1895 they humbled China in war. Europe was amazed and respectful, and the Japanese understood full well the source of this new respect: "When we sent you the beautiful products of our ancient arts and culture you despised and laughed at us; but since we have got a first-class Navy and Army with good weapons we are regarded as a highly civilized nation." Britain contracted a treaty of alliance. The Russians fought with the Japanese and for this effrontery saw their fleet destroyed by Admiral Togo in the 1905 Battle of Tsushima Strait. After that, in Africa and Asia, there were people who hardly even knew the names of Russia or Japan; but they knew that one of Europe's principal nations had been defeated in war by a nation of nonwhites.

Japan's pursuit of military power brought her cities to the flame and her Emperor to the status of a subordinate under General Douglas MacArthur. Japan then found herself almost a picture-perfect microcosm of the world as seen by some of our modern economists: constrained in size, poor in resources, deficient in agricultural land, lacking sources of energy, overpopulated, and barred from seeking to remedy these ills through military conquest. It is a tribute to her native spirit and tough-mindedness that she did not then collapse into a miasma of decline and divisiveness amid trendy slogans about limits to growth. Instead, she set out to become the first non-Western nation to rise to world economic power, with her industrialists and bankers gaining the sway denied her generals and admirals. With institutions and policies acting strongly to promote this growth, with rarely a question among her people or leaders as to its desirability or achievability, she called on her innate resourcefulness to carry her forward. She succeeded, and in the only way she could: through production of high-quality finished goods, often representing advanced technologies, produced from the raw materials she had to import.

Japan today has created nearly ideal conditions for economic growth. To begin, she has proceeded under the basic assumption that the objectives of government and industry are the same. Relations between industry and government have been very close, and in particular, the government has stood behind companies' bank loans. In Japan such bank loans have furnished much of the capital needed by industry; companies there typically carry far more debt than they own in assets or other equity. This practice would be unthinkable in the United States. In Japan it is a source of great strength, for it means that companies need not finance their growth out of current profits. Once they earn enough to cover the interest on their debts there are few financial constraints on growth. With needed capital provided not by stockholders but by the Bank of Japan, and with government guarantees for these loans, corporate directors need not focus on the bottom line, the current quarter's profits. Instead, they can build for the future, plan for the long haul. Throughout the 1960s, they set up distribution networks for their cars in the United States, often giving away dealerships at a cost of millions of dollars. They could afford to gamble that Americans in time would turn to their small, high-quality cars. Detroit automakers, preoccupied with immediate customer demand, could not take such a gamble, could not look so far ahead.

Their government has helped in other ways as well. It has often encouraged collaboration and joint ventures among companies. Thus, their automakers could get together with government officials and agree on a common design for antipollution equipment. In the United States that would have been a violation of the antitrust laws. In Japan, as one writer has put it, "anti-trust and other policies are a means to an economic end, and not part of a higher morality." In addition, the famed Ministry for International Trade and Industry works with other agencies to develop the broad outlines of industrial policy, setting new directions for growth and making major investment decisions. This treatment of industry thus has been compared to a giant garden where newly sprouted plants are given special support and protection, mature ones allowed to prove themselves, and declining ones carefully pruned. When the auto industry was trying to establish roots in the 1960s, it got a variety of tax breaks, financial incentives, tariffs on competing foreign imports, and exemption from import duties on raw materials. When that industry came of age in the 1970s, having proved and refined its products in domestic competition while laying the foundations for an export market, the government withdrew these subsidies and protective policies. The auto industry was left to shift for itself—and it promptly took 22 percent of the U.S. market.

These policies have given the Japanese economy something of the character of a conglomerate, or alliance of unrelated but mutually supporting industries. Indeed, this economy has often been called "Japan, Inc." They have been able to channel cash flows from low-growth to high-growth areas. They have applied the debt capacity of safe, mature businesses to capitalize rapidly growing but risky ventures. They have moved into electronics and other dynamic new industries, bringing financial power no competitor could match. They have acted to increase capacity quickly, thus rising to dominate the industries in which they have chosen to compete. The borrowing power of all Japan itself has been available to its growing industries. The resulting growth has indeed been astonishing. In the late 1960s, before the oil embargo, Japan was investing 34 percent of her gross national product in new capital. Her growth rate was some 12 percent, and her rate of return on investment thus was some 33 percent. There is a play by Stendhal where a father advises his son, "Remember that four percent is a good return on your money." Clearly the Japanese have done far better than that.

In addition to her industrial policies, Japan has had unusually good labor relations. The Japanese worker is employed for life. What is more, his salary is set almost entirely by his age, education, and length of employment. His union exists within the company rather than within a trade crossing company lines, encouraging labor and management alike to perceive common goals. Company loyalties run deep. Many Western travelers have noted how workers stand and sing a company song, with lyrics like, "A bright heart overflowing with life linked together, Matsushita Electric."

Strikes are almost unknown; in 1978 Japan lost 1.4 million workdays because of strikes, while the United States lost 39 million. In addition, since status and salary do not depend on job assignment, management has great flexibility in assigning personnel. There are almost no rigid work rules, no featherbedding or union-mandated make-work. The Japanese worker, with his lifetime job security, tends to see new production techniques not as threats to his job, but as useful tools to help his company. The Japanese thus have been free to explore the uses of robotics. At Nissan's body plant near Yokohama, workers have pasted photos of movie stars on the robots that make the cars. In this plant, 35 workers aided by robots produce 350 Datsun bodies every eight hours, seven times the productivity rate of competing U.S. automakers.

Permanence of employment and seniority-based wages act to channel labor into the most rapidly growing sectors while increasing pressure on the less dynamic ones. A fast-growing firm, recruiting directly out of the schools, has low labor costs but a highly skilled and youthful work force. Management responsibilities tend to go to the young and vigorous. The slow-growing firm, by contrast, hires few young graduates; its labor costs rise with age, and its management becomes older and less aggressive. Able and ambitious young people thus have strong incentives to seek out the companies and industries growing most rapidly.

Japan has also found merit in a suggestion embraced by Shakespeare's Jack Cade: "The first thing we do, let's kill all the lawyers." The whole country has only twelve thousand attorneys, compared with half a million in the United States. In Japan a typical contract between two parties may run to a paragraph or two. If problems arise, the people involved will sit down, pour some sake, and work out an equitable agreement. In the

United States a comparable matter might run to two hundred pages, in an attempt to foresee all possible contingencies, and a lawsuit is never far away. The Japanese simply do not believe in settling controversies by legal battle between adversaries. Their cultural homogeneity and long history as a crowded island people have given them ample experience in seeking and achieving consensus. Before making major decisions, government and industry leaders first consult extensively among interested parties. They thus take an unconscionably long time to decide about anything, but once a decision is made there are few difficulties in carrying it out.

In Japan, about one fifth of all bachelor's degrees and two fifths of the master's degrees are in engineering. In the United States the comparable figure, at each degree level, is around one twentieth. Between 1963 and 1977 Japan awarded about as many engineering degrees as did the United States, even though her population is only about half the size of ours. In Japan such engineering degrees offer a favorable route to business and social success. About half of these new-minted engineers actually enter the engineering professions; the rest become civil servants or industrial managers, and around half of her senior civil servants and industrial directors hold degrees in engineering or related subjects. In the United States, of course, it is often the law degree which leads to advancement. The difference is significant. Lawyers are concerned with dividing a pie; engineers seek to make it bigger.

The Japanese people do one other thing that helps their country. They save their money. On the average, a Japanese wage earner saves 26 percent of his income, compared with the average American who saves 5.5 percent. This makes all the difference between a nation of investors building for the future and a nation of consumers living only for the present. These massive savings represent the funds Japanese banks loan to industry, as well as to the government. Japan has one of the world's lowest inflation rates, 5.5 percent in 1981, little more than half that of the United States. It is not that she is free of dependence on OPEC oil and other imports; she is far more dependent than we are. Nor does her finance ministry run a balanced budget. In 1981 her national budget showed a $71 billion deficit out of total spending of $213 billion. America's deficit was $58 billion out of a total of $661 billion. Both governments finance these deficits largely by borrowing, but Japan has enough bank assets to permit the government to borrow without driving up interest rates or depriving industry of needed capital. In the United States such federal borrowing does put pressure on interest rates and on corporate borrowers; to keep these problems manageable, the Federal Reserve has often increased the money supply. That is why our deficits have been inflationary while Japan's have been much less so.

In addition to all this, Japan's products are remarkably free of defects and of unusually high quality. Knowledgeable U.S. auto buyers have long tried to avoid buying cars built on Mondays or Fridays, when absenteeism leads foremen and workers to cut corners on the assembly line. But few Monday or Friday cars are made in Japan. Indeed, U.S.-built autos which would be perfectly acceptable here are often returned from Japan as defective; at the least, such cars need an extra coat of paint to satisfy the demanding Japanese. To qualify for the export trade, Japan's autos and other products must survive fierce competition within her domestic market. Thus, long before Honda began selling motorcycles in the United States, its managers brought that company up from a distant Number Two position in Japan proper, beating out the leader, Tohatsu, and indeed dominating Tohatsu so thoroughly in domestic sales that that formerly leading firm largely gave up building motorcycles altogether.

The pursuit of quality is unstinting throughout Japanese industry. Groups of about ten workmen and a foreman form quality-control circles, which meet frequently and on their own time to find solutions to production problems, to seek ways to increase production and to raise standards. If a worker spots a faulty item on the production line, he can shut down the whole line to fix it. Until recently, an American worker who did that would be courting dismissal. Indeed, such quality-control circles would be easily sabotaged in the United States, where labor unions would often regard such notions as attempts to exploit the workers, while middle managers would see them as a threat to their authority. Nor would they be welcomed by higher management. As one senior U.S. auto official has candidly stated, "We wrote off the workers as contributors to the organization in the 1930s when they unionized."

Is Japan then a superstate, destined to sweep the world's markets? Some economists have worried that Japan might reduce the United States virtually to the status of a colony, exporting food and raw materials while importing manufactured goods from the mother country. But if this ever happens, it will be in the face of a vigorous American move to meet this overseas challenge, and on its own terms. The U.S. auto industry today has been buffeted severely by shocks brought on by its own complacency and lack of preparedness. It has been rushed through rapid federal timetables for enhanced safety, pollution control, and fuel economy. It has seen its sales and earnings fall from 9.3 million cars and a $3 billion profit in 1978, to 7 million cars and a $4 billion loss in 1980. At its peak, direct auto manufacturing employed a million workers; it now employs 600,000, and most of the lost jobs are gone for good. Gasping for breath, its sales at a twenty-year low, the industry nevertheless now is meeting head-on an era bringing the most rapid changes in its history. It is spending huge amounts of money for retooling and for building new plants. This investment program will cost $80 billion by 1985 and is the price of the industry's revival and of its future as a major force in the American economy.

In the midwestern heartland of the industry and in other states as well, spanking new production plants are rising, filled with the best equipment available today. Near Detroit and other old auto centers, aging plants are being gutted and stripped clean as the automakers rebuild them with new automated facilities. Thus, Chrysler has spent $100 million refurbishing its Jefferson Avenue plant in Detroit, whose ancient shell dates to 1907. There they have installed industrial robots to take over tasks previously done by assembly-line workers. Their spot-welding robots cost $60,000 each, but run for $6 an hour in operating expenses, compared with $20 an hour for a worker's wages and benefits. These robots operate under computer control, wielding welding torches as well as hooklike hands to lift heavy steel parts. Each auto body in Chrysler's K-car series, its Plymouth Reliant and Dodge Aries, has 3,000 spot welds; 98 percent of them are made by robots and other automatic welding machines. (This conjures up the vision of a flatbed truck pulling up to the Chrysler plant loaded with two dozen robots, whereupon the boss says to his men, "Okay, boys, we won't be needing you anymore." But it doesn't really happen that way; that's why there are union contracts.)

What sort of cars will these new plants be building? In the old days the ads used to boast that each new year's models were longer, lower, wider. During the 1980s, the new cars will be shorter, smaller, and much more expensive. Detroit has long subsidized its small cars with profits from its big ones, but that will stop. One way or another, the Big Three will have to get back their $80 billion investment, and the most likely way is via the price sticker on the side window. This process is known as downsizing and is the same with autos as with candy bars; the manufacturer shrinks the size while hiking the price. Inevitably, then, millions of would-be car buyers will balk at paying $10,000 for less car than could be had for $3,000 in 1970. The used-car market will boom, and some drivers will get many of the advantages of a new car, at one fifth the cost, simply by installing a brand-new engine.* Moreover, Detroit's new designs will often come equipped with bugs as factory-installed standard equipment, and it may take several model years to get them all out. Japan has much to teach us about quality control, and Chrysler's 1974 experience may well be repeated. To boost quarterly sales figures, Chrysler pushed questionable products onto the market on a near-panic basis, with the result that that year's Plymouth Volare and Dodge Aspen have the dubious distinction of being two of the most recalled cars in history.

* In 1973 I paid $1,500 for a used 1972 Pinto. In 1979 I paid $1,500 for a new engine with factory warranty, a new cooling system, and some transmission work. The car has over 200,000 miles on the odometer and runs like a charm.

Nevertheless, for those courageous souls who will brave the new-car showrooms, there will be new features aplenty. The auto of 1985 will be as different from that of 1980 as 1980 autos are from 1965 ones; fifteen years of innovations are being telescoped into five. Most of these changes have the aim of increasing gas mileage while retaining as much interior space and driver comfort as possible. Thus, more and more cars are getting front-wheel drive, with the engine mounted sideways. This eliminates the long, heavy drive shaft as well as the differential in the rear, while permitting a smaller engine compartment. Yet the passenger compartment stays as roomy as before. For example, the 1981 Plymouth Reliant cut 1,070 pounds or 30 percent from the weight of its predecessor, the 1980 Volare, yet sacrificed only 4 cubic feet (4 percent) of passenger space and 1 cubic foot (6 percent) of trunk space. This weight reduction allowed a cut in horsepower from 120 to 84 and a hike in gas mileage from 17 to 24 miles per gallon.

Weight reduction is very important; a weight loss of 1 percent can increase gas mileage by 0.7 percent. So automakers are increasingly turning to lightweight materials in place of the traditional alloy steels. Carbon steel is giving way to high-strength steel, and total auto industry demand for steel may be cut in half. Other substitutions are introducing aluminum, nylon, fiberglass, foam-filled structures, and reinforced plastics. In fact, plastic parts have a great future in the automotive world. They don't corrode and are cheap to replace. They are extraordinarily versatile; fenders made of certain plastics can absorb a dent and minutes later bounce back to their original shape. The list of metals that can do this is very small, like zero. Also, in the words of C. M. Heinen, a director of research, "We use plastics for beauty. The fineness of detail possible with plastics often cannot be achieved any other way. The breadth of beauty, of texture, feel, and comfort of body cloths possible with plastics cannot be matched by any other material, including fine Corinthian leather."

Other changes will run deeper. Borg-Warner is developing a continuously variable transmission, in effect making available an infinite number of gears. This will improve fuel economy by allowing the engine to run at the most efficient speed for a given power output. To choose this most efficient engine speed, the engine will be under the control of a microprocessor. Whether you are cruising on the freeway or maneuvering in traffic, this microprocessor will continually and automatically adjust transmission and engine together to give the best gas mileage. And this is only the beginning; as one auto company spokesman has said, "Once we get a computer on board, we'll be looking to give it more and more jobs to do." In 1981 some 6.5 million new U.S. cars were equipped with microprocessors. Already there are electronic carburetor controls and fuel-metering systems, while some 1983 Ford and Chrysler cars have microchip voice synthesizers to warn drivers: "Please turn off your lights"; "Your oil is low." Your auto's computer will not only help run it and talk to you as well; it will also communicate with your mechanic. It will offer self-diagnostic services, making it much easier to spot trouble in the making.

Will tomorrow's cars have futuristic engines to match their electronics? The answer is no, and it is well this is so; after all, these motors will have to be worked on by ordinary auto mechanics. The good old internal combustion engine has been around a long time, and it will be hard for any fancy new design to compete with it. Still, there will be changes. For one thing, we will be driving into the future on four cylinders. In its heyday, about 1970, the V-8 accounted for 90 percent of U.S. engine production; as late as 1979 this still stood above 60 percent. By 1990 the V-8 will be virtually extinct. Even the in-line six will be in only about 5 percent of U.S. cars. As early as 1985 the four-cylinder engine will be in half our new cars, up from 1 percent in 1969.

Also, quite possibly there's a diesel in your future; by 1985 they may account for 15 percent of U.S. auto production. The diesel boosts gas mileage by some 25 percent when compared to a gasoline engine of the same power, and diesel fuel is not only cheaper than gas, but gallon for gallon contains more energy. A diesel needs no spark plugs, carburetor, or distributor, and thus has fewer things to go wrong. What's more, they last practically forever. Diesels operate at much higher cylinder pressures than do gasoline engines and thus are heavier and more costly, but by the same token they are much more rugged and durable. Some truck diesels run half a million miles between major overhauls, and there are authenticated records of auto diesels lasting as long as 900,000 miles. This worry-free long life will be attractive to owners who want to hang onto their cars for a long time. In an age of five-digit sticker prices and of body styles staying nearly the same from year to year, surely many owners will do so.

Four-cylinder engines with front-wheel drive, lightweight bodies, computer control, diesels; with all these developments working together, the four-passenger car of 1985 may easily top 50 miles to the gallon. The mileage of diesels can be improved even more with a turbocharger, in which a turbine, spun by exhaust gases, runs a compressor to push extra air into the cylinders. Already the Volkswagen Rabbit with four-speed transmission offers 45 miles per gallon in city driving, 58 on the highway. An experimental turbocharged Rabbit raises these test mileages to 55 in the city, 69 on the highway. (Your own mileage may differ depending on the heaviness of your foot.) Nor are such cars underpowered; European drivers climb up the Alps in their four-cylinder cars, and our hills are hardly steeper.

Even this may not reach the outer limits of fuel economy. New from Japan, the two-seater Daihatsu Cuore has a length (not wheelbase!) of

126 inches. This is barely twice the length of an ordinary bicycle. Among autos, the Cuore thus is a foot shorter than the Honda Civic and over three feet shorter than the smallest of today's American cars, the Ford Escort. With a weight of just over a thousand pounds, it is about as small as a four-wheel car can be and still be street-legal. It may be just the thing for a commuter or for a housewife who puts groceries in the passenger seat. Certainly there will be others like it, two-seater microcars with the most advanced available auto technology. By 1995, according to automotive analysts Charles L. Gray and Frank von Hippel, such microcars may be reaching 113 miles per gallon.

This may prove too much of a good thing, even though Mae West said that too much of a good thing is wonderful. Today's small cars cost about twenty-five cents per mile to own and operate; this includes gas and oil, maintenance, insurance, and depreciation. The difference between a car getting 40 m.p.g. and one getting 60 is 0.0083 gallons per mile, and even when gas hits $2.00 per gallon this will be a difference of only 1.6 cents per mile. For that modest sum the auto buyer, choosing the lower-mileage car, may get a larger and more comfortable vehicle, a more powerful engine, even a less costly auto. After all, the added mileage probably will come from a turbocharger and from other advanced features appearing as options on the price sticker. Even in Europe in the mid-70s, where gas was already at $2.00 a gallon in today's dollars, average auto mileage rarely exceeded 25 miles per gallon. The day of the 100-m.p.g. car may be long in dawning.

We also should not assume the current rapid pace of change will continue for many years to come. It is quite possible that by 1990 the worldwide auto industry will have shaken down into its new patterns, and the autos of 1990, perhaps even of 1985, will be little different from those of 2010 or even 2030. During the 1980s we can count on the industry to offer a wide variety of designs and options, with different engines, transmissions, electronics packages, and types of body construction. Inevitably a few basic types will capture the greatest customer interest, and the industry will settle on these as its standard designs. Then, except for modest year-to-year changes, the auto companies will concentrate on these basic types, together with an assortment of variations or nonstandard types, to be built by specialist companies. The Daihatsu and Mercedes-Benz, representing the extremes of the automotive world, would be among these. Then, having gone through twenty years of rapid change, the industry of 1990 may well relax, catch its breath, and settle down to build a line of products which will change only modestly for at least several decades.

The auto will no longer be the Great American Dream Machine, but with its unparalleled flexibility of use it will still be the Great American Freedom Machine, and it will continue to dominate our transportation.

What the auto companies will be offering, both here and overseas, will be world cars. No longer will the American motorist drive a car that is different—larger, more powerful, and luxurious—than those of the rest of the world. No longer will U.S. firms build big cars like the Chevrolet Caprice for the domestic market and small ones like the Opel Rekord for the Europeans. Instead, the same cars will be seen on the streets of Liverpool and Lagos, of Kyoto and Kalamazoo. Indeed, such recent models as Ford's Escort and Mercury Lynx already count as such world cars. There will be a worldwide market, and it will be a rather homogeneous one, dominated by perhaps eight giant firms. In Europe will be Fiat, Renault, and Volkswagen; in the United States, Ford and General Motors; in Japan, Nissan, Toyota, and Mazda. All these will have overseas plants and subsidiaries. The next few years will see a great stirring and interpenetrating of national markets, together with a number of mergers and joint ventures. Thus, American Motors already is largely owned by Renault and probably will increasingly become a French subsidiary. In time Chrysler too may well enter into merger with an overseas partner, quite possibly Mitsubishi in Japan.

The auto industry of the year 2000 then will scarcely resemble the Big Three as we have known them. Instead of having our Datsuns and Volkswagens unloaded from ships at the docks, they will be built in this country, by those firms' U. S. affiliates. Of course, Ford and GM will be returning the favor, with their own plants in Europe and Japan. Parts and components increasingly will come from suppliers dispersed around the world, including developing countries. Final assembly plants may be located increasingly in countries where labor is less costly. These developments will be further hastened by "local content" laws in the Third World, requiring that a certain percentage of a car sold in a country has to be manufactured or assembled there. With Europe, America, and Japan having slow-growing populations and needing new cars mainly to replace the relatively few million each year that reach the end of their days, and with the Soviet Union still most likely closed to mass auto ownership, automakers indeed will be looking to the Third World as the next great untapped auto market. In the twenty-first century the people of Asia, Africa, and Latin America at last will gain the means to get behind the wheel. The world's Big Eight intend to be ready.

Not all countries will share in this widespread automotive uniformity. Brazil, for one, is finding its own path by going in heavily for alcohol as fuel; one could say the cars there will be drunk with power. They already have most of their cars running on gasohol, which in Brazil is one-fifth ethanol or grain alcohol. The alcohol is distilled from fermented sugar cane; in the midst of green cane fields, bright steel storage tanks have sprouted near the new distilleries, and Brazil is planting much new arable land to trap its sunshine in liquid form. They don't fool around in Brazil with finicky fuel formulas. They just pour in the alcohol till they cut the gasoline about one to four, then drive off. As a result, they'll blow out a million mufflers in the next two years. Cars coasting downhill rattle windows with exhaust explosions, and rush hour sounds like a World War II firefight. However, help is on the way. By planting still more cane, and by producing alcohol from the even more abundant cassava or manioc, Brazil hopes to convert to straight alcohol. By 1985 that country may have 3 million vehicles, about half its fleet, running on pure 200-proof. Such alcomobiles do not emit acrid gasoline fumes; instead, the autos give off an odor resembling vanilla.

In America the conversion of biomass to ethanol is a national folk art hallowed by tradition if not by law. Fuel alcohol also has a long history. Henry Ford built his Model T with an adjustable carburetor to burn either gasoline or alcohol, and he envisioned a world of alcohol cars. As he liked to say in the days before Prohibition, "There's more stills than service stations." However, he was opposed by the powerful Anti-Saloon League, one of whose main financial backers was John D. Rockefeller of Standard Oil, whose opposition to demon rum dovetailed neatly with his business interests. Recent years have seen increasing interest in gasohol, and quite possibly the gasolines of the future will be blended with at least a few percent of alcohol. However, we lack Brazil's virgin lands, and our agriculture relies far more heavily than Brazil's on tractors and other motorized farm equipment, as well as on fertilizers, pesticides, and irrigation —all of which rely upon or are derived from petroleum. As a result, in the United States fuel alcohol actually takes more petroleum to produce than it saves by the gasoline it replaces. What's more, most gasohol projects here rely on fermented corn, which raises the unhappy prospect of people seeking food competing with automobiles seeking fuel.

What fuel alcohol we have is heavily subsidized. In Iowa, for example, state and federal tax exemptions together give a subsidy amounting to $44 per barrel of fuel alcohol. However, farm-state Congressmen who have pushed heavily for gasohol have not really been concerned with oil imports; the real issue has been the price of corn. In Iowa the rule of thumb is that a 1 percent decrease in corn supply raises its price by 2 percent. If they could divert 500 million bushels of corn a year out of the normal production of 6 billion bushels into alcohol distilleries which they induce Washington to finance, they would raise the price of corn by one sixth. City people would be paying more for their meat from corn-fed cattle and more for their uncompetitive and tax-subsidized gasohol; but the farmers would be laughing all the way to the bank.

Nevertheless, fuel alcohol does have a future here for it provides a very useful solution to a messy problem. This is the problem of agricultural and municipal waste. Cheese whey, citrus fruit wastes, damaged crops, pulp mill residues from wood chips, bagasse from sugar cane, and ordinary city trash—all of these are produced each year in disconcerting quantities, and all of them are rich in cellulose or carbohydrates, which can readily be converted to alcohol. If gasohol in the United States is to be more than a scheme to prop up corn prices, most likely it will rely on these sources. During Prohibition in Chicago, when violin cases were known to conceal tommy guns, federal agents tried in vain to trace the illegal grain shipments from which Al Capone supposedly was distilling the city liquor supply. What they didn't know was that Capone was making his alcohol from a source over which he had better control: the city's garbage.

What sort of highways will our cars be running on? Here the situation is not so pleasant to contemplate. The 42,500-mile interstate highway system was authorized over a quarter-century ago, but little systematic provision has been made for its upkeep and maintenance. The interstates were designed to last twenty years, which to a politician is virtually eternity; but some stretches are already that old, and several thousand miles are already crumbling. Ironically, there still remain about 1,500 miles of authorized interstate highway still to be built, and there are places where potholes lie unfilled, virtually in the shadow of construction cranes. Thus, near San Jose, California a multilevel interchange now is nearing completion, designed to link together three freeways. Close by, Highway 101 is full of holes and according to Jon Carroll of *California* magazine, "There are many blood alleys in California, but this one leads the parade. An absolutely terrifying driving experience."

There are plenty of other roads like it. Pennsylvania just may be the pothole capital of the nation. On Interstate 90 near Erie, the road is in such bad shape that the speed limit is 30 miles per hour. On I-95 between Maryland and Virginia the Woodrow Wilson Bridge, built in 1961, has deteriorated so rapidly that it must be rebuilt at a cost of $60 million. Off the interstates, things are worse. In Louisiana, a former rodeo cowboy turned trucker had this to say about Highway 71 between Baton Rouge and Shreveport: "That road just tears the rig apart. It's like riding a bucking bronco." In eastern Kentucky, pockmarked roads with two-inch asphalt paving are being pulverized by overloaded heavy coal trucks, and drivers complain their tires blow out before they wear out. Where the roads don't stop you, the bridges just might. In Ohio alone over six hundred bridges are so dilapidated that they have been blocked off to traffic. Four thousand others show ominous signs of deterioration, yet are still in use. Overall, some one fifth of America's bridges need major repairs or refurbishment.

Moreover, some motorists have discovered that the open road may yet hold deeper delights in store. One fine spring day in 1981, some New York commuters awoke to the news that their trains into the city would be a half hour late. Fearing still longer delays, thousands of them took to their cars. But just as the morning rush hour reached its bumper-to-bumper peak, a four-square-foot section of concrete roadbed on Manhattan's elevated West Side Highway collapsed and fell to the ground below. A repair crew soon arrived to patch the hole with a metal plate, but the damage was done; traffic backed up for three hours. Some drivers did not reach their offices till noon.

With roads and bridges often in such poor condition, one would think the day would be at hand for alternatives to the automobile. Certainly there has been much talk of such alternatives, but the cities which have relied on them most extensively have often allowed them to fall into even worse disrepair. Particularly in the Northeast, much public transit is not merely in bad condition but stands literally as a menace to life and limb. Moreover, those subways and bus systems stand as a serious indictment of the misgovernment that mayors and administrators have visited upon their cities.

In Philadelphia, where rickety buses, subways, and trolleys are prone to breakdowns and accidents, transit riders filed suit in 1979 claiming they were "risking their lives" by riding. On the Broad Street subway line, the average car is 48.2 years old; some date back to the 1920s. On most days the Broad Street line has fewer than two thirds as many operable subway cars as it needs to operate at full service. As for its mechanics, when the city management decided to require maintenance workers to pass proficiency tests, the transit union went out on strike.

In Boston, subway service has deteriorated amid tight budgets, while militant labor unions have pushed costs sky-high. Two Boston bus drivers earned $55,000 in 1980, more than some experienced airline pilots. The people who clean the subway stations were earning $22,000. On four-car subway trains, union rules call for a driver, three carmen, and two "door guards" to open and close the car doors. The doors are fully automatic. No matter; the door guards are there, and they earn $33,000 a year. Moreover, they hold onto their jobs even while Boston has cut its engineering and maintenance work force by 16 percent. In Chicago the situation is much the same; labor costs account for 77 percent of that city's transit system operating expenses.

These problems pale, however, compared to those of New York, whose transit system serves five million daily riders, one sixth of the nation's users of public transit. The subways are filthy, dimly lit, Dantesque horrors resembling the sewers of Paris in Victor Hugo's *Les Misérables*. Graffiti covers subway cars both inside and out, thick as flies near a garbage can. This graffiti is not only a most unsightly vandalism, but a reminder to all that in 1980 there were twelve thousand muggings and other felonies on the system, including eighteen murders. The vandals of New York also break nearly a thousand subway windows each week. Some trains have been guarded by patrolling police with guard dogs, and there are martial-arts instructors who recommend that their advanced students ride the subways as part of their alertness training.

In a 1981 report, an advisory committee to New York's Metropolitan Transit Authority stated that "the picture that emerges is one of a system in physical collapse." Spare parts are scarce and cannot simply be purchased from the manufacturer; often they require the cumbersome procedure of competitive bidding. Thus, repair crews have resorted to cannibalizing some cars to fix others, and at any time over two thousand subway cars are out of service. Angry riders have revolted, refusing to get off trains being withdrawn from service because they were unsafe. Other passengers have endured being stranded for more than an hour in a broken-down subway trapped in a tunnel under the East River. Some tunnels in the system date to the nineteenth century and leak badly. Some maintenance facilities were built originally as horse barns. To put the system back into good order would cost over $1 billion a year for the next ten years; its managers are spending only $300 million a year.

If ever there were a situation made to order for new technology to come to the rescue, surely it is our aging transit systems. Yet quite often new technology has been part of the problem. Bus and subway manufacturers have often acted as if they were designing entirely new and revolutionary systems, rather than examining and improving on existing vehicles which have proved themselves in extended service. Such vehicles do exist. A streetcar designed in 1936 benefited at the time from years of practical experience in trolley manufacture. It is not only still in use but is a mainstay of trolley lines in San Francisco, Boston, Toronto, and Philadelphia. In 1914 the Brooklyn Rapid Transit Company introduced an all-steel subway car which saw successful use for fifty years. By contrast, when New York recently introduced a fleet of seven hundred shiny new R-46 subway cars, the cars soon had to be taken out of service due to cracked undercarriages. Returned to service, they had to be brought to the shops at least twice a week for inspection. New York won $72 million in damages in court, but the manufacturer still could not make permanent repairs.

The situation with buses is similar and would be tragic if it were not so laughable. In April 1978, Grumman began producing the Flxible 870. It had sleek sides, smoked-glass windows, computerized destination signs, cantilevered seats, push-button stop signals, and a low-slung body. With their dark-tinted front windows, New Yorkers called them Darth Vader buses. One art critic (doubtless a taxi rider) pronounced them as having a "cool, refined air." The first batch arrived in the summer of 1980 in Manhattan, city of killer potholes, and won immediate notoriety for breaking down. By December the city had to take all 637 of them out of service because of cracks in supporting frames. Grumman officials blamed New York's potholes. Grumman's critics responded that those bus frames had also cracked in Los Angeles, where the streets have plenty of smog but few potholes, and noted pointedly that it is standard practice to design a product for the conditions it will encounter in service. Grumman repair crews set out to strengthen each frame with two hundred pounds of welded steel plate, but in New York the repaired buses soon cracked again. Meanwhile, other cities were renovating their 1959-design buses from General Motors, which a Chicago transit manager described as "reliable, and pretty much all the bugs are out of them."

The federal government has provided operating subsidies as well as funds for the purchase of new equipment, but has attached long and very sticky strings to its aid. Thus, many transit systems have been under federal orders to provide "total access" to the handicapped, so people in wheelchairs can have unencumbered use of public transit. This means equipping stations with special ramps or elevators and vehicles with special doors and wheelchair lifts. The motivating spirit has been unexceptionable, the intentions most admirable; but these are the good intentions with which the road to hell is paved. In the San Francisco area, such costly changes in the Bay Area Rapid Transit system would serve about sixty wheelchair riders daily, out of a total ridership of 165,000. In New York these modifications would cost $1.4 billion, plus $100 million a year in operating expenses. Also, the Urban Mass Transportation Administration provides federal funds to purchase buses, but it requires the buses be air-conditioned. The assumption is the air-conditioning systems will work. They don't. They break down fast, rarely stay repaired, and when out of service leave little provision for fresh air. It would be perfectly possible to provide large windows that can easily open, but that would be *much* too simple.

No, the cure for these transit ills will lie with proper policies adopted by city and state governments. These ills have been long in the making and their solutions will not be quick. In New York the subways began as private companies early in this century, but over the years they went into bankruptcy, and by the time the city government inherited these systems they were already in serious disrepair. Unrealistically low fares did not help; as late as 1965 the price of the New York subway fare was fifteen cents. During the 1960s New York, like other Snow Belt cities, made a series of critical and ultimately disastrous blunders. They elected to follow "progressive" social policies, shifting resources away from transit systems, police and fire protection, school systems offering quality education, and city maintenance and sanitation. Instead they hiked welfare benefits, greatly expanded their public-assistance rolls, and caved in to unions of teachers and municipal employees by yielding to their demands for lavish wage and benefit packages. Predictably, tax-paying citizens of the middle class fled these cities in droves, while the liberal welfare benefits attracted hordes of ne'er-do-wells from out-of-state.

These policies were often accompanied by budgetary legerdemain. When the financier Felix Rohatyn began working to reform the finances of a nearly bankrupt New York in 1975, he confronted Mayor Abe Beame and Deputy Mayor James Cavanagh. They refused to admit that for years they had been cooking the books to hide billion dollar deficits, and Cavanagh chided him: "I can see you don't know much about municipal finance." Replied Rohatyn, "I know a lot about baloney, and we're going to have some serious discussions." During that fiscal crisis, however, the Metropolitan Transit Authority suspended routine maintenance of subways and buses and allowed the roster of skilled mechanics to dwindle by not replacing the ones who quit or retired.

For New York, as for many other cities, the road back lies in mayors and councilmen relearning the art of governing. They will have to run for office on platforms pledging serious efforts to improve the transit systems and offering visible measures of progress by which they can be judged. To do this, they will have to build coalitions among voters who regard transit as a first priority and will have to show themselves willing to confront the selfish unions. In Boston and elsewhere, they will have to appoint professional transit managers rather than treating transit as an exercise in patronage. As one former Boston executive director has said, "Transit is not the kind of place where you can hide your brother-in-law for four years. It is a sensitive profession. It is a very technical industry."

By telling the truth to the people, these leaders will face up to the painful task of building political support for fare hikes and perhaps for tax hikes dedicated to transit. They will have to resist the self-serving demands of individuals and groups claiming entitlement to special subsidies at the fare box. They will have to trust the public, who after all are not fools and who will be willing to pay more—in return for noticeably better service. These leaders will have to avoid the temptation to be all things to all people and to run to Washington whenever their receipts run short. Still, to govern is to choose. City administrators can choose the painful tasks which can bring renewal, or by default they can choose further rot and decay.

At the state level, crumbling roads and bridges can be patched with money, but again the need is for political leaders to tell the truth and to build coalitions which will support the required efforts. Most state highways are funded with gasoline taxes, and the interstates get support from a federal gas tax. In 1959 this federal tax was four cents a gallon. Today it is still four cents a gallon. In 1973 the average state gas tax was seven and a half cents; today it is only nine cents. Meanwhile, the cost of rebuilding a road is 166 percent higher than a decade ago. Some states have begun raising their gas taxes, but the moves in this direction have been few and halfhearted. Yet surely few motorists will wind up in the poorhouse if they pay a tax of an extra nickel a gallon; on the other hand, they may wind up in the cemetery if they get hit by a truck dodging a pothole. And there is a certain justice in the needed policies. The prosperous Sun Belt states have long highway mileages to maintain; the states of the Snow Belt are financially strapped, but geographically compact and with much less highway mileage per capita.

Some transportation systems work. The San Francisco Bay subway system (BART) was plagued with breakdowns, but its management has raised fares, improved maintenance, gotten trains to run on time—and increased ridership by 50 percent. New subway or rail systems sparkle in Washington, D.C., Baltimore, Atlanta, and Miami. Dallas in particular stands out as a model; it is the largest U.S. city to be run, not by a mayor, but by a professional administrator answering to the city council. Dallas keeps a computerized inventory of all street surfaces, curbs, gutters, sidewalks, and stop lights. Potholes are filled within three days; streets, bridges, and sewers are kept in fine condition. To help do this, Dallas assesses user fees, charging people for services. For instance, when streets are improved, the people who own houses or businesses on them pay part of the cost. This is in vivid contrast to the traditional practice of hiding the true cost of services from the people, by means of subsidies or outright budgetary tricks.

Transportation in the United States has long been a highly political topic, ever since the days of the Erie Canal and the National Road. In recent years the politics of transportation, as well as its public policy, have been inseparable from the name of Ralph Nader. So it is appropriate, in closing, to assess his influence and to suggest how he will be viewed in the future.

The American people have shown strong interest in the positions and policies advocated by Nader. They have supported improvements in auto safety, even while rarely purchasing the optional safety equipment items, or using the mandatory ones. They have endorsed environmental protection and auto emissions standards, while resolutely refusing to do anything about fume-belching older cars which continue to foul city air. They have supported the 55-mile-per-hour speed limit in the legislatures and ignored it on the highways. Truth-in-advertising, consumer protection, recalls of defective autos—all these measures have won wide support. Nader himself, however, has not; he has never sought public office. It is likely that public office will never seek him, for his humorlessness, his driven in-

tensity, his lack of the common touch are not the stuff of which successful political careers are made. He and his allies have operated at one
remove from politics, lobbying for new laws and insinuating their sup-
porters into the regulatory agencies. By these means they have gained
great influence, but it is an influence whose foundations are less than
secure. After all, in America power flows ultimately from the ballot box.
Courts and regulators may make law, but in the end only the voters will
affirm or reject these reforms. And Nader has never built a political force
with the strength and cohesiveness of, say, the labor movement. It is inconceivable that a president would appoint a Secretary of Labor who was
regarded as antilabor, but we have seen that it is quite possible to have a
Secretary of the Interior who is regarded as a foe of the environmentalists.

In the early years of this century the Socialist Party, led by Eugene
V. Debs, foreshadowed many of the century's most valuable social
reforms. It has been said that most of the programs of the New Deal appeared in the Socialist platforms of 1904 and 1908. Minimum wages,
eight-hour days, abolition of child labor, retirement benefits and pension
plans, workmen's compensation in case of accident or injury, legal protection for labor unions, government funding for housing and public works—
all these reforms and many more were sought by these Socialists. Yet that
party never gained power, never won the right to shape law and policy on
its own terms. Debs himself, though a frequent presidential candidate on
his party's ticket, never polled as many as a million votes. The American
people were willing to embrace social reform, but they balked at socialism.
The influence of Debs and his followers would be great, but it too would
be exercised at one remove from politics. For America was unwilling to
accept the rest of Debs's program: government ownership and control of
the means of production. Similarly, America has rejected Nader's demand
that its industries be subjected to control by representatives of the public
interest—that is, of the public interest as defined by Mr. Nader. Ralph
Nader thus stands as the Eugene V. Debs of the late twentieth century.

6

TOWARD NEW FRONTIERS

The people involved with the space program are a rather sentimental lot; they are forever recalling anniversaries. Thus, ever since 1969 space buffs have had a special affection for July 20, the date of the first lunar landing. Indeed, in 1976 this even was the date on which the first Viking spacecraft landed on Mars. More recently, the first flight of the space shuttle *Columbia* took place on April 12, 1981, twenty years to the day after Yuri Gagarin became the first man to orbit the earth. But probably a more significant anniversary is the year 1982. For this is the fiftieth anniversary of the year that astronautics achieved its modern character and form.

By 1932 there existed the American Rocket Society. It had actually been founded a couple of years earlier, but at first it called itself the American Interplanetary Society. The name change was significant, for it signaled a willingness to downplay wistful dreams of the distant future and to focus instead on the hard, necessary task of developing the rocket and its applications as technologies in being. In time this focus would make it the oldest professional society devoted to rocketry and astronautics. During the '30s it actually sponsored an Experimental Committee that did rocket research, initiating a line of rocket development which ultimately led to engines for the X-1 and X-15 research aircraft as well as for the Vanguard satellite-launching program of the 1950s. Beginning in the 1940s, however, the society took on the character of a professional engineering organization, providing a forum for meetings and publications to serve its growing community of experts. In 1963 it merged with the Institute of the Aeronautical Sciences to form the premier aerospace organization, the American Institute of Aeronautics and Astronautics, or AIAA. But even in the 1930s, the existence of the American Rocket Society meant that astronautics would not remain in the hands of amateur enthusiasts. Instead, it would be an activity for serious professionals, who would always be interested in the hopes of the far future but who in the meantime would emphasize the prospects of the near term, and the problems immediately at hand.

Also in 1932, rocketry for the first time passed into the hands of people who would have the money and the motivation to carry forward with rocket development in a major way. I refer, of course, to the military. In 1932 it was the German Army, and to be specific, the Ballistics and Munitions Branch of the Weapons Test Department. That was the year that this weapons group established a rocket research center in Kummersdorf, near Berlin. The group's technical leader was Walter Dornberger; in 1932 he hired Wernher von Braun. As Dornberger later wrote, "We wanted to have done once and for all with theory, unproved claims, and boastful fantasy, and to arrive at conclusions based on a sound scientific foundation. We were tired of imaginative projects concerning space travel . . . We wanted to advance the practice of rocket-building with scientific thoroughness." They did precisely that, and outside of the Soviet Union their work dominated the fields of rocketry and astronautics until the advent of the Space Shuttle in the 1970s.

So modern astronautics was conceived fifty years ago, but it was a long time gestating, and did not truly hatch till Sputnik in 1957. Nevertheless, in the first years after that, its progress was extraordinarily rapid. By 1960, the United States already had a space program featuring nearly all the principal activities which occupy it today. We were launching spacecraft with the aid of the Air Force's Thor and Atlas missiles, and the Titan was under development. Twenty years later, descendants of these rockets would still be boosting satellites to orbit. In 1960 we were pioneering the very important field of satellite communications, launching the Army's Courier satellite as well as the experimental Echo I of Bell Labs. Echo was simply a big metallized reflecting balloon and required ground antennas of such extreme sensitivity that its approach to satellite communications was soon abandoned. Still it had its uses. The sensitive Echo radio antenna at Holmdel, New Jersey, in time passed into the hands of the physicists Arno Penzias and Robert Wilson, who used it to discover the faint microwave hiss proving the universe had originated in an immense explosion some eighteen billion years ago.

The year 1960 was a good one for space in other ways as well. The Navy launched its first Transit navigational satellites. We also launched the first weather satellite, Tiros I, and thus opened up the immensely fruitful application of satellites to meteorology, which after 1970 would broaden into the more general topic of earth observations. (The name Tiros is Greek for "cheese," and indeed the early Tiros satellites resembled a great cheese wheel, but the program managers could not make the connection explicit, so they invented the acronym of Television and Infra Red Observation Satellite.) In that year we also laid the groundwork for the planetary program, launching Pioneer V which communicated to earth by a radio link from a distance of 22,462,740 miles. To be sure, twenty years later we would be getting high-quality color photos live from Saturn, a billion miles away, thus demonstrating an improvement in space communications by a factor of something like a million. But the whole field of deep-space probes started with Pioneer V.

What was more, we demonstrated the successful orbiting of a satellite and its return of a protected capsule safely to earth. That was Discoverer XIII, the lucky thirteen of an Air Force project with close involvement by the Central Intelligence Agency. What the Discoverers were out to discover, of course, were the military secrets of the Soviet Union. To this end they carried cameras and other equipment, returning the film in those capsules. As early as 1955, Lockheed had begun developing the Agena second stage for Discoverer (the first stage was the Thor), and test flights began early in 1959. The art of recovering a capsule from orbit took some doing to develop. Thus, Discoverer II by misfortune delayed releasing its capsule, which parachuted down onto the Arctic islands of Spitsbergen. A Norwegian ski patrol spotted its orange parachute. So, apparently, did residents of nearby Barentsburg, a town inhabited by a colony of expatriate Russians. Needless to say, the CIA never got its capsule back. The next ten Discoverers also malfunctioned in odd ways, but at last came success. This success not only opened the way for the very active Air Force field of satellite reconnaissance. It also was the first demonstration of what would soon become a familiar sight to TV viewers, that of a returning space capsule gently swaying below large red and white striped parachutes, slowly descending to the ocean where it would be recovered.

In fact, there are only two major space activities we undertake today that we did not do in 1960. We were not launching astronauts into space. To be sure, though, this was not for want of trying; we were pushing ahead vigorously with Project Mercury, and for a while there indeed were optimistic souls who believed the first Mercury astronaut would fly into space in 1960. (Alan Shepard finally made it in May 1961.) More significantly, we were not launching satellites to geosynchronous orbit, 22,300 miles up, where they would hover overhead and match earth's rotation with their motion. In the 1970s, geosynchronous orbit would become a well-traveled thoroughfare, but in 1960 all attention was on low orbits, no more than a few hundred miles up. A launch to geosynch then was just as demanding as a successful mission to the moon. Indeed, by 1960 we had already shot off several lunar missions, none of which were more than a partial success. The Soviets had successful moon shots in 1959, but we did not match them till 1964, a year after Syncom I and II pioneered the geosynchronous communications satellite.

In 1960 there was a vigorous ongoing debate about our space program, with the Democrats charging the Republicans with having done too little and promising to do more. We now can say that even by today's standards, our space activities in 1960 were certainly both vigorous and protean. The Democrats however, once in the White House, soon initiated the Apollo program, and midway between the twin foreign-policy debacles of the Bay of Pigs and the Berlin Wall, the United States suddenly found itself committed to landing a man on the moon by 1970. For space program advocates, Apollo was a dazzling triumph, but it was a triumph too easily won. Like an inexperienced ingenue suddenly finding herself the toast of three continents, the space community was dazzled with a shower of new programs which for several years were quite exciting, but which in the longer run could only feed delusions of grandeur. Apollo did not contribute to attitudes of space-program leadership which would confront the hard, necessary task of justifying its costly projects in the face of a skeptical Congress and public. Instead it spawned the legend of a benevolent prince, a new John F. Kennedy, who would save the space program from the grubby work of competing for funds in the face of other national concerns, who would bestow upon it the Cinderella slipper of funding which would turn its pumpkins into fiery coaches bound for the stars.

From the early days of astronautics, visionaries such as Von Braun had written of lunar flights, space stations, missions to Mars. Apollo spread the notion that simply because such projects were feasible and exciting, they therefore represented serious claims on the nation's priorities. Prior to Apollo, few were so bold as to predict an early manned lunar landing. Arthur C. Clarke, author of 2001: A *Space Odyssey* and surely among the most daring of the space visionaries, forecast in 1947 that the first moon rocket would fly in 1959. He was right. However, he also predicted that men would not orbit the earth till 1970 or land on the moon till 1978. Apollo thus encouraged a frame of mind which would smile at even an Arthur C. Clarke as being too conservative, lacking optimism, destined to be overtaken by rapid technical progress outstripping even the most visionary of dreams.

It is worth remembering, however, that until 1960 just about everyone who wrote of lunar flight did so with the idea that future manned missions would fly to the moon for reasons such as scientific research or exploration, reasons which would be sufficient in themselves to justify the flights, somewhat in the way that we send picked crews of scientists to Antarctica or to the depths of the ocean. Apollo, by contrast, was an exercise in national technical muscle-flexing. Its flights were affairs of state, complete with presidential visits. By 1972 the program had been milked dry of contributions to national prestige, and the question of its scientific value then became preeminent. What happened, naturally, was that with each flight costing $400 million, Apollo's contributions to lunar science, a restricted and specialized branch of geology, were costing as much as all the basic science funded by the National Science Foundation. On this basis Apollo simply could not compete; Apollo 17, the first flight to carry a professional geologist, was also the last flight in the program. Back in the '40s and '50s, anyone could have predicted that if the lunar landings were to be occasions of state, they would first take place a lot sooner than if they had to be a matter of routine space operations. But today we are still very far from going to the moon in the fashion that we go to the Antarctic. From this standpoint, then, Apollo's optimism stands as ill founded, and in a very real sense we still have not yet reached the moon.

In the mid-60s, during Apollo's heyday, NASA spending reached $6 billion a year, nearly 1 percent of the gross national product. A few years later, in the post-Apollo era, that annual sum stood at $3 billion in inflated dollars, or only about 0.2 percent of the GNP. For most of the past decade it has stayed at that level, with cost-of-living adjustments. Space advocates have roundly condemned this, declaring it to be the work of "bean-counters" lacking in vision for the future. Yet the post-Apollo cutbacks and the subsequently level NASA funding have reflected a broad and general national consensus about space, which since the early 1970s has produced remarkably little controversy. The demands of space advocates then appear not as visions for the future, but merely as the pleadings of special interests. We can say that after the early Kennedy burst of enthusiasm, space funding settled down to a level merited by the good it could do the country. Communications, navigation, meteorology, earth observation, military reconnaissance, space science, planetary exploration—these have been the meat and potatoes carrying the space program forward. Significantly, all of these existed in at least rudimentary form in 1960, prior to Apollo. We could say that Eisenhower planted a sturdy tree which would be nourished and would grow to serve real needs and ongoing applications. Kennedy, by contrast, gave the nation a hothouse plant which would wither as soon as the sustaining flows of money were withdrawn.

Nevertheless, that tree still stands. The space program of the 1980s is a collection of practical, useful activities standing on their merits, which have survived an often intense competition for funds. The rapid expansions of the 1960s are past, but a hectic and exuberant adolescence has given way to a solid and capable maturity. Its basic character is well established: that of launching unmanned satellites to provide services to users. This maturity of the space program does not mean stagnation, nor does it mean—not yet, at least—that its activities are as standardized as those of commercial aviation. Instead, it reflects a twist of irony which the pioneers of space flight, visionaries all, failed to foresee. They looked to astronautics to provide spacefarers with ships transporting them on journeys to distant planets. But astronautics today finds its greatest usefulness in building unmanned craft to sit in geosynchronous orbit, hovering over one spot on the equator without losing altitude, and which thus in effect go absolutely nowhere.

Probably the most significant of these are the communications satellites. The services they provide are valuable, widely used, and generally unobtrusive, which is the way an advanced technology should be. Indeed, we have become so accustomed to color TV transmissions live from anywhere in the world, and to direct-dial long-distance phone service, that it is not always easy to realize how recent they are. Old TV hands know. They recall how it was during the Korean War, when battle action was filmed in black and white. The film would then be processed and developed and had to be carried by plane on a 36-hour flight across the Pacific until it could be fed into a network affiliate on the West Coast. Anyone who was overseas in the early '60s probably recalls waiting three hours or more for a phone connection to the States via cable or shortwave radio. The quality of the connection was about on a par with tin cans and string, and you paid at least $10 (the equivalent of $30 today), probably a lot more, for three minutes of the privilege. If you had mentioned direct-dialing to the overseas operator, she would have had to politely dispel your illusions.

No, things are a lot different now, thanks to the satellites. What's more, they're getting better. Thanks to a Federal Communications Commission decision, you now can get cheap long-distance telephone service through companies like MCI and Sprint. You pay a subscription fee of $5 a month for evening and weekend service, or $10 if you also want service during the business day. Then, for example, a five-minute call from Boston to Los Angeles costs $0.70. A half-hour call from Houston to San Francisco runs to $4.01. The corresponding costs via Ma Bell are respectively $1.41 and $7.52, so if you make even a few long-distance calls per month, these new satellite phone services will save you money. In the world of satellite communications, though, this is just the beginning. In addition to the rapidly growing areas of television, Telex, and long-distance phone service, there is the field of data transmission. This includes

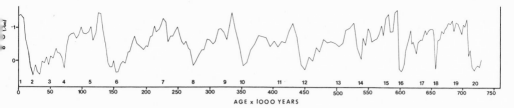

Fig. 17. *Emiliani's curve for temperature change during the past 700,000 years. Peak-to-peak change is about 10 Fahrenheit degrees.* (Courtesy Cesare Emiliani.)

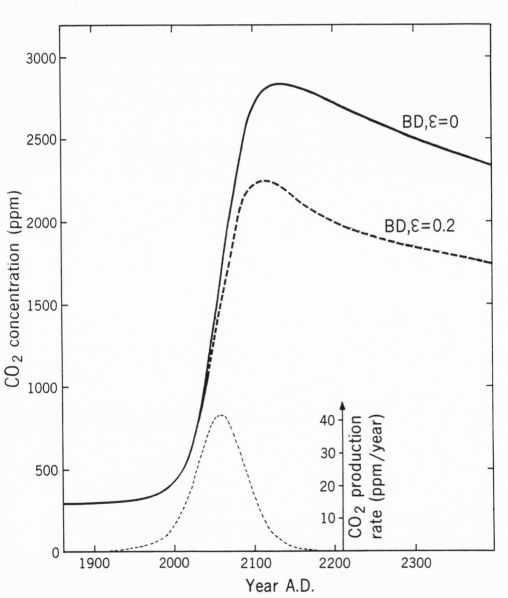

Fig. 18. *Predictions by Siegenthaler and Oeschger for the rise of carbon dioxide if all fossil fuel is burned. The two curves represent results from two different mathematical models. Bell-shaped curve at the bottom shows annual increase.* (Courtesy U. Siegenthaler.)

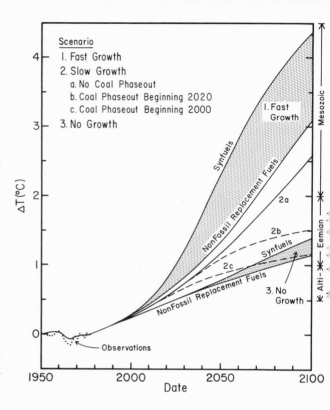

Fig. 19. *Predictions by James Hansen and associates for future temperature rises resulting from various scenarios for fossil-fuel use. "Fast Growth" is 3 percent per year. "Slow Growth" is half as fast. "Synfuels" replace oil and gas with synfuels as the former are depleted. "Coal Phaseout" scenarios diminish coal use over a 40-year period beginning in 2000 or 2020. "Altithermal" was the warm period 6,000 years ago. "Eemian Interglacial" was 124,000 years ago. "Mesozoic" was the very warm period in the age of the dinosaurs. It is likely that in 150 years, temperatures will be warmer than at any time since the dinosaurs died off. (Courtesy James Hansen.)*

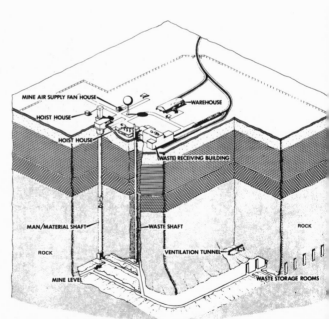

SIMPLIFIED CUTAWAY OF REPOSITORY

Fig. 20. *Proposal for disposal of nuclear waste. The waste would be stored thousands of feet underground, yet could be inspected and retrieved in case of need. (Courtesy Atomic Industrial Forum.)*

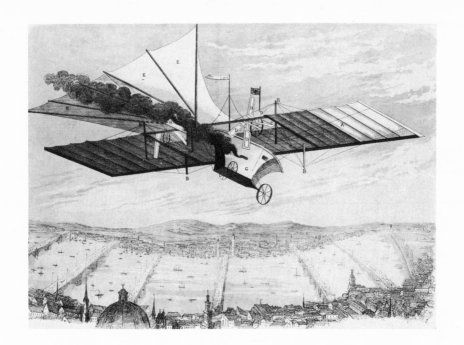

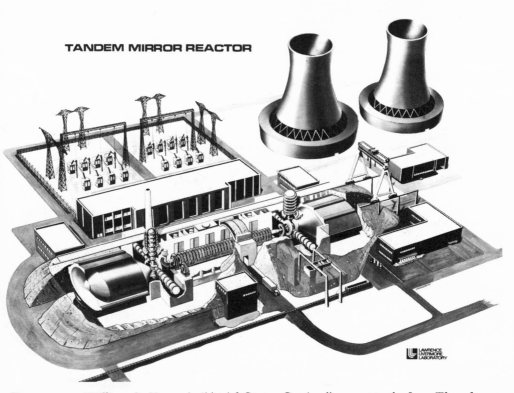

TANDEM MIRROR REACTOR

Figs. 21, 22. William S. Henson's "Aerial Steam Carriage" concept of 1843. Though eighty or more years ahead of its time, the design lacked ailerons and would have crashed if actually built and flown. Lawrence Livermore Laboratory concept for a future fusion plant based on the magnetic-mirror approach. A century from now such designs no doubt will show a similar mixture of prescience and egregious error. (Courtesy National Air and Space Museum and Lawrence Livermore Laboratory.)

Fig. 23. *The Japanese invade the West Coast: Datsun 280-Zs being off-loaded at Long Beach.* (Courtesy Nissan Motors.)

Fig. 24. *Meeting Japan on its own terms: robot welding systems, newly installed in the interior of the old Jefferson Avenue plant.* (Courtesy Chrysler Corp.)

Fig. 25. *Is this the Ford in your future? Artist's conception of the sort of small car Ford may be offering about 1990.* (Courtesy Ford Motor Co.)

Fig. 26. *New York's West Side Highway.* (Photo by Aryeh Brodsky.)

Fig. 27. *San Francisco's BART.* (Courtesy Bay Area Rapid Transit District.)

Fig. 28. *President Nixon greets returning Apollo 11 astronauts aboard the U.S.S. Hornet, July 24, 1969. The astronauts are in a mobile quarantine facility. The Apollo flights were affairs of state rather than exercises in science.* (Courtesy NASA.)

Fig. 29. *The Voyager spacecraft, taking close-up photos such as this one of Saturn's rings, have been among the triumphs of our "first look" strategy of planetary exploration.* (Courtesy Jet Propulsion Laboratory.)

Fig. 30. *Heavy Lift Launch Vehicle built using Shuttle systems. The airplane-like Orbiter has been replaced by a payload housing. The solid rocket boosters are dropping away at an altitude of forty miles.* (Courtesy Space Division, Rockwell International Corp.)

Figs. 31, 32. For *space buffs, lift-off of the Space Shuttle will be a familiar sight for years to come. It may be decades before the Shuttle gives way to some entirely new system such as Rockwell's proposed "Star-Raker."* (Courtesy NASA and Space Division, Rockwell International Corp.)

such business services as electronic mail and facsimile transmissions, as well as high-speed data links between computers.

International services are in the hands of Intelsat, the telecommunications consortium. It is an arrangement to give encouragement to all who hope for international cooperation. The highly publicized 1975 Apollo-Soyuz flight showed that two nations could cooperate briefly in space, at considerable expense. Intelsat has shown that its 106 member nations can cooperate in space, year after year, and earn a profit doing it. These nations run from Afghanistan to Zambia, and operate over 300 ground stations in some 150 countries. The countries include Libya and Israel, China and the Soviet Union, South Africa and Mozambique. The latest communications satellites, designated Intelsat V, have a capacity of 12,000 telephone circuits, fifty times more than on the first "Early Bird" (Intelsat I) of 1965. During those years the annual cost of a satellite telephone circuit has dropped from $32,500 to $800. The charge for an hour of prime-time color TV transmission has gone from $22,350 to $5,100. What's more, the growth of Intelsat and its services continues at an explosive pace. During the 1980s this growth is expected to double Intelsat's capacity every three years. The next satellite generation, Intelsat VI, will have a capacity of 40,000 telephone circuits on each spacecraft. The AIAA has projected that by the year 2000, some twenty Shuttle-loads of communications payloads will be headed for orbit each year, and this is just the traffic to be carried by U.S. launch vehicles. Already Intelsat has arranged to launch three of its Intelsat V craft using Shuttle's competitor, the European-built Ariane rocket. By 2000 the overall world demand for telecommunications may be five times larger than it is today. The growth in satellite communications will be even greater, perhaps by a factor of ten to thirty.

Such statistics, however, really deserve no more pizzazz than a report to Bell Telephone stockholders on new microwave repeater stations being set up along the highways. What counts are new services. These are on their way, and the most interesting of them will flow from the idea of complexity inversion. This idea harks back to the earliest days of satellite communications, when in order to reach orbit and function at all, the satellites had to be as lightweight and simple as possible. Thus, Echo I was just a big 136-pound balloon, entirely innocent of electronics; the ground stations, by contrast, were research-quality radio telescopes. Complexity inversion turns this around. It seeks to make the satellites big, powerful, capable, and complex, so the ground stations can be small, cheap, simple, frequently portable, and available to all. Already experiments have run satellite color TV through ground antennas as small as two feet across. Such an antenna would fit easily on your rooftop or in your backyard, and cost no more than a few hundred bucks at your nearby Radio Shack.

Direct TV broadcast will be especially easy to link in with existing cable television services. Direct broadcasting has faced legal obstacles from conventional broadcasters, but then so did cable TV, and this opposition delayed but did not defeat the cable services. When direct broadcast becomes available, at first few homes will have the necessary antennas, and the satellites will broadcast mostly to the cable services. In time more and more people will install home antennas, which will cost much less than a few years of cable subscription. Broadcasting directly to people's homes will be cheaper because of the high cost of adding to cable systems. In the long run the economic advantages of direct broadcast, with or without complementary cable TV, will prove overwhelming. Many local TV stations then will play a role similar to that of local AM and FM radio stations today, featuring news, movies, local events, and commercials. The networks will not rely on local affiliates for broadcasting, but with their satellites will offer a uniformly high quality of reception, if not of program content, everywhere in the country.

Mobile air and sea telephone and TV links will become common, just as has been the case with radar. Thirty years ago, ship owners often viewed radar as a costly and questionable solution to a problem they were not convinced they had. After all, for hundreds of years they had been navigating with reasonable accuracy while avoiding collisions, usually. Yet today even fishing boats carry radar. Similarly, we can expect them to carry satellite communications gear. Already there is the Marisat system to serve the ships at sea, as well as the international Inmarsat organization.

In the air, satellite communications links will enhance the pleasures of first class. More comfortable seats are all very well, as are better food and drink, but the airlines know full well that the back end of the airplane gets there the same time as the front end. However, most first-class travelers are business people, who often try to work while en route. Convenient phone service would certainly be attractive to them; already a few of United Airlines' aircraft have conventional phone equipment transmitting to the ground without satellites. As for the proletariat in economy class, they will be quite happy with a small TV screen on the back of each seat, which they can listen to with earphones. (That would still leave the forward wall available for an in-flight movie.)

Back on the ground, we can expect a great expansion of electronic mail, providing for rapid exchange of letters and documents among business and government users. For this service, each ground terminal would be a three-foot rooftop antenna linked to a facsimile machine, of the type operating over phone lines today to transmit a page of text every two to six minutes. Such systems would come in several sizes, from capacities of 31 letters per day up to 2,620 letters per day for large offices. There would be a "mail delivery" every ten minutes. Such systems would be cheap enough so that there could be half a million or more in use, serving offices with as few as five people. They would handle fifteen billion pieces of mail per year, and the cost per letter would be four cents.

By effectively bringing back the four-cent stamp, a company like AT & T, if sufficiently enterprising, could make the life of the Post Office just so much more difficult. So if we ever see such an electronic mail network, the government will probably run it as a legal monopoly, indeed as part of the postal system. To encourage its use, letters sent by electronic mail will cost rather less than those sent by ordinary first-class postage, though nowhere as low as four cents. The difference between the charged cost and the true cost then would serve to subsidize other Post Office services, and we could at least hope the cost of a first-class stamp would stop going up quite so fast.

All these services will call for ground or airborne antennas about the size of an umbrella. But there is no reason why satellite size and power cannot be boosted still further. A satellite of 54,000 pounds in orbit, with power of 280 kilowatts and antenna diameter of 220 feet, could communicate by telephone link with a ground antenna smaller than the one built into a pocket transistor radio. For those who like to be fancy, the ground receiver could actually be a wrist telephone resembling Dick Tracy's wrist radio. However, that would have a battery only large enough for about five minutes of use each day, between recharges. The true pocket telephone of the future would have the size and battery power of today's calculators and will be quite similar in appearance. It will have a digital display and a set of push buttons, as well as a speaker. A microprocessor within it would serve as your own personal switchboard. Like today's electronic pagers, beeping to let you know to call your office, it will travel anywhere: in your car, in an attaché case, even in a vinyl case looped onto your belt. It will not be the phone you will usually use to "reach out and touch someone." But it will be great during vacations, or to call ahead while driving, or for emergencies, or to keep in touch with it all while you're getting away from it all.

The success of satellite communications has long since brought them to the point where the spacy parts of the system fade into the background, leaving us free to make phone calls and watch TV. Another important space endeavor, the planetary program, has also shown dramatic success; but here the successes have gone far to leave it with only limited prospects for the future. Indeed, to a large degree the more it has accomplished, the less it has left to be done. It is a pity this is so, since this activity actually is the present-day fulfillment of the classical dream of exploring the planets by rocket. Just as Apollo has left us with little cause to send future missions to the moon, our planetary program has shown such brilliant accomplishments that it is well along toward working itself out of a job.

The U.S. planetary program began in the mid-60s with a rather uncertain focus. Its central point, so far as it had one, was the search for life on Mars, a strategy which emphasized one planet and one scientific field, biology. Bruce Murray, later to direct the Jet Propulsion Laboratory, proposed in 1968 that instead of this "Mars only" approach we should follow a "first look" strategy:

> In contrast, I advocate that we emphasize in our space strategy looking for the first time at new worlds. Go to each more-distant planet as soon as we possibly can . . . Since this is exploration, he who gets there firstest, gets the mostest. We must go for the most important, exciting scientific facts awaiting discovery at the distant new worlds . . . First of all—and this is the most important thing—we must be opportunistic, and maximize the exploratory return on a short-term basis. If there is a chance to go with something exciting now, do it. Don't worry about the future and the long-term evolution of spacecraft and so forth—just go . . . If a mission makes sense, particularly as exploration, if it is really going to increase our understanding or reduce in a dramatic way the enormous range of possibilities, go, go, don't wait.

Well, we didn't. In every year from 1964 to 1981, we launched or had a mission encounter at the moon or some planet. Often we had several such encounters in a year. We orbited Mars and mapped it with a camera, landed on its surface and looked for life. We flew past Venus, probed deep into its atmosphere, and by a neat trick of celestial mechanics flew by Mercury not once but three times. We discovered the volcanoes of Mars and of Io, satellite of Jupiter; viewed close up the rings of Saturn; were awed by the violence and the immensity of those giant planets. We took not only a first look, but a second and a third. From Venus to Saturn, every planet was the target for at least three visits by two or more distinct generations of spacecraft. We even took advantage of a rare alignment of the planets to send the highly advanced Voyager 2 on a planetary grand tour, visiting Jupiter in 1979 and Saturn in 1981, then flying on to planned encounters at Uranus in 1986 and at far-distant Neptune in 1989. As the poet Diane Ackerman has written,

> Neptune is
> elusive as a dappled horse in fog. Pulpy?
> Belted? Vapory? Frostbitten? What we know
> wouldn't fill
> a lemur's fist.

In a few years, however, what we know will fill at least a fair-size book.

With its astonishing successes, though, this "first look" strategy always carried with it the fact that sooner or later we would run short of interesting places to visit for the first time. That is the problem confronting the planetary program in the 1980s. The question is not whether we will send missions to explore the planets. We have already done so. The question is whether we can justify spending $500 million or more for a fourth or a fifth look at planets we have seen before. For example, when the Viking landers searched for life on Mars, they found no trace of carbon compounds, which are found in all forms of life as we know it. But these instruments also observed strange chemical reactions in the soil, which in some ways gave at least a superficial mimicry of life. It would evidently be interesting to learn more about this. It would also be interesting to return a sample of Mars' soil and rock for study in Earthside laboratories. But would it be two billion dollars' worth of interesting? That is how much such a sample-return mission would cost.

Planetary missions are funded out of public taxes, and the people who dispense this money know that to govern is to choose. To choose, in this situation, means deciding between programs which are vital or essential, those which are desirable, and those which would merely be nice to have. The planets have never had a firm hold even on the second category and increasingly have fallen into the third. That is why we will probably be launching only two new planetary missions during the whole decade of the '80s. In 1986 we will launch the Galileo orbiter mission to Jupiter. It will drop a probe into that turbulent and primeval atmosphere, studying that planet and its large satellites with cameras and sensors far better than the already very good ones aboard Voyager. Later, but no earlier than 1988, we will send a mission to Venus. It will orbit that planet with a very advanced radar system, peering beneath its dense clouds, and mapping its surface in enough detail to tell us a good deal about the geology of that enigmatic world.

During the late '60s and '70s, we were launching planetary missions at the rate of one or two per year. In the '80s and '90s, if we are very fortunate we will launch perhaps half a dozen missions in all. It would have been different if we had found something truly fundamental. If Mars was now teaching geophysicists how to predict earthquakes, if our studies of Venus' atmosphere had sharply illuminated the problem of climatic changes due to our carbon dioxide buildup, continued missions to the planets would stand today as projects of the first importance. As things stand, while the success of Intelsat will mean increasing traffic in orbit, the fulfillment of the long-held dream of exploring the solar system now may offer the planetary program no more than a genteel decline.

At the Jet Propulsion Laboratory and other planetary centers, it is fashionable to draw analogies between the discovery of America and our modern exploration of the planets. It is as much as to say that in space we will find the present-day equivalents of fertile land, productive mines, indigenous civilizations to be looted for gold, and natives who can be driven to the Cross. Everyone knows the names of Balboa, Cortez, Magellan; but a more pertinent comparison would be with explorers like Dirk Hartog, Jan Carstenz, François Pelsaert, Abel Tasman. These were the Dutch captains of the early seventeenth century, who discovered and explored the coasts of Australia. They described these lands in reports as an "accursed earth," full of flies that "perched on our mouths and crept into our eyes," and being "very dry and barren. We have not seen one fruit-bearing tree." Not till Captain Cook discovered the well-watered southeast coast a hundred and fifty years later did Australia begin to look attractive to Europeans.

Yet the most wasted and desolate of Australia's western deserts still will be far more clement than any environment found in space or on the planets, where even the oxygen must be manufactured. There has been some discussion of mining the moon for metals and other resources. But the moon has never undergone geochemical processes acting to concentrate metals into rich lodes of ore. Regarded as ore, lunar rocks or soils have about the resource value of ordinary clays or country rock. For a long time to come, the most important lunar resource will be the light it gives to lovers.

The lands of space are even more closely akin to those of Antarctica. No one is planning to establish colonies or cities in its frozen wastes, and while Antarctica has coal and other resources, they are insupportably costly to obtain. What counts in Antarctica is scientific research. That is why people go there, though they go not as settlers but as scientists, to live not in towns but in carefully prepared stations. The research results out of Antarctica are immensely valuable, and unlike oil or fertile farmland, the more widely they are shared and distributed, the more valuable they become. But Antarctica is not like India or America. No nation has risen to power by establishing an empire there, nor has any nation declined in power because of inattention to its glaciers and ice shelves. The same will probably be true of the moon and planets. Certainly, anyone who cherishes fond hopes for the future importance of these lands of space should first ask whether comparable hopes stand to be realized in Antarctica.

Nevertheless, in the 1980s we will indeed see new initiatives in space. The Shuttle has been widely touted as a means to open up entirely new and unprecedented space ventures, and to some degree this will happen. One area which will see activity is materials processing in space. In this area there have already been a number of predictions of commercial success. The basic idea is that in the weightlessness of space, processes become possible which are quite infeasible on the ground. Thus, it may be possible to grow exceptionally large crystals, or to produce alloys of metals which would separate out if mixed in gravity, or to achieve unusually high degrees of purity in materials. With characteristic enthusiasm, proponents of such space-processing ideas have drawn pictures of orbiting factories to "inaugurate a new age of technical civilization," or lead to a "third industrial revolution" in which space industries would become as pervasive and influential as the industries based on electronics and computers.

All this is far in the future, if it ever comes at all; but we can indeed look toward one goal. This is a program seeking to establish a national materials research center, based on the Spacelab manned laboratory which the Shuttle will be carrying to orbit. As the National Research Council noted in a 1978 report, the value of space for materials processing does not lie in commercial opportunity; at least, not in the near future. It lies in fundamental scientific research in such areas as the physics and chemistry of combustion, solidification, and convection, as well as phenomena occurring near the melting or boiling points of materials. However, such space research opportunities cannot stand alone. They must be guided by a strong program of ground-based research, not only to develop comparative data but also to guide the design of space experiments.

Should space manufacturing develop along these lines, we cannot soon look toward an era of space factories. But we will have a research facility resembling such astronomical spacecraft as the Space Telescope. As with such orbiting observatories, only the most carefully chosen problems, representing the highest scientific priorities, will gain experimental time. Again, as with these observatories, the choice of problems will be influenced strongly by the priority various problems will have received in ground-based work, and by the limitations on investigation which will have become apparent in such work. Because of their exceptional capabilities, the orbiting facilities will stand at the pinnacle of a hierarchy of laboratories to winnow and define the most exciting problems and observations to be studied in space. The resulting Shuttle-supported materials program will resemble other national research facilities, such as Palomar, Kitt Peak, Fermilab, the Francis Bitter National Magnet Laboratory, and Lawrence Livermore Lab.

To many people, a space materials lab would appear as a way station on the road to a space factory. This list of national laboratories was deliberately chosen to suggest this would not necessarily be the case. It is easy to envision that fundamental research at the frontiers of space will soon lead to new industries, but this connection is neither direct nor straightforward. When research points to something so basically new as computers or recombinant DNA, the new opportunities may develop with much encouragement and little competition. However, when research merely promises incremental improvements, the new prospects often will unfold only slowly, amid strong competition from existing industries. The slow growth of the solar-power and nuclear-power industries, in the face of continued availability of fossil fuels, is a case in point.

Space-based manufacturing may well be in the second category. As noted in that 1978 report, "When gravity has an adverse effect on a process, stratagems for dealing with it can usually be found on earth that are much easier and less expensive than recourse to space flight . . . The space environment usually contributes at least as many problems as it solves." Thus, in looking to the future of space manufacturing, two prospects may flow from the existence of a space materials laboratory program.

The problems addressed in space may become increasingly technical, remote from the real world of materials and processes. This would certainly be far from unprecedented. The Palomar and Kitt Peak observatories have nothing to do with the practical applications of astronomy in timekeeping and navigation. Nor do Fermilab or the other high-energy nuclear labs address issues touching fission power. A space materials laboratory could become a physics lab, aiding our basic understanding of matter rather than its practical application.

The second prospect is that a space materials lab could indeed contribute findings of practical importance—but these findings might find their greatest use in improving materials and processes in Earthside factories. New findings in materials science will far more likely lead to improvements in existing processes than will trigger the rise of industries in the entirely new milieu of space. If a space experiment produces a material having wonderful new properties, this does not mean people will rush to build an orbiting factory in order to produce it. More likely, the space results will be regarded as demonstrating a new standard, and scientists on the ground then will set themselves the goal of meeting this standard by the clever use of existing processes.

This issue has been put in particularly sharp focus by the recent invention of metallic glasses. These are produced when molten metal alloys flow through a slot onto a chilled metal wheel which rotates rapidly. When the molten alloy touches the rim, it "splat-cools" into a solid ribbon in about a thousandth of a second. This very rapid cooling leaves its molecules arranged in random, formless patterns like those in glass, rather than leaving the metal frozen into a mass of tiny interlaced crystals as in conventional alloys. The equipment for producing these metallic glasses is very cheap, and these ribbons can be extruded at 6,000 feet per minute. They are potentially the strongest, toughest, most corrosion-resistant, and most easily magnetized materials known to man. They have been produced with strengths far greater than commercial steel and have been made more corrosion resistant than stainless steel, lighter and stronger than aluminum. Such ribbons can be woven to make fabrics, braided to make tubes, wrapped in a helical pattern to make cylinders, or laminated to make plates. John B. Slaughter, former director of the National Science Foundation, has written that "the potential for these and other new glasses is enormous, perhaps rivalling the impact on modern society of the semiconductor technologies." Certainly space-produced materials will have far to go to compete with them.

The picture presented here is one of a mature space program, which during the rest of the century, and probably well into the next, will be making solid but often routine contributions to the work of the nation. Some areas such as communications will see rapid growth and expansion. In others such as planetary exploration, recent rates of activity will be cut back. In still others, such as materials processing, we will be gingerly feeling our way as we seek to discover what may be found. Yet all these efforts, and many more, will rely on the Space Shuttle. Its capacious cargo bay has from the start carried high hopes as well as payloads, and it will be the mainstay of our activities for at least the balance of the century. So it is appropriate to offer a few predictions as to what will become of it.

Its development will follow the well-trodden path of earlier launch vehicles. From the moment we could begin launching satellites with converted military missiles, about 1960, our rocket builders did not content themselves with what they had; they sought to make these rockets better. This meant developing advanced upper stages like the hydrogen-fueled Centaur, solid-fuel booster rockets, more powerful rocket engines, and larger tanks. By such means the Atlas, for example, originally an intercontinental missile not even intended to reach orbit, was developed into the Atlas-Centaur that could hoist Intelsat V, weighing more than a ton, to geosynchronous orbit. For the Shuttle, we can expect a collection of improvements offering advanced prospects for research in space, longer durations in orbit, heavier and more bulky payloads, and new upper stages.

Already nearing completion is Spacelab, a large cylindrical cabin which will be carried in the Shuttle's payload bay. It is being built by a European consortium as a research laboratory to be fitted out with a variety of items of equipment and will carry scientists who will work in orbit. The flights of Spacelab will begin about 1984 and will for the first time see Europeans flying into space aboard the Shuttle. However, in its basic design the Shuttle can stay in orbit for only a week before it begins to run low on supplies. For extended flights, there will be a 25-kilowatt solar array to generate electricity. This will be a large solar panel, over 200 feet long when deployed. With this extra electric power, it will be possible to add extra oxygen for the crew as well as extra groceries and toilet paper. Missions then will fly for a month and eventually may be up as long as four months.

A variety of improvements will accommodate payloads bulkier or heavier than the standard limit of 65,000 pounds for the basic design. By simply attaching small solid-fuel rockets near the back end of the Shuttle's booster rockets, the payload can be boosted to 75,000 to 90,000 pounds. The standard Shuttle relies on its large solid boosters during its initial ascent. Eventually these may be replaced by more powerful liquid-fuel boosters burning hydrogen and oxygen and incorporating the same powerful engines used in the Shuttle Orbiter. Such liquid boosters would bring the payload to a level of 100,000 pounds. With them, the Shuttle at launch would not produce its white-hot flame and heavy plume of smoke reaching from the launch pad far into the sky. The flame of a hydrogen rocket is not only smokeless but virtually invisible, and the Shuttle would appear almost to levitate itself into its ascent.

Other proposed developments get fancier. There is the idea of a Heavy Lift Launch Vehicle to be assembled from Shuttle components. It would keep the big fuel tank and booster rockets of the basic Shuttle, but would dispense with the airplane-like Orbiter. Instead, the Orbiter's three main engines along with its maneuvering rockets and its computers —that is, all the expensive stuff—would be assembled into a compact capsule or propulsion module. The Orbiter would be replaced with a simple cylindrical payload housing, and the payload could be as much as 185,000 pounds. Once in orbit, the propulsion module would be cast loose to maneuver back into the atmosphere to descend to the ground for recovery and reuse.

Particularly bulky payloads could ride on the front of the external tank, the large fuel tank carrying most of the Shuttle's propellants. This tank is 27.6 feet wide and could carry a payload compartment on its front end, holding payloads rather wider than the standard 15 feet carried in the Orbiter itself. For even larger payloads, this forward compartment on the tank could swell to a diameter of 37 feet, a configuration known as a hammerhead. The day may come when the Shuttle will look like an airplane attached to the stalk of a mushroom-shaped water tower.

Finally, there will be upper stages. The Centaur will be fitted into the payload bay, and this resulting craft, the Shuttle-Centaur, will launch the Galileo mission to Jupiter. An even more advanced upper stage will be the SEPS, Solar Electric Propulsion Stage, which will use the principle of ion propulsion. Rather than burn fuels in a rocket engine, SEPS will use electric fields to produce a beam of electrically accelerated ions of mercury. The thrust from the beam will be feeble, but this thrust will continue for months, allowing SEPS to reach very high velocities while using little propellant. SEPS will carry particularly large spacecraft to geosynchronous orbit. It also will be a mainstay of future planetary missions, and following the Venus radar orbiter, most of what missions we launch to the planets will rely on SEPS.

And who will benefit from the added power and flexibility these improvements will bring; who will pay for their development? The answer may well be the Air Force. To a degree far greater than is commonly realized, the Shuttle is a military spacecraft. When NASA initiated its Shuttle program in the early '70s, its managers were well aware that, as a National Academy of Sciences review board had written in 1971, "It is clear that space science and applications by themselves are insufficient to justify the cost of developing the shuttle." But when NASA sought Air Force support, the Air Force drove a hard bargain. NASA eventually agreed to absorb the cost of building two Orbiter vehicles for virtually exclusive Air Force use. They also built the Orbiters with a delta wing, larger and heavier than what NASA would have preferred, simply because the Air Force wanted it. What was more, at least one classified payload was so large and important that the Air Force was able to insist in 1971 that the Shuttle payload bay be 15 by 60 feet in size and pack 65,000 pounds, instead of the more modest numbers NASA wanted.

All this means the Air Force will be calling the tune in rocketry for a long time to come, just as various military departments have done so, both here and elsewhere, since 1932. The great flexibility of the Shuttle, its susceptibility to improvement and advance, means that it will be not years but decades before it is replaced by a brand-new launch vehicle employing future technology. In this respect the Shuttle will be much like the B-52 bomber. Designed in 1948, first flown in 1952, the last B-52s were built in 1962. Those postwar years saw many of the most rapid and dramatic advances in the history of aviation, and it appeared inconceivable that the B-52 should still be in service in the 1990s. It was almost as if the Sopwith Camel of World War I were to have served in the Korean War. Yet it will be 1985 at the earliest before the Air Force receives the first new aircraft of the generation that in time will finally retire those doughty old B-52s. The Shuttle will show similar longevity.

So we cannot expect the year 2001 to show a futuristic Pan Am spaceliner whisking passengers to a big space station amid the strains of the "Blue Danube Waltz." Nor will space flight be cheap. The Shuttle has been widely hailed as inaugurating a new age of low-cost space transportation. In reality it will not even offset the effects of inflation. In 1971, Martin Marietta was advertising a version of its successful Titan III launch vehicle which could carry payloads to orbit at a cost of $435 per pound, or $1,100 in 1982 dollars. Today NASA will rent the Shuttle's full payload bay capacity for $71 million a flight; with a payload of 65,000 pounds the launch cost will also be $1,100 per pound. That is a far cry from $5 a pound which in 1967 Barron Hilton, son of Conrad, stated would be adequate to allow him to build a hotel in orbit.

For those who wish to experience the thrill of spaceflight, not all is lost. There is always the Concorde for a transatlantic flight. Test pilots know that above 50,000 feet in altitude there is really very little difference from being in space. A Concorde passenger will board a sleek delta-winged jet looking remarkably like the Pan Am spaceliner of the movie 2001. There is the acceleration from the powerful Rolls-Royce Olympus engines, and the digital Machmeter on the forward wall will show your speed in multiples of the speed of sound. Out the window is the velvet-purple sky, brightening to a light-colored band above the horizon; if you look closely you will see the curvature of the earth. The coastlines of Cape Cod or of southwest England will be as distinct as on any map. Of course, the trip lasts less than four hours, and there is no weightlessness; you won't have the chance to make love in zero-g. But at a cost of only $2,000 or so, it's a real bargain. And when you get off the plane, you won't be in a cramped lunar base or an austere space station. You'll be in Paris.

Still, as Senator Edward Kennedy said on a different occasion, "The work goes on, the cause endures, the hope still lives, and the dream will never die." The legacy of space is more than applications programs, more than projects with greater or lesser degrees of usefulness. From its very inception, astronautics has been a spur to imagination, to hope, to endeavor. The plans and efforts put forth for space colonization, for extensions of human civilization into space, for interstellar voyages—these too are something significant. Then there is the enormous popular success of science fiction, of *Star Wars* and *Star Trek*, of generations of fictional heroes and villains from Buck Rogers to Darth Vader. In their own way these are consequences of astronautics just as real as spacecraft and satellites. They may not be up to the literary level of Ernest Hemingway or William Faulkner, but they have been a part of our lives. Space offers more than a collection of technical enterprises. It is a cultural endeavor, and no account of its prospects can overlook this. Space grips the popular imagination in ways no Antarctica, no Everest, no abyssal Challenger Deep can match.

There is a witchery to space, closely allied to that interest in celestial matters that is older than civilization. Astronomy is the most ancient of the sciences; for ten thousand years men and priests have declared their gods reside in the heavens. The flight of a rocket then readily conjures metaphors about mankind breaking its bonds, entering the realm of those gods. When we launch a rocket, we send fire leaping into the sky. We thus link two of man's strongest and deepest sources of awe, of ritual, of magic. The effect is enhanced by the rocket's ground-shaking thunder, of which only a pale echo reaches the TV viewer. The magic is strengthened even further when the rocket carries on its side the painted symbol of the nation.

Beyond this, space flight resonates peculiarly strongly with our national mythology. We use the word "myth" as a term of derision, meaning a superstition or falsehood which people should see through. But a mythology can as well sum up the character of a people, their view of themselves, and the folk history by which they define their background. In America, alongside our love of liberty and our belief in the perfectibility of societies and institutions, we cherish a sense of our growth as having been shaped by ships and railroads, and by the land. We see ourselves as having crossed the ocean to the New World, where from settlements on its eastern shore we set out into the West. Always there was the land, recently discovered and explored, extending toward the sunset, with a boundless promise of hope and new opportunity. America has always been a young country; we have had no history as the Europeans know it. Here are no cities founded during the Roman Empire, no landed barons whose titles date to the Norman Conquest. We have grown and prospered by looking to the future, by seeking vigorously the new territories disclosed by exploration, by breaking the land and making it fruitful with the aid of machines. Tom Paine wrote, two centuries ago, that in America there was the chance to "begin the world anew"; and so it has been.

Space flight offers hope that these centuries of progress will continue. Like America itself, astronautics has offered little history but great hope for the future; bright, brilliant prospects in the new lands of planets and space the explorers have disclosed; great rewards to be won in its vastness with the aid of technology. Significantly, in the Soviet Union this same mythology is cherished with at least an equal loyalty. Russia is an old land, but so far as has been possible its leaders have shut up its kaleidoscopic past in museums, have cultivated the fiction that its history began only in 1917, and have resolutely sought to turn the nation's face to the future. That nation, no less than ours, has been made great largely through the vastness of its land. What is more, since its Revolution it has seen a continued emphasis on technology as the key to a better life ahead, and a more powerful nation. In both the United States and the Soviet Union, astronautics is so in keeping with the national character that it is almost a surprise not to have found Alexis de Tocqueville predicting the space race.

Because of this, we will probably always be doing more in space than we would otherwise. We will keep alive our planetary explorations, if only at a modest level, since the alternative would be to stand with the gloomy St. Augustine in his later years and to declare, "I no longer dream of the stars." As we continue to launch satellites from Cape Canaveral, we will continue to launch interstellar starships in our movie theaters. The hope will endure that someday we will find our future hitched to a star. And who can say the hope is wrong? It is easy enough to project a straightforward, even routine array of space programs in the next twenty years. But there will be a twenty-first century, and a twenty-second century, a twenty-third, and twenty-fourth. They will likely differ from each other, and perhaps from our own, as much as did the sixteenth, seventeenth, eighteenth, and nineteenth. With space attainable, with bold undertakings achievable, it would be the height of folly to assert we will never build space colonies or send ships to the stars. As Senator Kennedy said, "the dream will never die."

7

THE ELECTRON
AND THE INDIVIDUAL

On maps it is called the Santa Clara Valley. It extends north from San Jose, along the San Francisco Peninsula, and is a new kind of industrial region. Its importance in the world is rising rapidly to challenge that of such long-established areas as Germany's Ruhr Valley, with its extensive coal and steel industries, but here are no fume-belching smokestacks, no dreary coal-laden railyards. Nor are there the grimy, ancient factories one sees in almost every town or city when traveling through the Northeast. Instead there are low, neat company buildings showing attractive architecture and landscaping, with plenty of windows, looking for all the world like so many college campuses. Their people appreciate this, for many of them have advanced degrees. This California valley is the nation's leading center for electronics, and in time even the mapmakers will probably change its name to conform with local practice. For to its people, and throughout the world, it is known as Silicon Valley.

Silicon Valley furnishes an excellent example of how a first-class university, supporting research in advanced areas of science, can serve as a seed which will bring a wide surrounding area into economic bloom. The story of Silicon Valley actually began in the late 1930s at Stanford University in Palo Alto. There Fred Terman, a professor of electrical engineering, was in the thick of research on radar and other advanced topics of the day. He encouraged some of his most promising students to establish electronics firms locally, among them Russel and Sigurd Varian,

William R. Hewlett, and David Packard. After World War II, Terman and the university administration persuaded the trustees to establish an industrial park on land owned by the university. Thus began Stanford Industrial Park, which became the nucleus of electronics in Palo Alto. The first industrial tenants were Varian Associates and Hewlett-Packard, and in time nearly seventy other firms would take the opportunity to establish facilities in that park, where they could be close to the talent of Stanford.

The real founder of Silicon Valley, though, was William Shockley. He had been raised in Palo Alto, and called it his hometown. In 1928, age eighteen, he left to go to college, winning a B.S. from Caltech and a Ph.D from Massachusetts Institute of Technology. He then joined Bell Labs, directing research efforts that in 1947 brought the invention of the transistor, and which ultimately brought him a share of the Nobel Prize in physics. Finally, nearly thirty years after leaving home on the southbound train, he returned to Palo Alto. In 1956 he set up Shockley Semiconductor Laboratory, staffing it with an exceptional group of engineers, including such future industry leaders as Robert Noyce and Gordon Moore. Within two years, eight of Shockley's key people, including Noyce and Moore, left to found their own firm. This was Fairchild Semiconductors, which they set up with support from the East Coast firm of Fairchild Camera. Fairchild set up shop in Mountain View, just a few miles down the road from Stanford. It was an almost immediate success, but hardly was this company well launched when some of its bright people started a trend by leaving the parent firm to set up their own enterprises. In 1959 several Fairchild alumni set up Rheem; in 1961 and 1962 others started Signetics, Amelco, and Molectro. Meanwhile, five people from Varian were setting up Spectra-Physics to develop applications of lasers, and Memorex was getting started in Mountain View. All through the 1960s Silicon Valley kept on growing, and then between 1967 and 1969 came a bumper crop of new companies. Some of the most significant were National Semiconductor, Precision Monolithics, Computer Micro-Technology, Qualidyne, Advanced Memory Systems, and Four-Phase—all founded by Fairchild graduates. The most important of these new firms got its start in 1968, when Noyce and Moore left Fairchild with $500,000 of their own money and $2.5 million in venture capital. The firm they founded was Intel, which soon was introducing the first microprocessors. Those early efforts, the Intel 4004 and 8008, were among the first to pack the power of a computer onto a single small silicon chip, thus opening the way to small calculators, digital watches, electronic cash registers, personal computers, video games, automated bank tellers, and much else.

Things kept right on rolling in the 1970s. At IBM in Menlo Park, Gene Amdahl was director of the Advanced Computing Systems Laboratory. He urged his company management to back him in a large new computer he wanted to build, but they declined. So Amdahl left IBM in 1970 and set up his own company in Santa Clara. Six years and $40 million later, the first Amdahl computers went into use. Meanwhile, Atari was introducing Pong and its other video games. Steven Jobs, a young Atari engineer, believed it was possible to introduce a personal computer based on the chips Atari was using in its games. As at IBM, though, he could not persuade his management to back him. So along with his friend Stephen Wozniak of Hewlett-Packard, he too hit the road, but they didn't go far —only to nearby Cupertino. There they founded Apple Computer. Their hunch was right, of course, and a few years later their firm was worth $600 million.

Silicon Valley is far from being the nation's only advanced center for electronics. Around Dallas is a similar region, centered on Texas Instruments. Near Phoenix, Motorola plays a similar role. Outside Boston, Route 128 was for many years the nation's leading electronics center, with Raytheon and MIT playing the role of Hewlett-Packard and Stanford. But nowadays everyone looks to Silicon Valley. It is home to some 1,000 electronics companies, and these and other high-technology firms employ 630,000 people in the region. The entrepreneurial climate is unmatched; four out of five new companies succeed. However, the place is getting crowded. Although 40,000 to 50,000 new jobs are being created annually, only 12,000 new housing units are available each year, and the price of an average house in Palo Alto shot up from $56,000 in 1973 to $150,000 in 1979. So it should be no surprise that just as Fairchild spawned a host of newer firms, Silicon Valley itself may give rise to new electronics centers in other states. In April 1980, Maryland's Governor Harry Hughes hosted a luncheon for industry leaders at the Santa Clara Marriott, hoping to lure some of this growth eastward. The menu, titled "Tastes of the Chesapeake," featured oysters on the half shell, sautéed oysters, cream of crab soup shipped from Maryland, and crab cakes prepared according to a special Maryland recipe. Said James Belch, Maryland's director of industrial development, "Silicon Valley is similar to the Chesapeake Bay if you turn it upside down."

What all this activity is concerned with, of course, is microelectronics. It is fashionable to talk of a microelectronics revolution, and within its scientific circles, claims of revolution are nearly as common as within left-wing political circles. But just as not much is truly new under the left-wing sun, so claims made by electronics buffs must be taken with a sizable grain of silicon. Thus, here is a prediction made in *Time* magazine some years ago:

The house was like none ever built before. Its roof was a honeycomb of tiny solar cells that used the sun's rays to heat the house, furnish all the electric power. Doors and windows opened in response to hand signals; they closed automatically when it rained. The TV set hung like a picture, flat against the wall—so did the heating and air-conditioning panels. The radio was only as big as a golf ball. The telephone was a movielike screen, which projected both the caller's image and voice. In the kitchen the range broiled thick steaks in barely two minutes. Dishes and clothes were cleaned without soap or water. The house had no electrical outlets; invisible radio beams ran all appliances. At night, the walls and ceilings glowed softly with glass-encased "light sandwiches," which changed color at the twirl of a dial. And throughout the house, tiny, unblinking bulbs of a strange reddish hue sterilized the air and removed all bacteria.

You will readily appreciate that this was not written last week, but you can be forgiven if you date it to 1970 or so. Actually, it led off a cover story on electronics titled "The New Age," in the issue of April 29, 1957. A quarter-century later we do have the microwave ovens, but that's about it. Flat-screen TV is still far away, as is ultrasonic dishwashing and laundry. The beaming of electric power was a laboratory experiment in 1957, but will hardly replace the good old reliable plug in a wall socket. More interesting are predictions like the Picturephone and the golf ball-size radio. The Picturephone is thoroughly feasible and has been for some time. The only problem is, it's impossible to transmit a decent TV picture over ordinary telephone lines. The advent of light-wave communications, featuring lasers and optical-fiber cables, today gives new promise to the Picturephone, but it still will be some time before this service is available in ordinary households. As for the radio, nowadays one finds both AM and FM in a set of headphones which can be worn while jogging or at the beach. Also there are sunglasses which actually contain a radio with an earphone. But in a living room, "*the* radio," meaning the main one in the house, is usually the AM and FM bands of a stereo receiver, and this has hardly shrunk at all since 1957. Conspicuous by their absence are predictions of calculators and home computers. In 1957 a computer was a roomful of racks of vacuum tubes and other equipment; only major companies or the Air Force had them. Today there is more power in the desk-top systems you can buy at your local computer store, while programmable hand calculators, computers in themselves, are sold in stationery stores.

Then there are the photovoltaics, the solar cells. It still is not possible to phone a local building-supplies company and order a photovoltaic roof, but here at least there are strong grounds for optimism. Solar cells were invented at Bell Labs in 1953, while integrated circuits were invented at Texas Instruments in 1958; perhaps that is why *Time's* 1957 predictions mention the one and not the other. But the two technologies share many similar features, and the next few years may well see the rise of a true photovoltaics industry. Those who know this field assert the basic methods and techniques are in hand, and as markets expand and production experience increases, the costs and performance of solar cells will improve apace and will open the way to still wider applications for them.

This is reminiscent of the situation with integrated circuits in the early 1960s. As Robert Noyce has noted, by 1960 the basic methods were already available for designing and fabricating integrated circuits on silicon microchips. It then was merely (merely!) a matter of driving down costs through competition and production experience, while designing the circuits with increasing cleverness to give steadily better performance. The solar-cell industry thus may follow the integrated-circuit industry, with a lag of twenty years; about 1990 we may see a real commercial market for these cells. Thus far they have been bought for use in remote areas, serving radio repeater stations and signal lights in places where the cost of installing a power line would be prohibitive. The Department of Energy has actively nurtured this industry, sponsoring such installations as a 3.5-kilowatt solar array serving the Indian village of Schuchuli on the Papago Reservation in Arizona. Larger installations are on their way. Already there is a 100-kilowatt array providing all the electricity for the buildings at Natural Bridges National Monument, Utah.

There are also houses with photovoltaic roofs. The Department of Energy has built them in Arizona, Florida, Massachusetts, and Hawaii. These count as experiments, since photovoltaics today cost about $5 per watt when in full sunlight, about ten times more than electricity from your nearby utility. Still, this is a good improvement over the $15 per peak-watt of 1977, and a vast improvement over the $1,000 or so per peak-watt in the 1950s.* When photovoltaic roofs do become available, they will offer a number of advantages. They can serve as total home energy systems, since the sunlight that does not convert to electricity—about 85 percent—can heat water or warm air. Since homeowners will own them, they will actually add to the value of the house. Most homeowners will still need tie-ins to the district electric grid, in case of cloudy weather and at night, but by law, people will be entitled to sell their surplus electricity to the local utility and thereby run their electric meters backward. They can also cut their electric bills by scheduling some activities for the sunny part of the day. For thousands of years, housewives waited for a sunny day to do the laundry. Soon they may do so again, but for a different reason.

Overall, what electronics is bringing is not a revolution so much as a New Deal. Franklin Roosevelt's administration brought sweeping change and reform to many institutions and policies, but he did not turn the nation upside down. Rather, his New Deal was carried out well within the

* These costs refer to capacity; e.g., an array capable of generating 1,000 watts today might cost $5,000. The actual electricity it generates, even though made from free sunlight, has a price resulting from this cost.

confines of the Constitution and of our system of laws. He strengthened and buttressed some aspects of the nation, introduced innovations into others, but he steered clear of a wholesale overturning. Rather, his was the creative use of existing laws and institutions.

Similarly, electronics today stands as the consequence of a century and a half of development and change, dating to Samuel Morse and his telegraph. Electronics today calls for the inventive use of existing physical laws and industrial or commercial institutions, but the resulting changes are less a radical shift than an extension and continuation of trends long established. Better and more varied communications, more diverse forms of entertainment, new energy sources, new or more sensitive instruments, improved control systems, higher productivity in manufacturing and in the service industries, new ways to handle and process information, better ways of doing calculations—all these are trends long predating the microchip. Electronics today brings us, not a new revolution, but an extension of an old one. It brings a continuation of the Industrial Revolution by other means.

Viewed as an enterprise in itself, however, modern electronics stands as a monument to the creative turning of necessities into virtues. Microchips, integrated circuits, are rich in transistors, but silicon is not necessarily the best transistor material. Silicon is indeed only a so-so semiconductor; in some ways a better one is germanium, and indeed the early transistors were made from germanium. What gives silicon its value is not the silicon itself, but its oxide. When manufacturing a microchip, the oxide readily forms on the chip surface as a dense, impervious film, adhering tenaciously to the underlying silicon and providing exceptional electrical insulation even in thin layers. It thus becomes possible to prepare the chip as a finely detailed array of conducting and insulating zones, which is the basic approach to chip fabrication.

However, only a few electrical circuit elements can be fabricated in silicon. If you look inside a stereo or TV set, and understand what you are seeing, you will find the circuitry is made up of transistors, diodes, capacitors, resistors, and inductors, all strung together with wires or conductors. On a silicon chip it is easy to make transistors and diodes, and thin pathways of aluminum can be laid down to serve as conductors. But it is difficult to make resistors. Fortunately, it is possible to make transistors serve most of their uses. Only very small and weak capacitors can be fabricated in silicon, however, and inductors are out of the question. In practice, then, a microchip often turns out to be mainly an intricate array of transistors. This would represent a very severe limitation in ordinary practice, since the capacitors and inductors serve to ensure the faithful reproduction of the complex wave forms they receive. An example of such a wave form is the squiggly groove on a stereo record. Transistors, by contrast, act mainly as switches, permitting small flows of current either to stop or go. But they are exceedingly good switches, and a circuitry based on them can operate very capably if it deals with binary digits—strings of 1's and 0's, representing switches which are either on or off. That is why modern electronics shows a heavy bias toward digital operation. It is an elaborate stratagem to avoid the need to deal with continuously varying wave forms, the stock-in-trade of more traditional approaches to electronics.

So we wind up using silicon not so much for itself as for its oxide, and we go to digital operation because it is almost the only thing circuits fabricated in silicon can do well. Nevertheless, the resulting technology has proved so widely useful, so susceptible to innovation and improvement, as to stand as a wonder of the age. The first integrated circuits, in 1959, featured only a single transistor. In 1964 Gordon Moore, then director of research at Fairchild, pointed out that the number of elements in advanced integrated circuits had been doubling each year. He set forth "Moore's law," stating that this trend, the doubling in complexity each year, would continue. By 1977 circuits containing 2^{18} or 262,144 elements were available, just as Moore had predicted. Since then things have slowed down a little; early in 1981, Hewlett-Packard announced a chip containing some 600,000 elements. But the industry today is full of preparations for the chips of the later 1980s, which will contain millions of transistors and other elements.

What has driven these advances? It is a common trend in an industry that the more experience it has the more efficient it becomes, so its prices (in constant or uninflated dollars) can drop. Most industries reduce their constant-dollar costs by 20 to 30 percent each time their cumulative production doubles. Data from the semiconductor industry shows integrated-circuit costs have declined 28 percent with each doubling of the industry's experience. However, most industries serve markets which grow in pace only with the population or the gross national product. The proliferating electronics markets, by contrast, gave rise to a two thousandfold increase in annual production of transistors (mostly on chips), representing eleven doublings, in the seventeen years from 1960 to 1977. This means that during those years, the price of the product was dropping by close to 20 percent per year, in constant dollars. This naturally gave an enormous impetus to research and development. A year's lead in introducing a new product could give a company a 20 percent cost advantage over the competition; by contrast, a year's lag would mean a comparable disadvantage.

For several years the more advanced chips have contained more circuitry than the most advanced electronics systems that could have been built in the 1950s. However, few chips have cost as much as $100; most have cost far less. There are a variety of ways to visualize this complexity and low price. The first large electronic computer, ENIAC, was built in the 1940s as a wartime crash program. As early as 1976, the Fairchild F8 microcomputer had somewhat better performance, but was 300,000 times smaller, consumed 56,000 times less power, was more than 10,000 times more reliable, and cost 10,000 times less. Put another way, the modern microcomputer can do everything its vacuum-tube grandfather can do, but it fills a briefcase rather than a meeting room, and uses the power of a light bulb rather than of a locomotive. In terms of cost, the early transistors cost $200 each; since at least 1977 a jelly bean has been more expensive than one transistor on a complex chip. In terms of complexity, a microchip with its transistors and aluminum conducting paths can be likened to a street map of a city. In the 1960s, chips were about as complex as a small town. Circuit designers could know their way around the chip as they knew their way around their hometowns or neighborhoods. Today's advanced chips have the complexity of the Dallas-Fort Worth metropolitan area. By 1990, chips may be as intricate as a complete street map of the United States. What's more, all this detail will have to fit onto a chip area the size of a postage stamp. Moreover, the standards for perfection are stringent. A typical calculator chip today may have 6,000 transistors, somewhat more than the number of city blocks in Manhattan. Each one must be interconnected by a thin aluminum path, and the entire chip is useless if there is a fracture in the metal at any point. That would be like shutting down Manhattan completely if any street had a crack across it more than a foot wide.

The ordinary hand-held calculator is a typical product of these advances, and it is appropriate to give a few words in its praise. In the scientific world, a standard book of mathematical tables is Jahnke and Emde's *Tables of Functions*. In its 1933 edition, Fritz Emde had this to write:

> The Table of Powers was calculated on a Brunsviga nova calculating machine in which the resulting product can with a single turn be put back into the machine as a factor for further multiplication. I must express my gratitude to Herr Otto Hess, the Stuttgart agent for the Brunsviga machines, who kindly placed the machine at my disposal. A machine with the following properties would be very useful: (1) one should be able to calculate on it from left to right without striving after an absolute accuracy . . . (2) one should be able to throw back into the machine with a single turn not only products but also quotients; (3) a storage mechanism should enable one to sum a number of terms having either all the same or alternate signs without having to write down intermediate results.

Well, today even the simplest El Cheapo calculator can do that and more, which means that anyone with $7 or $10 can buy more calculating power than was available in the 1930s to even the most indefatigable compilers of math tables. This means much more than accurate tax returns, more than taking the terror out of such chores as balancing the checkbook. The basic technology of the calculator has made possible language-translating devices, operating at least at the level of handy phrases for the tourist. If you are lost on the streets of Fritz Emde's Stuttgart, not knowing German but having a translator, you simply key in the phrase "Where is the railroad station?" and its display will show the translation, "Wo ist der Bahnhof?" Anyone who wishes to help you can turn the translator around and key in a German phrase which you will read in English. This type of communication still is a bit cumbersome, but it beats thumbing through a dictionary, and it is a vast advance over people shouting at each other in hope of making themselves understood.

A calculator is built around a microprocessor and there are many more ways microprocessors will be turning up in our lives. What these chips actually mean is the disappearance of the computer; that is, its disappearance into the inner works of the equipment it controls. They offer compact, distributed power for computing, controlling, or handling data. To appreciate their future significance, begin by recalling that a century ago there were no such things as engines or motors for household use. Now take a walk through your home and garage, and try to find all the electric motors in the various items you possess. The big ones in an air-conditioner or clothes dryer are obvious, but don't forget blow dryers, cassette tape drives, electric typewriters, and the small motors that drive your auto windshield wipers. Count them all up, and if the total is much below two dozen, you're probably living close to the poverty line.

Microprocessors will become similarly ubiquitous. Think of having them in appliances, replacing the complex cams and rotating control parts of dishwashers and washing machines. Look ahead to "smart" air-conditioners or home thermostats, continually adjusting the temperature setting throughout the day or night. Think of a smart oven with cooking recipes stored in its memory; you will put a roast in the oven, punch the words 5-LB. STANDING RIB ROAST onto a calculatorlike display, and the oven will know how long to cook it and at what temperature. Imagine a home security system which will not only turn lights and TV on and off while you are away but will be linked to stereo speakers to simulate the sounds of a family at home—the kids yelling, the dog barking, the parents arguing. Other chips will be more subtle, providing such services as automatic frequency control to give the clearest signal reception in a radio, or automatically tuning the TV for the sharpest picture.

The telephone may give some of the most useful applications for microchips. Equipped with an electronic memory, it can store frequently called numbers, which can be dialed by pushing just one button. A special button marked "Last number dialed" helps out when you get a busy signal; simply pushing that button will automatically redial that number. Similarly, the number of the police or fire department would be stored and dialed with a single push button. All this would be in addition to such currently available services as call forwarding (forwarding incoming calls to another phone number) and call waiting (placing a call on "hold" so you can handle another incoming call). Even more advances will come when the phone is equipped with a microprocessor and digital display. It will time long-distance calls, display the phone number you are dialing, and even perform calculations with what would amount to a built-in calculator. Most important of all, while other people are talking to you, the microprocessor would allow the phone company's computer to talk to your telephone. An incoming call would display the calling phone number, as well as storing it in memory, and you could recognize who was calling even before you picked up the receiver. If you wished, you then could simply let it ring.

This alone could bring a considerable change in the way we use telephones. The anonymous phone call would become much more difficult. A woman alone at night, tormented by an obscene caller, would know where the call came from, and she could set up a three-way phone conversation with the police listening in. Even if the call came from a pay phone, she could try to keep the pervert on the line long enough for the cops to get to him. If the phone were to ring while you were in the shower or outside the house, there would be no need for a mad rush to get to it; the caller's number would be there on the display for you to redial. The microchip would act as an answering machine, recording the numbers of incoming calls while you were away, or while you were asleep and had the telephone turned off. (Yes, Virginia, it will be possible to do that.) It will even be possible to bring to telephoning something of the leisure of letter writing. We do not accept a tyranny of the mailbox, in the way we often allow a phone call to intrude on our daily activities; instead we open letters and answer them at our leisure. Similarly, if we wish we will be able to take our time in reviewing the record of incoming calls, and to take even more time in answering them.

The auto will see the most widespread introduction of electronics. Already some sixteen electronic systems are available:

Headlight control	Antiwheel-lock braking control
Alternator rectifier	Clock
Voltage regulator	Intrusion alarm
Tachometer	Crash sensor, air-bag inflation control
Cruise control	Electronic fuel injection
Electronic ignition	Lamp timing control
Climate control	Spark plug timing control
Windshield wiper control	Electronic digital displays

In the future many more will be available. Among the proposed systems, many of which have already been built as prototypes, are the following:

Automatic cruise control	Vehicle locator
Automatic radar brakes	Automotive diagnostic systems
Radar crash sensors	Vehicle blind spot detectors
Crash recorders	Emergency location transmitter
Electrocardiographs	Multiplex wiring systems
Sleep detectors	Low tire pressure indicators
Alcohol ignition interlocks	Vehicle load weighing systems
Automatic vehicle guidance	Ultrasonic fuel injectors
Engine-knock limiting control	Controlled exhaust-gas regulation
Continuous transmission control	Dual-displacement engine control

Automotive radar systems would offer obvious safety advantages, as would devices to detect alcohol or sleep, or for heart patients a wrist cuff to detect pulse patterns and serve as an electrocardiograph. Other systems would improve engine performance. The continuous transmission was discussed in Chapter 5. An engine-knock control would retard spark timing to allow the motor to use low-octane fuels less costly than high-octane gas. Exhaust-gas control systems would monitor the engine exhaust and regulate the air-fuel mixture to give the best efficiency for emission control. A dual-displacement control would shut down some cylinders when the car is cruising along. Multiplex wiring systems would simplify the extensive array of cables and switches in an auto's electrical system. All in all, by 1985 electronics may account for 10 percent of an auto's cost, or close to $1,000. This fraction of the cost will likely increase in later years, and will increase still more with stereo and tape-deck systems. Tomorrow's cars may resemble General Motors' Alpha V, which has thirty-four automotive functions under the control of a central microprocessor.

All these uses of electronics, in the home or telephone or car, represent applications using a microprocessor in a straightforward way to control a process or store data, or at best to perform the routine chores of a simple calculator. More interesting possibilities emerge when we look to the home computer. Since the mid-70s people have been buying these systems, paying $3,000 to $5,000 and sometimes more, for their entry into the world of bytes and bauds, RAMs and ROMs, floppies and PROMs.†
These home computers have more power than the computers used by the

† For those who don't speak computerese, a byte is a standard length for words or numbers stored in a computer. Baud is a data transfer rate. Random-access memory (RAM) and read-only memory (ROM) are data storage arrays that respectively hold a computer's program or data, and its operating instructions. A floppy or flexible disk vaguely resembles a phonograph record and provides permanent storage for programs or data. A PROM has nothing to do with dances or corsages, but is a programmable read-only memory.

Air Force in the 1950s to control air-defense systems and more power than the large commercial computers of the early '60s used to design nuclear reactors. Originally it was mostly computer buffs or hobbyists who bought them; some even went to the trouble of building them from kits. In recent years, though, these microcomputers have found a solid niche among professionals and small businessmen. They can be bought with programs allowing a businessman to keep accounts, quickly calling up data on any phase of the business. Naturally, they can be particularly helpful at tax time. Account-keeping has always been an onerous chore, and thirty years ago large corporations were glad to order the big IBM computers of the day to help out; now small companies can get the same sort of electronic bookkeeping. Many small computers come equipped with graphics packages, allowing the user to prepare graphs and charts on a TV screen, often in color. What's more, they can store an amazing amount of business data. One Minnesota gas station owner uses his computer to send out postcards to his customers, reminding them when it's time to come in for an oil change.

A particularly useful version of the microcomputer is the word processor. This allows a writer or secretary to type a draft of text on a typewriter keyboard, the text being stored on a floppy disk and displayed on a screen. The writer then can edit the text at the keyboard, adding or deleting words or sentences, correcting spelling, even shifting whole paragraphs. At the end, when the draft is satisfactory, he trips a key and the computer automatically types the text with an electric typewriter, letter-perfect. Many writers like the word processor because of the freedom it gives them in editing or changing text. The permanency of floppy-disk storage means that even months-old drafts can be readily reedited. Legal secretaries and real estate agents, among others, like it even more. These people are forever being faced with the need to prepare documents differing only slightly from day to day, but which must be finished quickly and with no errors. With a word processor they can store their basic texts of legalese on a floppy, then easily modify their documents to fit the case in hand.

Will the day come when there is a computer in every home? There is an interesting parallel between the rise of auto ownership, early in this century, and the anticipated rise of the home computer. The auto and oil industries grew up together and greatly encouraged each other's growth. The availability of oil made possible the production of automobiles; the rise of mass auto ownership meant a vast new market for oil products. But the auto could never have risen to prominence without the widespread network of paved roads it called into being. Similarly, the existence of the microprocessor has made possible the home computer, and the mass production of small computers has contributed to the widening markets which have made microprocessors steadily cheaper and more capable. But the home computer still needs its "roads"; it still needs places to go. In their most immediately useful form, these will come in the form of data tie-ins, linking computers by means of telephone lines to a wide variety of data.

When these links are built—fortunately, a task more like extending telephone services than like paving America's roads—you will be able to sit with your computer and do far more than you can conveniently do today with a telephone. Planning an evening out? A central data bank will have information on current movies or activities around town, which you can call up on your display screen and review. (Perhaps high school students will maintain this electronic community billboard, thus serving their community while they are learning about computers.) Planning a vacation? You will be able to check airline schedules, motel accommodations and rates, trip itineraries, and much more; you will even be able to order tickets to sports events or to a Broadway play. Need to pay for them? Electronic funds transfer services will be available via your bank, using secure codes to protect your account from theft. Thinking about investments? A data line to Merrill Lynch will give you up-to-the-minute information. Planning for college? Your kids can sit at their leisure and review the contents of university catalogs. Going shopping? Other catalogs, from Sears to Gucci, will be there for you to call up and review. Did your newspaper get caught in the rain? A news-ticker service will give you stories direct from the wires of the Associated Press.

In addition, the computer color graphics will advance to the point of offering the kind of animation we now see in Hollywood special effects. There will be a wide variety of electronic games, often set against highly detailed and realistic backgrounds. This will amount virtually to fantasy TV, simulating just what it would be like to fly a spaceship past Jupiter along a course you select. Or to approach the speed of light in a starship. Or fly and maneuver at will through the Grand Canyon. Or simulate the ship traffic, under your control, maneuvering near you as you steer a supertanker into New York Harbor. Or shrink yourself to atomic size and fly down the corridors of carbon atoms in a diamond crystal. Or build the world's highest building or most advanced monorail line within your city. For generations, kids of all ages have been enthralled with model railroads and Erector Sets. Their future electronic counterparts will be far more vivid, colorful, detailed, and realistic. For those whose tastes run to more conventional games, there will be computer chess, poker, or backgammon, which for brave souls will be available at world-class championship levels of play. These will all be examples of packaged computer programs, available from a catalog. Already a two-page ad from Apple Computer today lists (in very fine type) over a thousand such programs for nearly any imaginable application. The experienced computer programmer, capable of writing his own, could add even more.

In the business world, these capabilities have already brought the advent of the electronic office. R. J. Spinrad, vice-president for research at the Xerox Palo Alto Research Center, has described how he handles his daily correspondence and paper work:

> When I arrive at my office in the morning and begin my work I flip a switch and press a few keys and an index of incoming messages is displayed on the screen before me. Since I am pressed for time I read only one relatively urgent message. Moving a pointer, I position an arrow on the screen and press a button. Instantly, the message I have selected is displayed . . . After some thought I send a message to a colleague asking for help in organizing a somewhat lengthy response. Then I set about the principal work of the morning, completing some equipment layout by directly manipulating the on-screen diagrams. Meanwhile, my colleague is structuring an answer to the urgent inquiry. He calls to the screen an old, but relevant report, composes an appropriate "cover letter," and sends it on to me. Later, after I have added some new information to the response, I send it via the network linking the North American parts of our organization. The message is automatically stored for future reference and, if a printed copy is desired, it is available in seconds from the nearby copier-printer.

It is easy, too easy, to see in such prospects the vision of a brave new electronic world very much at variance with the one we know today. So it is useful to note that in many ways this future world will still be very much the world we live in. For instance, it would not be difficult to equip every gas station or retail chain store with a microchip terminal that would record the details of every credit-card transaction. At the end of each day the recorded data would be transmitted rapidly, in code, over an ordinary telephone line to computers in the credit-card companies' central offices. This would speed by up to several weeks the billings on millions of dollars' worth of sales. However, such systems would threaten the existence of entire credit-card company divisions: the optical-character-recognition departments that transcribe into computer-usable form the transaction data from millions of separate pieces of paper filled out and mailed in by service-station attendants and store clerks. Entire divisions of companies do not willingly disappear. Similarly, home computers with data links would challenge airline and hotel reservation divisions, ticket and travel agencies, credit-card authorization and account-information divisions, and some bank and brokerage activities. There are plenty of service agencies and organizations whose people will be more than glad to see the small computer with its video terminal, as long as it stays in their hands. They can be counted on to resist offering data-link services that would put their services in the hands of ordinary customers with their own computers.

Still, if these people will be unwilling to let ordinary computer owners do their jobs, there will be plenty of professionals who will welcome the chance to do their work on their own home computers. This will be the widely predicted telecommuting or teleconferencing, in which professionals will work at home rather than commute to computer terminals at a central office, using their home systems to communicate with clients or associates and to serve their customers. It is hard to say how widespread this will become; the pattern of people working in offices or company buildings is quite deeply ingrained. Nevertheless, some people will have the chance to opt for this type of self-employment, and it is a safe bet that some people will find they married their spouses not only for better or worse, but also for lunch.

We certainly should not expect that when catalogs can be called up and displayed on video terminals, the end of books and magazines will be in sight. It will be perfectly possible to put the contents of a book onto a small computer-readable card or disk to be displayed page by page. But for some decades it has been possible as well to do precisely this, using microfilm or microfiche, in which pages are photographically reduced so that a hundred of them fit on a piece of film the size of a postcard. A microfiche reader, featuring a bright light and a display screen, then enlarges each page to full size. Interestingly, such readers look rather like computer video terminals. Yet microfilm has never caught on outside of libraries. Indeed, we could pose a challenge to computer scientists: "Devise a means of storing hundreds of pages of text at a cost of only $3.00 or so. The invention must be lightweight and compact, and require zero power to store or recover the text. It must require no special electronic equipment to make the text available to the reader. The invention must store its text for years or decades without fading or loss, and must be completely portable, capable of being used at home, aboard an airplane, or at the beach. The user should be able to begin reading at any place in the text, finding his desired place almost immediately, and must be able to read the text at his own pace, skipping material or even re-reading parts as he may wish. The invention should use only inexpensive and commonplace materials, and pose little or no environmental hazard when being produced or when it is finally disposed of."

This of course would be an ordinary paperback book. We tend not to appreciate it because it has been around so long. But it is safe to say that if Gutenberg had invented the microchip back about 1450, while Robert Noyce and his associates had invented the art of printing books with movable type only within the past dozen years, we would today be hailing Noyce's invention as ushering in a new age.

Video games deserve a word or two. By now everyone has seen the vividly colorful games like Asteroids and Space Invaders that keep their fans going late into the night, happily blasting away at enemy spaceships. These are basically sophisticated versions of the good old pinball machine. They can be much more enthralling, however, perhaps even addictive. So far there have been few if any verified accounts of Pac-Man addicts stealing the family jewelry or TV set to pawn in exchange for rolls of quarters with which to pursue their passion; but give it time, give it time. In any case, though, if a choice is to be made, such an addiction is certainly to be preferred to cocaine or angel dust. From now on, amusement parks and similar places will feature arcades garishly lit with naked light bulbs, full of high school students and other slightly seedy characters putting quarters into game machines having more electronics than the best scientific labs of thirty years ago. But in their essentials—their smoke and noise, their beer and raucous talk—they will be the same old pinball arcades we have always known.

Then there is television. The technology of TV certainly is changing, what with UHF channels, cable services, direct-broadcast satellites, video cassette recorders, and video disks. The disks in particular look like crystalline treasures from the year 3000, shimmering and glistening with rainbows of bright colors. This is an optical effect resulting from light reflecting from their closely spaced tracks or grooves. For over a century astronomers have used gratings ruled with finely spaced lines to break light into a spectrum of its component colors, and today's videodisk surfaces do the same thing. Many of these disk systems are so advanced they use a laser. Yet what is the program content on the disks? It may be *Doctor Zhivago* or a concert by Neil Diamond, but a popular choice is pornography. It is ironic, though hardly unexpected, that these precision instruments and almost magical technologies should be used for such programs as *Deep Throat* or *The Devil in Miss Jones*, but that is the nature of television.

The 1960s saw much learned talk about the influence of TV, and in some quarters it was even held up as a topic fit for intellectuals. Marshall McLuhan assured us that television had made us all residents of a global village, which if true would necessarily imply the existence of global village idiots. The tube gained an even greater mystique when its news reporting helped spark opposition to, and an earlier end of, the Vietnam War. Yet in fact it is not so hard to understand the role of TV in society, with its fads, its sensationalism, its cheap entertainment, and oversimplifications. One need only recall the Hearst press. (Which has often been called the "yellow Hearst press," and not only from the color of its newsprint.) In the heyday of press baron William Randolph Hearst, a popular newspaper game was the crime wave. In a city where circulation was flagging, news editors would play up and focus attention on stories of crime, particularly murders and rapes. Screaming headlines would proclaim the public danger. People all over the city would rush to buy the latest editions to learn of their peril. Then, with circulation nicely hiked, the editors would bring the crime wave to an end, let tempers cool down, and return to a slightly less dramatic form of news emphasis.

Did TV help end the Vietnam War? Well, the Hearst press helped mightily to start one. In the late 1890s, Cuban patriots under José Marti were struggling for independence from Spain, in virtually the last outpost of Spain's once great New World empire. Hearst took a personal interest. He had his editors play up and emphasize reports of Spanish atrocities in Cuba, at the same time editorially urging America to come to Cuba's aid. The resulting press coverage inflamed public opinion. President McKinley sent the battleship *Maine* to Havana harbor to help safeguard American interests. When it blew up and sank with heavy loss of life, no one could think of anything so commonplace as a boiler explosion touching off the ship's ammunition; no, it had to be those dastardly Spaniards with a submarine mine. Soon the cry of "Remember the *Maine!*" was on everyone's lips, and Hearst had his Spanish-American War. The influence of TV has been rather similar.

Nevertheless, for all its banality, television will likely continue to stand as the most obviously evident and pervasive form of electronics. Photovoltaic panels will sit on the rooftop like so many shingles. Microprocessors will work silently and unobtrusively, and we will take them as much for granted as electric lights; they will simply be there when we need them. Even home computers, however important or dramatic, will hardly engage a hundred million people for five hours each day. There is a considerable subtlety to many new forms of electronics, in that they do their work quietly and with no jarring intrusions into our lives. We could think of the uses of electronics in aviation, for reservation services, in aircraft check-out and maintenance, as well as for in-flight navigation and air-traffic control. But we do not dwell on these things; what we think about is that it is easy to reserve a flight and that we know the plane will get us there in one piece. Modern forms of electronics are not intrusive like television; they are not like George Orwell's loudspeakers blaring the words of Big Brother on every street corner. They are much more like the convenience of an automated bank teller, which gives you cash out of your account at any time, night or day and even on holidays, if you insert the appropriate plastic card—and who would object if one of those were on every street? Still, other electronic developments have at least the potential for considerable controversy and debate. These fall into the categories of artificial intelligence and of robots.

Robots, like astronautics, illustrate the point that science-fiction

writers may successfully predict that an important new enterprise will take
form, while completely missing a successful prediction of the actual form
it will take. Thus, the classic prediction was that astronautics would give
us spaceships to take people to distant planets, but its actual character has
been to build packages of instruments or equipment which are put into
orbit and just sit there. In robotics, the standard prediction has been for a
robot to be a mechanical or electronic man, perhaps a polite and almost
diplomatic one like C3Po in *Star Wars*. What robots actually are today,
however, are electronic workers. The difference is important since, as
many industrial workers will attest, their jobs call for almost nothing of
what we regard as characteristically human. An assembly line is a deaden-
ing place for a man, but it's perfect for a robot.

In Danbury, Connecticut, is the nation's biggest robot builder,
Unimation Inc. It builds the large Unimate robots as well as a smaller
version known as PUMA (for Programmable Universal Machine for Assembly), sending some 55 robots per month into the world. Only slightly
less busy is Cincinnati Milacron, builder of the sophisticated T-3 robot.
These devices look nothing at all like the Tin Woodman of Oz, though
they perhaps do have a passing resemblance to his arm. It is the arm, after
all, that is important in a robot. Its brains (computer) can be in a box located almost anywhere, and it has no need for legs or a body, let alone the
ability to sing "If I Only Had a Heart." What makes the arm useful is its
programmability, which allows it to adapt in minutes to a new job. To
program a robot, an operator with a hand-held controller guides the arm,
and its wrist and handlike gripper, through the desired sequence of movements. Following each motion, the operator instructs the computer to
store it in memory. A robot then acts as a machine which plays back and
executes recorded motions, rather as a programmable calculator plays back
and executes recorded steps of mathematical computation.

Robots cost typically $40,000 to $100,000 each, but for that and their
operating expenses, they average $6.00 per hour. By contrast, an assembly-
line worker costs $20 per hour, including fringe benefits. Robots don't get
bored, take vacations, qualify for pensions, or leave soft-drink cans rattling
around inside the assembled products. They cheerfully work round the
clock and will accept heat, radioactivity, poisonous fumes, or loud noise,
all without filing a grievance. They are at their best in dangerous jobs like
mining coal or handling irradiated parts of a nuclear reactor, and in
stupefying jobs like auto welding or painting—both of which are best
done at temperatures hotter than a worker can stand. At Chrysler's Jefferson Avenue plant, mentioned in Chapter 5, 50 such robots, craning forward and emitting sparks, work two shifts and do the work of 200 welders
who formerly worked there. At a Ford plant in Wixom, Michigan, robots

measure openings for windows, doors, and lights, working ten times faster than humans.

Robots also are valued outside the auto industry. At a Pratt & Whitney plant in Middletown, Connecticut, ten Unimates build ceramic molds for the manufacture of jet-engine turbine blades. The molds are helping to increase production from 50,000 to 90,000 blades per year, and the robot-made molds are so much more uniform, their blades last twice as long as blades molded by humans. At the General Dynamics plant in Fort Worth, a T-3 selects bits from a tool rack, drills a set of holes to a tolerance of 0.005 inches, and machines the periphery of 250 types of parts. The robot makes parts four or five times faster than a human, with zero rejections; it cost over $60,000 and paid for itself in eight months of operation. Joseph Engleberger, president of Unimation, tells of getting a call from some Australians who thought robots would be just the thing for shearing sheep. He politely told them they were crazy. Later they explained that Australia has 14 million people but over 130 million sheep, most of which require an annual barbering, and to replace their expensive sheepshearers they might need as many as 130,000 robots. That would be nearly 200 years of Engleberger's production, and he suddenly saw the underlying wisdom of the idea. The Aussies are now working on these robots. They have found that clipping a sheep's back and sides is not too hard, but have reported "significant difficulties" in finishing up the neck and head.

All this might raise the specter of angry workers wielding clubs to bash the sources of their unemployment. However, most industries where they would be used are strongly unionized, and the unions can and do ensure that robots merely replace men who quit or are transferred, rather than forcing layoffs. The attitude of union leaders often is, "No firing of our people, and give us some of the goodies that come from increased productivity." As one New Jersey president of an electrical workers' local has said, "If we can bring in a robot here to do, say, the painting that a man does for $7.00, then we can move him to another job at $7.50 an hour. We say, 'Train our people for the skilled jobs that are in today's market.'" On the shop floor, however, robots often go through their mechanical calisthenics not only unattended but hardly noticed by their human colleagues. The arms and boxes are, after all, just robots. When they are noticed, it is often with favor. Some industrial observers report that working with a robot seems to confer status. Some of them even inspire affection. When one machine known as Clyde the Claw broke down at a Ford stamping plant in Chicago, its human co-workers gave it a get-well party.

A 1980 world census of robots, taken by Bache Halsey Stuart Shields, Inc., showed that the United States had 3,000 of them. In Europe, West

Germany led with 850. Sweden had 600, Italy 500, and France, Norway, and England each had about 200. The whole Soviet Union had only 25, evidently experimental devices for the most part; but Poland had 360. Czechoslovakia, home of Karel Čapek who coined the word in 1921, was not listed. This state of affairs is ironic in that "robot" is of Slavic origin, being derived from words like the Russian *rabota*, work. But the true homeland of the robot today is Japan, as might be expected, with 10,000 in the census, or more than the rest of the world combined. The Japanese have done this in a characteristic way, deliberately setting out to make it easy for industrial managers to use robots. Robots are expensive, and many managers would be reluctant to buy them without first having proven them out with on-the-job experience. The problem has been how to let businesses get this experience without incurring high costs. The Japanese solution has been to set up a robot-leasing company to buy the robots from the manufacturers in large numbers. User companies, able to lease robots without having to buy them, have been willing to take on many robots, since they have known they could return them to the leasing company if things did not work out.

The Japanese in 1980 were producing some 7,500 robots per year, compared with 1,500 in the United States. Early in 1981 their firm of Fujitsu Fanuc opened a $38 million plant with robots working round the clock to produce—more robots. At a Matsushita color TV plant near Osaka, about 80 percent of the parts for a set are put in place by robots. Describing his visit to this plant, Philip Abelson, editor of *Science*, wrote that "At the end of the tour, I realized that I had seen only one inspection station—at the end of the line. I asked the official accompanying me why there were no inspection stations at intermediate points. He replied that until a few months ago there had been such stations, but they never found any defects and so they were scrapped. When I returned to the United States, I tried to arrange to see a comparable plant here. I was told that none existed."

Some writers have prophesied that robots would usher in a new cornucopian age, with goods so abundant and cheap that they would have virtually no intrinsic value. Givenchy gowns might be tossed away after one wearing; Mercedes-Benzes might be disposed of once the ashtrays were full. But in the coming decades, what we can hope is for robots to help hike our industrial productivity. In the late 1970s it was rising at less than 1 percent a year and in 1980 was declining at a rate of —0.9 percent. If robots could only help boost this to Japan's average level of 7.3 percent, what a welcome change it would be! Inflation would ease, since workers' pay increases, running around 10 percent a year in recent years, would be matched by comparable gains in productivity. Interest rates would fall, the dollar would rise against foreign currencies, our exports would increase, the stock market would rally, and the price of gold would go down. What's more, unemployment would probably drop and employment would rise, for a booming economy employs more people. Thus, paradoxically, robots would not displace workers from jobs but would actually create more jobs. Automation and increased productivity have always worked to do this. One need only compare the high unemployment in 1980 (11.5 percent) in the anemic, labor-intensive industries of Britain with the low unemployment (2.1 percent) in Japan.

Certainly the Japanese will be ferocious competitors, in robots and in other advanced areas of electronics no less than in automobiles. Some of their electronics firms have branches in the States and rely on American workers for some manual assembly jobs—a startling turnabout of the American practice of using Asian labor for such tasks. Between 1977 and 1979, U.S. manufacturers of computer memory chips failed to expand rapidly enough to fill the demand. Hewlett-Packard, one of the largest users of such chips, turned to Japan to fill the gap and the Japanese quickly moved in, grabbing 40 percent of the market for these chips, key components in computers. Fortunately, there is business for all. Some financial analysts predict the growth of the U.S. electronics industry will be limited mainly by how fast companies can add new production capacity, because demand will exceed supply for many more years. By 1990 electronics could be a $400 billion worldwide industry, putting it in the same league with such heavyweights as automobiles and even oil.

Robots will be only a small part of this, but they will surely play their role. What's more, they will be getting smarter. Today's robots are essentially blind grabbers. For instance, Unimation's PUMA can screw a light bulb into a socket, but bulb and socket must be precisely where they are supposed to be. If the bulb is upside down in its tray, PUMA will contentedly try to screw it in upside down. If the socket is a quarter-inch out of position, PUMA will jam the bulb against its rim and twist away. If the bulb isn't there, the robot will close its gripper on empty air and try to screw it in. None of this is particularly useful, and as a result a lot of effort and expense goes into making sure that parts and assemblies are aligned properly and will be where a robot expects to find them.

The problem of making robots smarter is one phase of artificial intelligence. One topic receiving attention is that of giving vision to a robot. It is easy to form an image with a TV camera, but it is not so easy for a robot brain to interpret what it sees. The image must be broken into small elements, converted into numbers, and interpreted in terms of patterns the computer finds in the numbers. Still, there has been progress. At SRI International near Stanford, a robot named Shakey could be given tasks that required it to be able to recognize boxes, doorways, and corners within various rooms through which it could move on wheels. In one experiment, Shakey was placed in a room with a box on a platform, and a ramp some distance away. It was given the question, "Can you knock over the box?" Shakey's computer was able to figure out that by pushing the ramp against the platform, it could wheel itself up the ramp like *Star Wars'* R2D2 and knock down the box.

Such a feat is reminiscent of the intellectual power of a monkey. A chimpanzee in a cage can learn to fit two short sticks together to make one long enough to reach a banana; Shakey displays similarly emergent powers of perception, foresight, and what psychologists would call goal-oriented behavior. In most ways, to be sure, monkeys are much smarter than a computer or robot. Computers have a terrible time recognizing anything more complex than simple shapes. For example, it is difficult for them to recognize faces. In Zurich the zoo master Heini Hediger would approach through a crowd of people; a monkey would spot him a hundred feet away and raise a cheerful commotion. Current computers, by contrast, would be as undiscerning as Hediger's ring-tailed lemurs, who even after a number of years never learned to recognize him. But there is one area where artificial intelligence studies are truly beginning to teach computers something new and useful. That area is understanding human speech.

Generating speech today is no problem. Computers today have recourse to programs which synthesize speech in rather the same way we do, by producing an appropriate flow of sounds. Thus, for $50 or so there is Texas Instruments' Speak and Spell game, which speaks in response to questions typed in from a keyboard. Also there is the microchip voice synthesizer in today's Fords and Chryslers, which warns you, "Please turn off your lights." But to actually understand spoken speech, with its grammar and syntax, its similar-sounding words and its differences peculiar to individual speakers—that has not been easy. Nevertheless, it has been possible to make a good start. At Bell Labs, an experimental system can not only listen to a speaker but process information extracted from his speech. It then goes on to compute an appropriate response, encode the response in a proper English sentence, and deliver a reply using synthesized speech. The system has a vocabulary of 127 words and acts as an airline ticket agent. Here is a sample conversation carried out over a telephone between man and machine:

MACHINE: This is the Bell Laboratories flight information system. May I help you?

MAN: I want one first-class seat on flight number three one to Denver on Sunday.

MACHINE: I am sorry the flight is not available on Sunday.

MAN: I would like to leave on Saturday.

MACHINE: Flight number three one leaves New York at nine forty-five A.M., arrives in Denver at one-ten P.M.

MAN: What is the flight time?

MACHINE: The flight time is five hours and twenty-five minutes.

MAN: How many stops are there on the flight?

MACHINE: The flight makes one stop.

MAN: I want a nonstop flight.

MACHINE: I am sorry there are no nine A.M. nonstop flights.

MAN: I want to go at four P.M.

MACHINE: Flight number one six leaves New York at four-fifty P.M., arrives in Denver at seven P.M.

MAN: What stops are there on the flight?

MACHINE: This flight makes no stops.

MAN: Is a meal served on the flight?

MACHINE: Dinner is served.

MAN: How much is the fare?

MACHINE: The round-trip first-class fare is eight hundred sixty dollars.

MAN: What plane is on the flight?

MACHINE: The aircraft is a Douglas DC-ten.

MAN: I prefer the Boeing seven four seven.

MACHINE: The aircraft is a Douglas DC-ten.

MAN: Please repeat the departure time.

MACHINE: The departure time is four-fifty P.M.

MAN: I will take flight number six five.

MACHINE: I am sorry there is no flight six five to Denver.

MAN: Please repeat the flight number.

MACHINE: The flight number is one six.

MAN: I want flight number one six.

MACHINE: Your reservation on flight number one six to Denver on Saturday is confirmed. Thank you.

Anyone who has studied a foreign language will recall how the early lessons are quite simplified, often featuring conversations in a highly limited vocabulary keyed to some stock situation such as a brief conversation in an airline terminal. The gap between such elementary lessons and the ability to read Anatole France or Alexander Solzhenitsyn in the original then illustrates how far computer speech recognition has to go to become truly fluent. Still, such speech systems will certainly be valuable when they enter common use. For instance, when you call your credit-card company to ask about your account, you are actually talking to someone who works in what amounts to an electronic sweatshop. A roomful of people, mostly young women, sit at computer video screens taking phone calls, and often they face a quota: take care of, say, 300 phone calls from customers in a day. Yet the data and the questions they handle are quite stereotyped: "What is your account number?" "What is my current balance?" Have you received my last payment?" A computer voice-recognition system could save these people from one of the few truly tedious jobs the electronic age has spawned, while leaving human employees free to deal with such questions as, "What do I need to do to get you to increase my credit limit?" Voice recognition will also be useful in dealing with robots and other smart machines, and science-fiction writers should take note: a command to a computer such as "Open the pod bay doors, Hal" will be possible well before the year 2001.

Artificial intelligence can also turn computers into superior game players. Computer chess has been stirring interest for a number of years, and recently there have been prizes to encourage further activity. The Dutch computer firm, Volmac, in 1979 offered $50,000 to anyone whose program could beat the Netherlands' Max Euwe, former world chess champion, who died undefeated in 1982. More recently Edward Fredkin, an MIT professor, has offered the Fredkin Prize of $100,000 for the first computer program to win the world chess championship. The best chess programs can beat some 99.5 percent of human players, but the current computer chess champion, a program from Bell Labs named Belle, does not rely on the subtlety and deep chess insight of a world titleholder such as the current champion, Anatoly Karpov. Instead it uses special-purpose computer circuits to execute its program rapidly. At any point of a chess game a large number of moves and countermoves are possible, and Belle is fast enough to examine thirty million such combinations during the three minutes allowed for a move in tournament chess. This is really just brute-force number-crunching and comparing, with the computer picking its move not by insight but by mechanically searching through the possibilities. Certainly few people believe that Karpov soon must fear for his crown.

Yet in a different game, backgammon, the contest is already over. In July 1979 Luigi Villa of Italy won the world title in that game, in a competition in Monte Carlo, Monaco. It was the summit of his career as a backgammon player. Earlier it had been arranged for the champion to play the winner of a competition among backgammon computer programs, this being the program BKG 9.8 written by Hans Berliner of Carnegie-Mellon University in Pittsburgh. Both sides put up $2,500, winner take all. The program was run on a PDP-10 minicomputer in Pittsburgh, and the computer was connected via communications satellite to a robot in Monte Carlo, dubbed Gammonoid, which made the moves. BKG 9.8 got off to a good start, winning the first game 2–0 and going on to win the second and third, making its score 5–0. (In backgammon it is possible to gain what amounts to a double win.) Then Villa struck back and won a victory, making the tournament score 5–1. The crowd watching in Monaco's Winter Sports Palace was ready to see Villa stage a stirring comeback. But in the fifth game, BKG 9.8 again triumphed and won the match by the score of 7–1. It was the first time a computer program had beaten a world champion at any board or card game.

Will computer intelligence continue to rise, perhaps in time outstripping man's? This is not really a proper question, for computer and human intelligence stem from different sources. In both species (using the term loosely) intelligence is connected with the ability to evaluate data and make decisions. In a computer, the basic element of decision-making is the same as in 1946, when John von Neumann was developing the theory of computers. It is the ability of its electronic logic to compare two numbers and determine which is the larger. On this all of artificial intelligence has been built, including the successes of Shakey, of Bell Labs' airline ticket agent, and of BKG 9.8. By contrast, C. R. Carpenter of Pennsylvania State University, a longtime observer of the lives of monkeys in the wild, has stated what sort of decision-making would have been common among even our prehuman ancestors:

> You are a monkey, and you're running along a path past a rock and unexpectedly meet face to face another animal. Now, before you know whether to attack it, to flee it, or to ignore it, you must make a series of decisions. Is it monkey or non-monkey? If non-monkey, is it pro-monkey or anti-monkey? If monkey, is it male or female? If female, is she interested? If male, is it adult or juvenile? If adult, is it of my group or some other? If it is of my group, then what is its rank, above or below me? You have about one-fifth of a second to make all these decisions, or you could be attacked.

Computers of course are very good at logic, at mathematics and data handling. They are quite poor at speech recognition or vision, at exercising judgment or common sense. Also they have barely the foggiest notions of learning from experience, of emotions, or of a human society—matters with which virtually everything significant in life must be closely associated. Human beings can with difficulty be taught to do mathematics or to reason by logic, though the resulting conclusions are always subject to being overridden, as in the Yiddish proverb: "When the penis stands up, the brains lie down in the ground." But any human alive can see, speak, feel, and live with fellow humans. In addition, many people can learn to deal with material of tremendous richness and variety, material touching the human experience, with which computers deal only crudely if at all. A computer can analyze the solar spectrum, but can it describe a sunset?

We two species, human and computer, complement each other. We are strong where they are weak, and vice versa. They need us, and we stand to gain enormous benefits from them. Yet they will not compete with us, for as a naturalist would say, we inhabit different ecological niches. They are not like intelligent horses or dogs, who might someday threaten us; they are more like intelligent corals or sponges, happily content to live in the sea. Indeed, virtually the only way they could compete with us is on the highly structured and artificial field of a game board. In many ways computers, and their associated electronics, will be making our lives brighter. Certainly they will make them more varied. If there is one theme that echoes throughout electronics, it is that it expands the choices and possibilities available to each of us. It is, preeminently, a technology from which everyone can pick and choose according to his liking. It is, first and foremost, a technology of the individual.

8

A CODE FOR GENES

All living things are made up of cells, and to understand something of how a cell works, a good way to start is to think of it as similar to an automated chemical factory completely under robot control. Like a chemical plant, a cell takes in raw materials (oxygen and food compounds), synthesizes them into useful products, and produces wastes which must be disposed of. Elaborate systems allow cell and factory alike to change the mix of products being produced, or to modify their activities so as to cooperate with adjoining cells or factories. And at the core of both cell and factory are lengthy and complex sets of operating instructions which guide all activities, storing the information drawn on for day-to-day work. The factory might encode these instructions on reels of magnetic tape, to be read by a computer, a master controller over the robots. In the cell, the information is encoded in the genes. These genes are made of DNA, deoxyribonucleic acid, and there is a most remarkable similarity between a cell's DNA and a robot's programmed magnetic tape. Indeed, the easiest way to understand DNA is to think of it as a computer tape operating on the molecular level.

A cell is far more intricate and complex than a chemical plant. A large plant, like South Africa's Sasol complex, may cover a square mile, and in its construction and activities it draws on the resources not only of chemistry but also of mechanical and electrical engineering. A cell is so small that thousands of them wash off your body every time you take a shower, and its inner workings can involve only chemistry, with no opportunity for installing mechanical pumps, blowers, evaporators, or motors. Yet this limitation is actually an advantage, for the cell not only can operate from day to day, it can reproduce itself. Its DNA encodes not only operating instructions, but a complete set of what amount to working blueprints, which is more than is stored in a robot's computer tape. But the principle of coded information is similar in both cases. Run a section of tape through the computer; the factory equipment will execute certain well-defined actions, produce certain resulting products. Run a length of DNA through the cell's chemical machinery, in particular, through substructures within the cell known as mitochondria; the mitochondria will produce specific proteins.

Nowadays, computer tape consists of iron oxide on plastic film and resembles the tape of a cassette recorder, but in early computers it was paper tape punched with holes in a coded sequence. DNA is like that; a very long molecule strung with a coded sequence of specific submolecules. In punched paper tape, the important part is the holes; we could say the paper serves to hold the holes and keep them in proper sequence. Similarly, in DNA the important part is the submolecules, and the long DNA molecular "tape" acts mainly to hold them in their own sequence. These submolecules, known as bases, are called by the letters A, T, C, and G, or by the corresponding names adenine, thymine, cytosine, and guanine. These bases have flat molecular structures resembling tiles with specific shapes, and along the DNA molecule one always finds A joined to T and G to C. But while each base has its own shape, the shape of the combination A with T is virtually the same as that of G with C. The overall DNA molecule has two long parallel strands built with molecules of the sugar deoxyribose, which gives DNA its name, and which serve as DNA's basic frame. Within this frame may be fitted any specified sequence of paired bases, which fit like identical rungs along a ladder. Indeed, the DNA molecule resembles such a ladder except that it is twisted along its length, one twist or turn for every ten of its "rungs."

Suppose we could examine closely a short length of DNA, following along one of its deoxyribose strands. A sequence of the attached bases might run GAGCATCCT. Then by the base-pairing rules, the corresponding sequence on the other strand would have to read CTCGTAGGA, and the complete length of DNA would have the base pairs $\frac{GAGCATCCT}{CTCGTAGGA}$. When the DNA molecule reproduces, its twin strands separate, each carrying its attached bases. The separated bases then pair up with their corresponding mates, which the cell itself supplies, and new deoxyribose strands grow to hold everything in place. There thus arise two DNA molecules where before there was only one. This happens when a cell reproduces, and it is DNA that transmits heredity, the inherited characteristics passed on from generation to generation.

A gene then is a particular length of DNA, coding for a particular protein. As DNA can be described as a specific sequence of bases or base pairs, so proteins can be described as specific sequences of amino acids, which are proteins' particular constituents. There are four DNA bases, but there are twenty amino acids, so one base does not code for one distinct amino acid. However, the bases are grouped in sets of three, and each triplet is called a codon. There are sixty-four possible codons, and it is these that correspond to the amino acids. With sixty-four codons coding for twenty amino acids, there is naturally considerable redundancy. However, the resulting genetic code is as complete and explicit as the correspondence between Morse code and standard text, or between a computer's binary digits and ordinary numbers.

Thus, in Morse code · · · − − − · · · is SOS, the mariner's distress signal. In binary code, 11110111100 is 1980. And in the genetic code, GAGCATCCT can only be the amino acid sequence, glutamic acid-histidine-proline, which is the basic structure of the hormone TRF. TRF is secreted by the hypothalamus of the brain and is the master hormone regulating the action of the thyroid gland.

All this was known by the late 1960s. In addition there was the very important finding that this genetic code is universal, holding from everything from bacteria to people. This raised the possibility that useful or valuable proteins for man might be encoded as genes of DNA and made to work in bacteria, thus turning the bacteria into factories for these proteins. Such proteins could include vaccines, hormones such as insulin, antibiotics, contraceptives, enzymes like urokinase which dissolves blood clots, and agents useful in fighting cancer. The particular bacterium to be put to work on such tasks had long been known. This was a strain of *Escherichia coli*, a tiny single-cell creature shaped like a short rod with rounded ends, only about a ten thousandth of an inch long. *E. coli*, as it is commonly called, lives in the intestines of many species, including humans; the species name *coli* means "of the colon" or lower intestine. In the laboratory, *E. coli* had for decades been the microbiologists' guinea pig, because it was quite easily grown and adapted readily to life in the lab, without being virulent or requiring exceptional precautions lest it cause disease. Indeed, so many microbiologists had studied and worked with *E. coli* that in its inner workings it could be described as the best-understood species in the world, more so even than man. This laboratory workhorse was to receive the genes to turn its cells into protein factories.

There was no hope of introducing designer genes into bacteria en masse, but the hope was that a very few bacteria, out of a large number, could be so treated. *E. coli* reproduces very quickly, however, about once every twenty minutes. Thus a single such bacterium with appropriate genes, selected out and allowed to grow, would multiply into a billion in only ten hours, each one identical with the founding father, carrying its genes and producing its proteins. However, to produce even one such bacterium, four developments were necessary. A method was needed for breaking or cleaving lengths of DNA obtained from different sources, then splicing them together. There was need for a gene carrier, a loop or other form of DNA to which foreign DNA could be spliced, and which would then reproduce or replicate both itself and the foreign DNA. Also necessary was a way to introduce the composite or recombinant DNA of this gene carrier into *E. coli*, where it would reproduce along with the multiplying bacteria themselves. Finally, a technique was needed to select from a large population of bacteria those few, and only those, which had acquired the recombinant DNA.

Between 1967 and 1973, a series of rapid developments fulfilled all these requirements. The first was the discovery of restriction enzymes. These would cleave a strand of DNA at particular base-pair sequences, somewhat like snipping a coded computer tape wherever a particular combination of letters would appear. Thus, a particularly useful restriction enzyme designated Eco RI would cleave DNA where and only where it found the sequence $\frac{GAATTC}{CTTAAG}$ along the molecule. The ends of the two cleaved lengths would have the sequences $\frac{G}{CTTAA}$ and $\frac{AATTC}{G}$, which overlap like mortise-and-tenon joints. Such overlapping ends are "sticky," and when two DNA strands are cleaved with Eco RI, their overlapping ends can easily be joined together, with the help of another enzyme called DNA ligase.

This solved the first problem. Stanley Cohen of Stanford solved the second, finding an appropriate gene carrier. Within *E. coli* cells, most of the DNA exists as a single intricately tangled mass. However, small amounts of the DNA exist as plasmids, small loops carrying only three or four genes each, out of some 4,000 in the whole bacterium. Such plasmid genes had long been known to code for proteins making microorganisms resistant to antibiotics such as penicillin. Cohen's contribution was to find a way to isolate the plasmids. Then, at a 1972 conference in Hawaii, Cohen got together with Herbert Boyer of the University of California at San Francisco, discoverer of Eco RI. Launching a new series of experiments, Cohen and Boyer used their Eco RI to cleave plasmids that Cohen had isolated from *E. coli*, and to cut snippets of DNA from other bacterial plasmids. The original *E. coli* plasmids contained genes conferring resistance to the antibiotic tetracycline. The other plasmids had made their bacteria resistant to penicillin. The scientists then mixed together the cleaved lengths of DNA and added ligase to seal the splices which they expected to occur. They thus used the *E. coli* plasmids as their gene carriers, into which they added the new genes from the other bacteria.

To solve the third problem, to get the recombinant plasmids back into the bacteria, they treated the *E. coli* cells with dilute calcium chloride. This made the cell walls permeable, so a few bacteria, perhaps one in a million, would take up plasmids. It was entirely a matter of hit-or-miss whether the DNA ligase would glue the original *E. coli* plasmids back together, repairing the original cleavings; or glue two cleaved plasmids into a single larger one; or—and this was what they wanted—splice the new genes into the cleaved plasmids and then seal the spliced DNA into loops. However, the *E. coli* they were using would be killed by tetracycline, allowing a sensitive test. The normal or unspliced *E. coli* plasmids conferred resistance to tetracycline. Only the new recombinant plasmids, with the new genes, would confer resistance to both tetracycline and penicillin. Thus, they put the bacteria on nutrient plates where they could grow and multiply, but spiked the plates with a mixture of tetracycline and penicillin. Thus, only bacteria that had taken up the new recombinant plasmids would survive the antibiotics. All others, in particular the vast majority which had either taken up no plasmid or had taken up a plasmid lacking the gene for penicillin resistance, would be killed. This solved the fourth problem.

The experiment worked. Not only did they succeed in producing spliced plasmids containing the new genes, but a few of the *E. coli* took up the recombinant plasmids. They then survived and grew, forming characteristic spots or colonies on the nutrient plates, each colony being composed of millions of identical cells. Then, just to show their technique was strong, Boyer and Cohen transplanted into *E. coli* genes taken from a higher animal—a toad. Theirs was no magic wand to turn a toad into a princess, or a bacterium into a toad; but they did succeed in verifying that their bacteria were producing proteins characteristic of toads. With these successes, recombinant DNA was born.

Soon, however, controversy was in the wind. In 1971, even before Boyer and Cohen had gained their successes, Paul Berg of Stanford had been first to produce molecules of recombinant DNA—an achievement which in 1980 won him a Nobel Prize. Berg also used Eco RI to cleave two DNA loops, but he did not know the cleaved ends would be "sticky," so he stuck the cleaved ends together using a more laborious method, employing the enzyme terminal transferase to build up overlapping mortise-and-tenon ends. He got his loops from two viruses, a common lab virus known as SV 40 and another called lambda phage. A phage or bacteriophage preys on bacteria, injecting its DNA into the host like a malaria mosquito infecting someone's arm. Having joined his SV 40 genes to those of the phage, Berg next proposed to use the phage to inject the recombinant DNA into *E. coli*. He met shocked criticism from some of his colleagues. They knew *E. coli* lives in the human intestine. They also knew that SV 40 causes tumors in mice and can turn human cells cancerous. What would happen if *E. coli* bearing recombinant SV 40 genes escaped from Berg's lab? Would these altered bugs cause a massive outbreak of intestinal cancer? Berg could not rule this out.

Thus, hardly had recombinant DNA been launched, with all its high promise of useful proteins and new research knowledge, than it attracted close and critical scrutiny. There was no proof or even good evidence that genetically altered bacteria would pose new risks of infection or act as the germs of new diseases. Nevertheless, the 1970s were a time when hypothetical or speculative risks were often given more weight than proven benefits. Paul Berg started by abandoning his proposed experiment, pouring the contents of his test tubes down the laboratory drain. He then set out to warn his fellow biochemists of the possible hazards of recombinant DNA. At the Gordon Research Conference in 1973, the attendees raised this issue with a letter to the National Academy of Sciences, suggesting that the Academy establish research guidelines for safety. The Academy set up a committee headed by Berg, which in 1974 came out with a list of possibly hazardous experiments, requesting that his fellow scientists avoid them. The list was actually quite extensive, and Berg's proposals amounted to a moratorium on much of the most interesting DNA research.

Nevertheless, the reputation of Berg and his committee was such that gene researchers obeyed this call for a moratorium. Also influential was the fact that this DNA story had moved from the cloisters of the Gordon Conference into the Sunday supplements, and editors were writing that even the scientists themselves were warning of perils from mutant bugs escaping from the lab. In short, everyone was treading very carefully. In February 1975, the gene-splicing world gathered at the Asilomar conference center in Pacific Grove, California, to write formal guidelines for the research. To England's Sydney Brenner, a conference leader, a successful guideline would be one that in the future would be made weaker. Brenner emphasized the need for safety procedures so tight that no one could accuse the scientists of being self-serving. The director of the Cold Spring Harbor Laboratory, James Watson, complained there was no proof of risk; "they want to put me out of business for something you can't measure." But others felt the hot breath of public opinion. As Stanley Cohen put it, "If the collective wisdom of this group doesn't result in recommendations, the recommendations may come from other groups less well qualified."

The issue was far broader than DNA research, for the Asilomar scientists were the first to face some vexing modern questions. To what degree should we guard against hazards in new technology before, rather than after, accidents and even disasters occur? To what degree can we foresee risks in advance, rather than wait to learn from harsh experience? Must we forgo promised benefits to avoid apparent risks? These questions will recur time and again, as future advances unfold, but DNA research was the battleground over which they were first fought. The outcome of that controversy thus would touch issues ranging well beyond DNA, so it is worth recalling just what happened. To some, recombinant DNA work was to be banned altogether. Among these was Erwin Chargaff, who in 1949 had done an important DNA experiment proving that A and C existed in equal amounts with T and G. By 1975 he was highly critical:

> Knowing that the desire to improve mankind has led to some of the most horrible atrocities recorded by history, it was with a feeling of deep melancholy that I read about the peculiar conference that took place recently in the neighborhood of Palo Alto. At this Council of Asilomar there congregated the molecular bishops and church fathers from all over the world, in order to condemn the heresies of which they themselves had been the first and the principal perpetrators. This was probably the first time in history that the incendiaries formed their own fire brigade. The edict published in due course, which lists the various forbidden items, reads like a combined [record of published research] of the conveners of the conference . . . Is there anything more far-reaching than the creation of new forms of life? . . . Have we the right to counteract, irreversibly, the evolutionary wisdom of millions of years, in order to satisfy the ambition and the curiosity of a few scientists? . . . My generation . . . has been the first to engage, under the leadership of the exact sciences, in a destructive colonial warfare against nature. The future will curse us for it.

To which Stanley Cohen responded,

> It is this so-called evolutionary wisdom that gave us the gene combinations for bubonic plague, smallpox, yellow fever, typhoid, polio, diabetes, and cancer. It is this wisdom that continues to give us uncontrollable diseases such as Lassa fever, Marburg virus, and very recently the Marburg-related hemorrhagic fever virus, which has resulted in nearly 100 percent mortality in infected individuals in Zaire and the Sudan. The acquisition and use of all biological and medical knowledge constitutes an intentional and continuing assault on evolutionary wisdom. Is this the "warfare against nature" that some critics fear from recombinant DNA?

Lewis Thomas, president of the Memorial Sloan-Kettering Cancer Center and author of *The Lives of a Cell*, perhaps best put matters into perspective:

> Considering the vast number of microbial species on this planet, the property of causing disease by infection is excessively rare, almost freakish.

Most of the bacteria and fungi make their living by browsing, reducing dead matter to reusable organic forms. The few microbes that have evolved as infectious agents have done so only after millions of years of adaptation and interliving. Most of them are equipped with elaborate signalling systems, special markers at their membranes, and bizarre products that imitate enzyme reactants in certain cells of their hosts. Organisms like these have to have multiple guidance mechanisms before they can even approach the tissues of a host. Pathogenicity is a highly skilled trade. It takes a kind of arrogance to assert that microbiologists can manufacture complicated creatures like these, by choice or by chance. On the other hand, the pure research potential of the recombinant DNA technique is simply tremendous . . . Deep questions can now be asked about chromosomes and genes and about the most fundamental processes of living cells . . . Our greatest handicap in coping with human disease has always been our ignorance of how the organism really works.

Time clarified the controversy. By 1977, several years' experience with recombinant genes in *E. coli* had convinced leading scientists, including the directors of the American Society of Microbiology, that the particular *E. coli* strain in laboratory use not only could not live in the human intestine, but could not be made pathogenic, not even when scientists deliberately tried to introduce genes known to give other bacteria their virulence. In one important series of experiments, researchers introduced genes from a cancer virus into *E. coli* and injected the transformed bugs into mice. The virus itself, polyoma, was highly infectious in mice. But the transformed *E. coli* were either totally noninfectious or, at worst, no more than one billionth as strong as the naked virus. Since such deliberate efforts to create dangerous germs had failed, it could hardly be imagined that random accidents or acts of ignorance would somehow produce new and lethal bugs. These and other results marshaled the scientists into a solid phalanx in favor of loosening the research guidelines.

Environmental groups opposed this, continuing to urge tightening of the rules. One such group, Friends of the Earth, went so far as to challenge the legality of DNA research if not accompanied by a full environmental impact statement. These groups met widespread and strong opposition from their scientific advisors. Such eminent scientists as Paul Ehrlich, René Dubos and Lewis Thomas publicly broke with their environmental groups over this issue, and the environmentalists found themselves out on a limb, bereft of technical support. Amid this changed climate, it passed almost unnoticed in 1979 that the rules were substantially relaxed. At the height of the restrictions, entirely straightforward DNA experiments were required by law to take place under safety precautions more stringent than the sealed isolation chambers used by the U.S. Army in germ-warfare experiments. Following their relaxation, experiments using standard lab *E. coli* could proceed with only the ordinary care of standard microbiology technique. The outcome of the DNA debate thus gave a strong rebuff to those who had been seeking exaggerated levels of freedom from risk, or who had argued there was knowledge man was not meant to have, areas of study best left untouched.

Meanwhile, the recombinant-DNA techniques were becoming more and more straightforward. In 1981 a branch of the University of Maryland set up the nation's first program to train genetic engineers. The program sprang from the realization that most of the techniques for recombinant DNA research were already in place, so that specific projects could be carried out by technicians. The industry that would employ these new graduates was already rising. In fact, as the fears and doubts of an earlier era evaporated in the late '70s, they were replaced in some quarters by an almost equally exaggerated euphoria. We could say that the Fear-and-Worry Era gave way to the Bullish-on-Bacteria Era. No, no Wall Street firm ever issued a statement such as "Merrill Lynch is bullish on bacteria"; but for a while they might as well have done just that.

On October 14, 1980, the firm of Genentech offered the first public sale of gene-splicing stock to trade on the over-the-counter market. The listed price was $35 a share. Forty-five minutes later it was at $89. When it closed the day's trading at $71.25, its market valuation (the value of all its stock) stood at $529 million, one twelfth the value of the chemical giant Du Pont. Yet Genentech was purely a research firm, years away from marketing a commercial product. It had been founded in 1976 by Herbert Boyer, discoverer of Eco RI, in partnership with the venture capitalist Robert Swanson. A series of research successes soon built for it a strong reputation. In 1977 it announced a bacterial process to produce the brain hormone somatostatin, which prevents people from growing too large or tall. In 1978 it came out with a process for human insulin; in 1979, for human growth hormone which prevents dwarfism; and in 1980 for the virus-fighting agent interferon. In 1981 Genentech would go on to develop a gene-spliced vaccine against a form of foot-and-mouth disease in cattle. These advances had whetted Wall Street's appetite for its stock offering, but even veteran brokers were slightly dazed by this performance.

The science writer Nicholas Wade explained what was happening early in 1981, in an article in *Science* titled, "How to Keep Your Shirt—If You Put It in Genes": "Stock prices . . . relate to perceived worth, an evanescent quality that differs from actual worth by the levitational factor known as hype . . . The hype factor, well understood to the professionals of Wall Street, often goes unrecognized by small investors, such as the widows and orphans who are left holding the baby when stock prices collapse. For the latter class of investors, the following advice is humbly offered. A common route to commercial viability is to produce a product. Although much of the gene-splicing industry seems intent on eschewing this well-trampled path to success, signs of a well-defined intent to manufacture something should not be regarded as an outright handicap. Which companies have definite plans to produce products in the United States?" He then went on to list the companies and their products.

Among the most significant of these were two forms of human interferon. Interferon was discovered in 1957, as the body's first line of defense against viruses. When a virus enters a cell, the cell secretes interferon, which passes quickly to nearby cells and stimulates them to mobilize their defenses by producing appropriate antibodies against the virus. Interferon acts against a very broad range of viruses, and from the start the scientists hoped it could be an important antibiotic, working against viruses just as penicillin and other antibiotics work against a broad range of disease germs. Interferon proved to be exceptionally potent in very small quantities and could be produced by cells and harvested. What's more, as a substance produced naturally by cells it promised to produce few side effects. But to fulfill its promise proved difficult. The cells secreted it in such minuscule quantities that very little could be garnered for clinical use. Thanks to recombinant DNA, however, it is now being produced by the gallon. It promises to be very useful against hepatitis, influenza, and even the herpesvirus which causes a particularly intractable venereal disease. What's more, it appears to cure the common cold, and we may be seeing interferon nasal sprays.

Interferon is only one of a variety of highly valuable substances which the body manufactures in such extraordinarily small quantities that up to now there have only been enough of them to tantalize researchers, but far from enough to use as pharmaceuticals. All are candidates for manufacture by the bacteria in the vats of Genentech and the other gene-splicers. Among them are the powerful brain hormones somatostatin and human growth hormone, which help us to grow to normal height but no further. These are already in production. There are other substances as well within the brain, linked to such disorders as schizophrenia and drug addiction. Surely the ability to produce them with DNA techniques will open new vistas for treatment. Then there are what are known as the angiogenesis inhibitors, which may have a very useful property indeed. They may cure cancer.

There is really no such disease as cancer; there are cancers, plural, a great many of them. Considering the different tissues and various types of malignancies which can be involved, cancers come in over a hundred varieties. The different kinds of cells in the body come equipped with genes that prevent them from becoming cancerous, but these genes can mutate one by one during a person's life. When enough of them mutate, a cancer cell is born; for instance, intestinal cancer requires about five such mutations in a cell. The causes of the mutations lie in the world outside the cell: cigarette smoke, cancer-causing chemicals, too much sunlight, nuclear radiation from rocks such as granite, even our choice of foods. (A diet rich in meat stimulates the flow of bile, which can be broken down to produce cancer-causing agents by bacteria living in the intestine.) With so many producers of mutations all around, it is little wonder that so many of us will find that our protector genes will all have been picked off.

A cancer tumor starts small, and as long as its cells stay in one place the tumor usually does little harm. They don't stay in one place, however; they metastasize. They break loose and travel through the bloodstream to distant parts of the body to grow there and form new tumors, which continue the process. The cancer spreads; soon the body is riddled with the sickness, and death is not far away. This tragedy strikes millions of people a year, but it may be we can interrupt it as it unfolds. It is important, of course, to prevent mutations of the anticancer genes. That is why we discourage smoking and watch with great care the chemicals to which people are exposed in their food and while at work. But once a cancerous tumor is well started, it will stop dead in its tracks, no matter what kind of cancer it is, if we interrupt the one thing it needs: a blood supply.

When a cancer tumor begins its growth, it is a small round mass about the size of the head of a thumbtack. At this stage the tumor has no blood vessels. To grow larger, it must develop such vessels by causing capillaries to penetrate and interweave its volume, a process known as angiogenesis. The tumor does this by secreting a substance called TAF, tumor angiogenesis factor, which causes nearby capillaries to branch and grow into its volume. As this blood supply develops, the tumor can grow and become malignant, while the metastasizing cells establish new tumors elsewhere which also secrete TAF and develop their own blood supply. The process feeds on itself, the tumors becoming ever more virulent as they develop blood vessels more and more rapidly. The process does not stop till the patient is dead.

However, no one ever died of a tumor in its early stages, before it begins to release TAF. If the effect of TAF can be blocked, metastasizing cells will be unable to grow into large tumors. In fact, it has been found that TAF not only gives rise to a tumor's blood supply, it also maintains that supply once it has developed. This means that when TAF is blocked, not only are new tumor blood vessels prevented from growing, but even in a large and well-established tumor the existing vessels will die away. When the vessels go, so does the tumor, and even a large and malignant mass will wither and shrink. The substance that blocks TAF is called angiogenesis inhibitor or AI and is found in cartilage. A small piece of cartilage placed between an early tumor and a nearby artery will prevent capillaries, branching from that artery, from reaching the tumor.

Just as interferon exists only in minuscule amounts in cells, AI is in similarly small supply within cartilage. But it is an attractive candidate for the gene-splicing trade. If it can be so produced, and if it fulfills its promise, it would stand as a true cancer cure. Just as interferon works against a broad range of viruses, so AI would work against any tumor and nearly all cancers, shrinking the most virulent and invasive masses of contagion back to thumbtack size. AI would also resemble interferon in that both are natural substances which should produce few side effects. There would also be similarities in their operation. During aeons of evolution, viruses have been quite unable to develop resistance to interferon. Similarly, tumors would have no defense against AI, for a tumor deprived of its blood supply would be as dead as a mosquito larva smothered by a film of oil floating on its stagnant pool of water.

When this sort of cancer therapy becomes routine, we will see that cancer treatments will have passed through three levels of understanding. As outlined by Lewis Thomas, the first or earliest of these is basically a laying on of hands. This phase of medicine is highly valued since it makes the patient as comfortable as possible, but it offers nothing to change the course of a disease, or affect its outcome. When a patient has terminal cancer, this today is all anyone can do. Many malignant cancers, however, can be treated today at Thomas' second level of understanding, which he calls "halfway technology." These are therapies that deal with cancer after the fact, to manage the disease in the face of ignorance of the more fundamental knowledge that would achieve an actual cure. Chemotherapy, radiation therapy, surgical operations which are often painful and disfiguring—that is what we do today to manage our patients' malignant cancers.

But there is a third level of understanding, giving therapies so simple and effective that the public comes to take them for granted. The treatment of polio will illustrate the difference between these levels. In the early 1950s, polio could merely be managed, not cured or prevented. Iron lungs, leg braces, injections of gamma globulin, a blood fraction rich in antibodies—all were halfway technology. The Salk vaccine succeeded so brilliantly that polio rapidly vanished. An AI type of therapy would similarly be so simple, so unobtrusive, safe and effective as to quickly send cancer to join leprosy, smallpox, and plague among the dreaded diseases of the past.

Should cancer thus be vanquished, its name will still be feared, but its legacy may in time grow dim. This has already happened with another old terror, which has left a trace everyone recognizes and almost no one understands. Everyone knows a nursery rhyme which in its original version goes

> *Ring-a-ring o' roses,*
> *A pocketful of posies,*
> *A-tishoo! A-tishoo!*
> *We all fall down.*

This bit of apparent nonsense is about bubonic plague. The lines refer to the rosy mark on the chest of a plague victim, the bunches of flowers advertised as warding off infection, the convulsive sneezing in the victim, and for the afflicted communities, mass death. We can only hope that one day children will be free to sing with similar carefree innocence about what we would recognize as our own plague, cancer.

If we can sketch serious and plausible scenarios for the conquest of viruses and even of cancer, can we speak of overcoming death itself? Can we think of immortality not of our souls but of our bodies? This question is not frivolous, for when presented with a research technique as powerful as recombinant DNA, questions like this help us understand the limitations as well as the promise of the technique. Aging and death, of course, are not diseases like cancer, even though these are often closely related. It appears, however, that aging and death represent a wearing out and using up of the body at the level of individual cells.

Many human cells can be grown in tissue culture, where they grow and multiply in glass bottles when supplied with nutrient. In the 1960s, in a significant series of experiments, Stanford's Leonard Hayflick overturned the prevailing belief that such cultured cells could continue to grow and multiply forever. He worked with human fibroblasts, cells found in skin, lungs, and other tissues. His technique was to grow cells in flat bottles and let them expand to the periphery, then divide the resulting cell masses in two and transplant half to a new bottle, let them grow to cover its surface, and continue dividing and transplanting. What he found was the "Hayflick limit": the cells could not be made to divide more than about fifty times.

This limit proved to show up in many ways. Fibroblasts could be made to divide a few times, then be frozen and kept in cold storage, perhaps for years. When thawed, they would "remember" their age, and the total of divisions from before and after the freezing would stay near fifty. Cells taken from embryos divided the full fifty times, but cells taken from children and teenagers divided only about thirty times; those from adult donors made only about twenty divisions. Fibroblasts from the embryos of chickens, rats, mice, hamsters, and guinea pigs—all with short lives— made only about fifteen divisions, and the number was considerably less when the cells were from adult animals. Most remarkable was an experiment done with male and female cells, which carry the X and Y chromosomes and thus can be distinguished. Hayflick seeded a bottle with cells from a male population that had undergone forty doublings, together with other cells from a female population that had had only ten doublings. If the Hayflick limit resulted from some problem within the bottles, male and female cells should succumb equally. But no; the males died off after only ten more divisions, while the females lasted for forty more.

There were almost as many explanations for this state of affairs as there were scientists trying to do the explaining. But a popular explanation was that as cells reproduce generation by generation, errors would creep into the copies of the cells' DNA, and eventually the errors would accumulate to the point of being lethal. This theory put the blame on deficiencies in the processes of DNA repair, by which cells heal nicks in the DNA molecule or correct faulty transcription of its data. In one experiment, fibroblasts from various species had their DNA damaged by exposure to ultraviolet light, and the rate of DNA repair was compared. Human cells were twice as active in this as were cells from chimpanzees, who live only half as long as we do. Fibroblasts from humans, elephants, and cows repaired their DNA five times faster than fibroblasts from rats, mice, and shrews.

This meant that chemical compounds resulting from processes in the cell, tending to cause DNA damage, would contribute to aging. One group of experimenters added vitamin E, which controls the buildup of such chemicals, to Hayflick's fibroblasts. They found the Hayflick limit rose sharply, from 50 divisions to 120. Other investigators showed that the drug L-dopa, used to treat Parkinson's disease, had an effect. High concentrations of L-dopa in the diet of mice raised their life-spans 50 percent. Such experiments raise the prospect of spry and active octogenarians busily gulping their vitamin E, or whatnot, in a novel version of macrobiotic diets. But the human gene-repair mechanism is certainly one of the world's fastest, and it is not clear that we could coax it to greater speed. Thus, the best hope these research findings offer is that people will live to age eighty or so, then fall apart somewhat like Oliver Wendell Holmes' wonderful "one-hoss shay." It lasted a hundred years, then went to pieces "all at once, and nothing first, just as bubbles do when they burst." Still, even if DNA research offers little immediate hope of cheating death, that is nevertheless far from the worst way to go.

As long as we are alive, however, we will be needing fuel and food. In these areas gene research will help, and we can look forward to developing microbes to put DNA power in our tank. (There may be an advertising slogan in that somewhere.) Already there are petrobugs to turn air and sunlight into oil. At the University of Toronto, the chemists Morris Wayman and Allan Jenkins have found combinations of different microbes which give very interesting results. The most efficient pairs the alga *Chlorella*, which has green chlorophyll to produce starches by photosynthesis, with either of two oil bugs which eat the starches. These bugs then grow fat with an oil resembling that from the tar sands of Canada's Athabasca region. The oil bugs include the bacterium *Arthrobacter*, which looks like a plastic bag bulging with oil drops, and the yeast *Candida*, which resembles a bean pod filled with petroleum lumps. Other petrobugs may also be in store. Only a few months after the Iranian oil crisis, as Nicholas Wade wrote in *Science*,

> *Alga ayatollahphobera** is a remarkable organism. It secretes a fine emulsion of hydrocarbons that is convertible to high-octane gasoline at $2.50 a barrel. It grows in seawater, a medium useless for food crops. It forms its hydrocarbons from carbon dioxide, thus helping to retard the worrisome buildup of the gas in the atmosphere. From nothing more than sea, sunlight and waste gas, the industrious microorganism produces gasoline almost too cheap to meter.

A. *ayatollahphobera* has yet to be discovered or invented, but it is the sort of thing the gene-splicers will surely be trying to make. In addition, given any petrobug that we find promising, we surely would seek improvement. This has been done quite successfully in the case of penicillin produced from mold. The original penicillin culture in 1941 produced about four units per tank of growing mold. After a worldwide search to find the best penicillin producer in nature and then after some mutation programs, yields increased to around a thousand units. Since then, further programs of mutation and selection have boosted the yield to over ten thousand units per tank.

These mutation programs involved nothing so sophisticated as recombinant DNA. They were done by a very slow, arduous, and imprecise technique. The molds were exposed to ultraviolet light or to other forms of radiation, or to chemicals causing mutations in the DNA. There was no way to know which genes would mutate or how the mutations would affect penicillin production; in fact, most of the mutants simply died. However, this scattershot approach did produce a few cells that not only survived but produced more penicillin. This amounted to Darwinian evolution in the laboratory. Penicillin molds could survive if they were naturally fitted to their environment, and they were fit if they produced lots of penicillin. The ones that didn't got dumped down the sink. Meanwhile, the flurry of induced mutations was mimicking, on a greatly speeded-up scale, the random mutations of nature which in wild populations produce variation of characteristics within a species, among which will be those whose variations will have made them most fit. From this view, we could say that the difference between these mutation programs and recombinant DNA is the difference between evolving the human eye and inventing the camera.

If we can do to the oil bugs as we did for penicillin, then the future for oil may echo Obed Macy's *History of Nantucket:* "In the year 1690

* "The alga that brings fear to the ayatollah."

some persons were on a high hill observing the whales spouting and sport-
ing with each other, when one observed: there—pointing to the sea—is a
green pasture where our children's grandchildren will go for bread." In
remote but sunny regions of the ocean, large areas of sea may be fenced
off by lightweight side structures floating on pontoons, thus converting
the enclosed waters into enormous sea farms. These may be the places
where *our* children's grandchildren will go for their fuel.

The flexible floating sides would act simply to demarcate the area and
prevent currents from scattering the enclosed oil bugs. While they would
carry warning lights and radar reflectors, if a ship struck this boundary it
would easily pass through what would essentially be a seagoing chain link
fence. Within these fences would be the fields of algae, turning the sea a
bright green with their chlorophyll, stirred by waves that pass through the
boundary and simply lift the pontoons. Crossing and recrossing these sea
ponds would be the skimmers, stimulating the growth with flows of fertilizer pumped into the sea, scooping up the water, straining the algae,
pumping them to onboard processing plants to separate out the oil, then
docking at a central station to transfer the oil from onboard holding tanks
into a floating tank farm. The main storage tanks would be huge steel
spheres and ellipsoids, resembling in size and shape those which oil
refineries build. They would be weighted to float nearly submerged to
avoid being damaged by storms or high waves. Tankers would come there
to pick up their oil, and the supertanker traffic in the vicinity would
resemble that near any oil port.

Other life-forms may find their way to our gas tanks, for unlike
money, diesel fuel grows on trees. In 1979, following a trip to Brazil, the
Nobel-winning chemist Melvin Calvin announced he had observed a tree
in the jungle that produces virtually pure diesel fuel. The natives have
known of it for a long time. They drill a two-inch hole into its three-foot
trunk and put in a bung to seal it. Six months later they remove the bung
and tap off four or five gallons of the stuff. Calvin has shown that it can
be used directly in a Volkswagen Rabbit. Certainly this tree will not soon
serve to prop up the fortunes of Exxon, but it does show something of the
amazing diversity of nature.

Not all of nature is so amenable to gene manipulation as are the
simple microbes. If we wish to improve the output of the diesel tree, or to
do such things as add genes to allow plants to fertilize themselves by fixing
their own nitrogen, we will find ourselves dealing with whole complexes
of genes rather than with just one or two. We are making progress, slowly
advancing up the scale of complexity of life. Thus, whereas the bacterium
E. coli has been the mainstay of most work to date, we are beginning to
see good use made of a somewhat more advanced one-celled organism. In
March 1981, Genentech scientists announced they had coaxed yeast

cells to make interferon (and thus boosted their company's stock another $7.00 a share). Such advances are rather modest, but they have certainly followed one another rapidly. If we can advance up the evolutionary scale at all, even merely in going from bacteria to yeast, there is the obvious question of whether we will stop, or whether instead we will progress to those most advanced animals of all, the higher mammals, and in particular to ourselves.

This question can be posed in many ways. One of the first things we would try to do, in applying genetic manipulations to humans, would be to cure inherited diseases. The simplest of these would be illnesses controlled by a single gene or at most a small cluster of them. The basic procedure would be similar to that of the 1973 experiments of Boyer and Cohen with E. coli. It would be necessary to isolate the desired gene, to transfer it into appropriate human cells, to get these cells to grow and multiply within a tissue, and to ensure that the new gene would be expressed; that is, that it would actually work. Some of these steps are easier than others. We are beginning to get a reasonably good grip on the problem of finding specific genes the lack of which give particular diseases. As one scientist has said, "Today it is easier to work with DNA than with any other macromolecule, which is the exact reverse of the situation ten years ago."

To get the gene into the cell nucleus, it is possible to use special viruses similar to the lambda phage that injects DNA into E. coli, but it is also possible to be more direct in the matter. Today there are glass micropipettes, miniature hypodermic needles if you will, with a tip diameter of only about 0.5 micron, less than a hundredth that of a human hair. With these it is possible to actually inject genes directly into the nucleus. From the cell's viewpoint the operation is about on a par with a person being speared by a telephone pole, but the cells are durable; about three fourths of them survive the treatment. Then it is necessary to put them in the body and try to have them multiply. In experiments done to date, an important goal has been to try to treat blood diseases such as thalassemia and sickle-cell anemia, in which the blood lacks proper forms of hemoglobin, its oxygen carrier. Since blood cells are manufactured in bone marrow, that is where the transformed cells have been injected. Then, in a neat turnabout on the idea of nontransformed E. coli being killed by tetracycline, it is possible to provide the transformed marrow cells in advance with an extra gene to give them a survival advantage. For instance, in earlier experiments in mice, the extra gene conferred resistance to the drug methotrexate. Ordinary marrow cells thus could be inhibited in their growth, but the new cells would proliferate. Eventually the marrow might be entirely of new cells, and if they did their job the patient would be cured of his genetic disease.

The real problem today is getting them to do their job; that is, getting the new genes to express themselves, as in producing the needed forms of hemoglobin. It is not enough to transfer the gene and to select the cells to carry it. The gene product must be made only in the appropriate cells, at the right time, and in the proper amounts. If the gene is not properly controlled it may not help the patient and might even hurt him. Unfortunately, the regulation of gene expression is one of those problems which today is not understood. In the words of Richard Axel, a leading specialist on gene transfer, as of 1981 "I know of no one who has been able to insert a tissue-specific gene into an appropriate cell line and have its expression regulated in the normal way."

The problem of gene expression in man is fearfully complicated for our genes are strung together on chromosomes. It may be necessary to put a transferred gene into its normal chromosomal site to get normal control, rather than simply injecting it loosely into the cell nucleus. Yet we will see this problem solved, at least in part. Genes indeed will be transferred and expressed, and bit by bit we will make progress against at least some genetic diseases. It may also be possible to correct the gene defect during conception. Today egg cells or ova can be fertilized in a laboratory dish and transferred to a mother's womb to grow into a normal baby; these are the famous test-tube babies. In the laboratory, this process is routine in dealing with mice. Already, functioning genes have been introduced into mice by injecting their DNA into mouse ova during the moment of conception, a privileged time when it is normal for the future embryo to accept new genetic material. In principle, this can also be done in humans.

Such gene-transplant babies would be the future counterparts of the heart-transplant patients. The operation would be splashed across the covers of newsmagazines and would be hailed as a dramatic and exciting advance in medical science. But like heart transplants, gene transplants would be exceptional procedures for use when all else fails. The more usual gene therapies, like the more usual heart therapies, would be much less dramatic. It is possible in many cases to know in advance that two parents have a high risk of producing a child with a genetic defect. For instance, the folk singer Woody Guthrie died of the inherited disorder known as Huntington's chorea, and his son Arlo has a fifty-fifty chance of having the same disease. Parents facing such risks might opt for a test-tube fertilization. The protoembryos, growing in glass dishes, would have a few cells removed and tested for the presence of the bad gene, and only an embryo which passed this test would go to the womb for normal growth. But where both parents carry genes for an inherited disease, the risk could run to 100 percent. In those cases, if the needed gene can be added at conception, the disorder might be prevented not only in the new baby, but also in his children. Genes, after all, are inherited.

Nevertheless, if it is so difficult to devise means to treat genetic disorders where there is a real need, it will be far harder to manipulate the genes controlling intelligence, body build, strength and dexterity, or other interesting features. These are all polygenic traits; that is, they result from the interacting work of many genes. There is a vast difference between them and the genetic diseases we may treat. A person either has sickle-cell anemia or he doesn't; there is no halfway point, and nothing he does in life will change his situation. But there is a wide range for people's intelligence or body build, with a continuous variation of intermediates. Moreover, people can get smarter by learning and can build their bodies by exercising; the genes' action is influenced by the environment. Furthermore, we have not identified even one gene or protein whose variation contributes to the normal range of behavior or intelligence, but we would need such information for many genes before we could try to modify people's basic natures by manipulating DNA.

Still, let us leap lightly over these difficulties and envision a time in the future when parents will be able to exercise a large measure of control over what kind of children they will have. No doubt many will choose to have their kids in the good old-fashioned way, with perhaps no more than an amniocentesis. This is a simple procedure which extracts a bit of fluid from the amniotic sac so that the cells within it can be studied and the fetus thus checked for genetic problems. (It is already becoming a common procedure.) But also there will be many who will want to have their say in deciding what the baby will be like. What kind of children will they choose to have?

Even if it becomes possible to do so, I very much doubt that parents would routinely choose to bring into the world children combining the beauty and grace of Miss America with the brains of Einstein, the temperament of a diplomat, and the artistic talent of Michelangelo. Parents usually want children who are pretty much like they are, rather than being unusual in brains or talent or physical perfection. Prodigies, by and large, are disturbing. What the parents actually would want, indeed, would be that the children be a little bit better than they are in those departments, but not so outstandingly so that at an early age, mom and dad will experience the mixed hope and doubt that comes from having a little Mozart or Brooke Shields or Bobby Fischer, the chess prodigy. What the parents would get, however, would ordinarily be rather different.

Because traits such as beauty, intellect, talent, healthfulness are controlled by many genes, children show a characteristic that statisticians call regression to the mean. This means a tending toward the average, a tendency toward averaging out whatever exceptional characteristics the parents may have had. The children of geniuses are themselves often quite intelligent as well, but somewhat less so. In the general population, children may be a little better off or a little worse off than their parents, in brains, health, and the like, but they show no progressive trend. In this regard, having a child is something of a shot in the dark. But if these traits can be controlled, parents will choose to have children who are at least as smart or healthy as they; maybe a bit more so, but certainly not less.

The average of the next generation, then, would show improvement by having been skewed. The same would be true of succeeding generations. By preventing children from sliding backward and having less of a given trait than their parents, the normal and reasonable wishes of parents then would actually work to improve the human race. In a thousand years or so, almost everyone you might meet on the street would, in today's society, stand out as truly unique.

What would such future societies be like? We already have small-scale societies of exceptional people: Hollywood for the beautiful and talented, Wall Street for the shrewd, university campuses for the bright. Significantly, they do not appear to run particularly better or worse than ordinary towns in the Midwest. Hollywood is famous for divorces and other instabilities (though the rest of the nation has largely caught up in that area), Wall Street for nervous tension and tranquilizers, campuses for loud idealism and an inability to take firm stands or decisions except when engaged in moral posturing. Midwest towns, on the other hand, tend to be quite narrow and provincial, as when their local book-banning committees treat *The Catcher in the Rye* as though it were the magazine *Hustler*. All this proves there is no such thing as a perfect society because societies are made up of human beings, and humans aren't perfect. Genetic engineering will not change that. The most it may do is reshuffle the imperfections.

There is another area ripe for speculations: the cloning of people. This means making a copy or copies of a person using the genetic information stored in any of the many trillions of his body cells. Cloning got its start in 1952, when Robert Briggs and Thomas King transplanted the nuclei of frog embryos into frog eggs from which the nuclei had been removed, and got copies of frogs. The current status of frog cloning is that when the nuclei are taken from cells at the earliest stage (the blastula) of frog embryonic development, 75 percent of the transplants give tadpoles or adult frogs. By the next embryonic stage this figure falls to 15 to 20 percent. No one has been able to grow adult animals with nuclei transplanted from adult cells, though eggs carrying such transplants may develop through several embryonic stages. Biologists know the frog is peculiarly suited for cloning. Its egg is large, develops in the water and outside the frog's body, and is so easily triggered that it will begin dividing if merely pricked with a pin.

More recently mice have been cloned. Again this has been a matter of taking cell nuclei from embryos at the earliest stage of development and transplanting them into recently fertilized mouse eggs whose own nuclei had been removed. One round of this work produced 363 such transplants. Only 48 went on to advance to the next embryonic stage, the blastocyst, in which the embryo exists as a hollow ball of cells. And not all of these blastocysts appeared normal. Thus, even in the earliest embryonic stages, the genetic material is apparently altered, probably irreversibly, with genetic information being switched off for good. All this is hardly the stuff to justify a protest demonstration outside the biology lab, with demonstrators shouting, "Stop Xeroxing people."

In going from the stage of a fertilized egg to that of a muscle or gland cell in an adult human, virtually all the genetic information is blocked. The egg has it all. The mature cell apparently has a full complement of DNA, but switches on only the very limited amount—perhaps four parts in a thousand—that it needs to do its job. The associated process of transition is called differentiation, and persuading an ordinary body cell nucleus to fertilize an enucleated ovum would be dedifferentiation. If the fertilized egg is like the Encyclopaedia Britannica, the mature differentiated cell is like that set of volumes with virtually all its text heavily marked out with India ink, leaving only those few articles dealing with some specific topic like metalworking. Dedifferentiation then amounts to turning the cells' lost information back on, removing the "India ink"—without ripping the pages or damaging the underlying text. Perhaps someday there will be an enzyme, dedifferentiase, to do the job. Right now that is like prophesying bridges across the oceans made of the miracle metal Unobtanium.

Again, though, let us leap lightly over these appalling difficulties and imagine cloning becomes a reality. Some horror stories picture a 1930s-type dictator proceeding to turn out identical clones en masse as cannon fodder for his army. Such a dramatic fancy raises an obvious question: where would this dictator get the wombs? Even Hitler, no neophyte at coercion or persuasion, could not convince his nation's Frauen to raise significantly their baby production. No, if cloning ever comes to be, this will not be its use. Cloning will produce multiple copies of prize cattle or horses; the technique may even be developed for this purpose. But for human beings it will be a more personal affair.

Imagine a woman in her early thirties, no longer quite young, but wanting a baby. Imagine further that she has no significant lover or fiancé, but she wants a child just the same. She could go down to her local fertility clinic and fertilize a clone of herself, using one of her own ova and body cells. The baby then would actually be her identical twin sister, born thirty or so years later.

What would the life of mother and daughter be like? We can get some fascinating glimpses of the possibilities by considering studies of identical twins separated at an early age and reared apart. A study at the University of Minnesota has looked at a number of such twins and has found that the scores of such identical twins on many psychological and ability tests, including the IQ test, are closer than would be expected for the same person taking the same test twice. To the psychologist David Lykken, a study leader, "the most important thing to come out of this study is a strong sense that vastly more of human behavior is genetically determined or influenced than we ever supposed."

Suggestive anecdotes abound, as in the case of the British housewives Bridget and Dorothy. They had been separated during World War II while in infancy, one being raised in modest circumstances, the other one's family being well-off. When they met, their manicured hands each bore seven rings. Each also wore two bracelets on one wrist and a watch and bracelet on the other. They each had a son and a daughter. One had named her children Richard Andrew and Catherine Louise; the other, Andrew Richard and Karen Louise.

In Ohio, Jim Springer and Jim Lewis were adopted as infants into working-class families and did not meet till they were thirty-nine years old. As Constance Holden wrote in *Science 80*, "Both had law enforcement training and worked part time as deputy sheriffs. Both vacationed in Florida; both drove Chevrolets . . . Both had dogs named Toy. They married and divorced women named Linda and remarried women named Betty. They named their sons James Allan and James Alan . . . In school both twins liked math but not spelling. They currently enjoy mechanical drawing and carpentry. They have almost identical drinking and smoking patterns, and they chew their fingernails down to the nubs. Investigators thought [the similarities of their] medical histories were astounding."

Advocates of the brotherhood of man would find food for thought in the twins who became a Nazi and a Jew. Oskar Stöhr and Jack Yufe were born in Trinidad in 1933 and separated soon after birth. Oskar went with his mother back to Germany, her home, where his grandmother raised him as a Catholic and a Nazi. Jack stayed in the Caribbean with his father, a Jew, and spent part of his life on a kibbutz in Israel. Yet when reunited these twins also showed noteworthy similarities in manner, personal habits, dress, and the way they do things.

So what would a mother-daughter relation be like if the baby were a clone? As she grew up, the mother would see the little girl reliving her own childhood in often poignant detail. The daughter, for her part, would learn in time that what her mother is today, she would be herself when grown. They would be unusually close; for who could be closer than a daughter who not only is like you, but actually *is* you? As more and more women (though perhaps never very many) would have such daughters, it would become widely appreciated that these were very special family groups, indeed. Then, among husbands and wives, a woman who loved her man to an exceptional degree might show her love by bearing, not merely his baby, but his clone.

What will we do, then, with the godlike power science promises, in controlling genes and manipulating the very stuff of life? We will do with it as in earlier decades we have done with the godlike power to fly, to communicate instantly, or to uproot mountains. We will weave it into the fabric of life and society, putting it to our advantage. A century ago, flight was the realm of angels and crackpot inventors, though railroads were familiar and in common use. Today we freely use aircraft and trains alike, often during the same trip, and no one finds this unusual. The reservation systems for Amtrak are the same as those for the airlines, while an Amtrak coach looks like a sawed-off airliner fuselage. Similarly, tomorrow we will accept the miracles of DNA in the same way that today we accept hybrid corn and antibiotics.

Perhaps one of the most influential things these godlike powers will give us, then, will be an easing of the belief that they are godlike. In centuries past, blood was regarded as ineffable and mysterious, a portion of nature's design in which man should not meddle. Today some people react with the same superstitious awe to the word "gene." But genes are made of DNA, a kind of cellular computer tape carrying information, and DNA is becoming well understood. Surely, if one more aspect of the world around us is stripped of its aura of magic and mystery, it can only mean we will have taken one more step away from ignorance. The human race as a whole can only gain for that.

9

WAR AND DETERRENCE

On the cover of each issue of the *Bulletin of the Atomic Scientists* is the doomsday clock, its hands poised at a few minutes to nuclear midnight. Every few years, as international tensions increase or ease, the minute hand moves up or back; yet never will it stand at more than fifteen minutes to midnight. The clock has been there since the founding of the *Bulletin* in 1947. Clearly, however, something is amiss. If we are presented with such a graphic warning of imminent disaster, yet the disaster fails to arrive for half a human lifetime, then we are justified in concluding the danger is not quite so imminent.

Certainly we can continue to expect warnings of nuclear war. There are numerous writers and academics whose pens show an almost seismographic sensitivity, fluttering and vibrating in response to even the slightest of tremors on the international scene. But the war itself, the Armageddon which will reduce the world to radioactive rubble—that war will probably never arrive. It appears quite possible to say, even today, that we have made it through the night, we have gone past the time of danger. Indeed, it is possible to say even more. The time of danger we passed was over thirty years ago.

The great wars of our century did not come by surprise. They were preceded by several years of crises, of diplomatic confrontations and local wars or invasions, which like warning earthquakes led up to the massive earthshakings of the World Wars. The countdown to 1914 began in 1905, with the Kaiser throwing the gage of war before France, personally visiting Tangier and challenging French colonial interests in Morocco. This was only three weeks after the defeat of France's ally Russia in the Russo-Japanese War. France, unable to fight, backed down. In 1911 the Kaiser sent the gunboat *Panther* to Morocco, making his challenge to France even more explicit, and in all the chancelleries of Europe was the whispered monosyllable, "War." This time, however, France had the backing of Great Britain and it was the Kaiser's turn to retreat. These Moroccan crises greatly heightened tensions in Europe. Still, as Chancellor Bismarck had predicted before the turn of the century, "some damned foolish thing in the Balkans" would ignite the next war.

Between 1908 and 1913 the rivalries and ambitions of Russia, Turkey, Serbia, and the Austro-Hungarian Empire led to a succession of crises, beginning when Austria annexed a Turkish province which Serbia coveted. Serbia, backed by Russia, then challenged Austria by launching the Balkan Wars of 1912 and 1913, which conquered most of Turkey's holdings west of Constantinople. When Serbian nationalists assassinated Austria's Archduke Franz Ferdinand at Sarajevo in 1914, Europe's cup of peace was drained of its last drop.

In the 1930s the descent into the maelstrom took only three years. In 1936 Hitler challenged France by sending troops into the demilitarized zone along the Rhine, which had been established as a buffer region. In 1937 he repudiated the Treaty of Versailles, the instrument of German surrender in 1919. In March 1938 he invaded Austria and annexed it to his Reich. That September he humiliated British Prime Minister Neville Chamberlain at their conference near Munich, forcing Britain and France to acquiesce in his seizure of part of Czechoslovakia. The following March he took over most of the rest of that country. Then he demanded part of Poland, signed the Stalin-Hitler Pact to give himself a free hand, and launched his invasion. This time France and Britain were prepared to fight, and on the clocks of the world it was wartime once more.

Just such a succession of crises, confrontations, and local invasions took place between 1948 and 1950, bringing the world closer to nuclear war than it ever has been, before or since. Following several years of deteriorating relations between the United States and the Soviets, in 1948 Stalin's agents seized control of Czechoslovakia, which had reestablished itself under a democratic government, and installed a communist puppet regime. That same year Stalin struck at Berlin, blockading land and canal routes to that city in a direct challenge to the British, French, and Americans who shared in its control. America responded by setting up the Berlin airlift, successfully bringing supplies to that beleaguered city, and by organizing NATO as an anti-Soviet military alliance. American power rested on our possession of the atomic bomb, but in 1949 the Soviets detonated their own bomb and shattered our monopoly. At nearly the same time, Mao Tse-tung's army gained final victory over the U.S.-backed Nationalist Chinese, and proclaimed the People's Republic of China. At the time this looked as though Stalin had again extended his power, this time to encompass the world's most populous nation. That impression was reinforced in February 1950, when Stalin and Mao signed a treaty of military alliance and proclaimed the Sino-Soviet bloc.

The following June China's ally, North Korea, struck across the 38th parallel, the international boundary at 38 degrees north latitude, and invaded U.S.-backed South Korea. President Truman rushed U.S. forces to the defense, but the North Koreans continued to advance and soon pinned the defending forces into a small corner of southeast Korea. In September the U.S. commander, Douglas MacArthur, launched his master stroke. With strong naval support he invaded Korea at Inchon, midway up its length, and struck across the North Korean supply lines. The North Korean offensive collapsed. MacArthur himself took the offensive, crossing the 38th parallel and seizing the capital, Pyongyang—the only instance thus far of a communist capital falling to advancing American troops. By Thanksgiving his army was approaching the Yalu River, the boundary between North Korea and China. With this, Chairman Mao unleashed his shock troops. They crossed the Yalu and in a matter of days struck suddenly and with great force at the Yanks, turning their triumphant advance into a virtual rout. Almost overnight it was a new war, and it looked as though the Americans would be swept from the Korean peninsula.

On November 30, at the height of the Chinese advance, Truman held a press conference. In his prepared statement he warned that his forces, fighting under United Nations authorization, would strike directly at China if the UN should vote to brand China an aggressor. (In those days the UN was largely controlled by the United States and its allies.) He then engaged in an exchange with newsmen:

TRUMAN: All necessary steps to meet the military situation would be taken just as we always have.

NEWSMAN: Will that include the atomic bomb?

TRUMAN: That would include every weapon that we have.

NEWSMAN: Mr. President, you said every weapon that we have. Does that mean that there is active consideration of the use of the atomic bomb?

TRUMAN: There has always been active consideration of its use. But I do not want to see it used. It is a terrible weapon that should not be used on innocent men, women, and children who have nothing to do with this military aggression.* That was what happened when the bomb was used.

NEWSMAN: Mr. President, I wonder if we could retrace that reference to the atomic bomb? Did we understand you clearly that the use of the atomic bomb is under active consideration?

TRUMAN: It always has been. It is one of our weapons.

NEWSMAN: Does that mean, Mr. President, use against military objectives or civilian?

TRUMAN: Selection of the objectives or targets is for the military authorities to decide. I am not the military authority that passes on those things.

NEWSMAN: Mr. President, you said this depends on the United Nations action. Does that mean that we would not use the atomic bomb except on a United Nations authorization?

TRUMAN: No, it does not mean that at all. The action against Communist China depends on the action of the United Nations. The military commander in the field would have charge of the use of the weapons, as always.

It has often been written that war between the superpowers might result from miscalculation, from overreaction, from callousness as to the effects of the bomb, or as an act of madness. All these elements were present in December 1950; rarely have any of them been present since. It was by miscalculation, by gambling that MacArthur could safely push north of Pyongyang toward the Yalu, that the Chinese entered the war. Their stunning successes were, for a few days at least, nearly as shocking to America as the Japanese attack on Pearl Harbor nine years earlier. Harry Truman had ordered the bomb used against Japan, and some historians argue that when he struck at Nagasaki, it was less to speed the Japanese surrender than to impress upon Stalin that we had the weapon and were willing to use it, even if its use was not entirely necessary. Finally, Stalin had long established himself as a ruthless and possibly mad leader willing to kill by the tens of millions to achieve his aims, which included massive extensions of communism and of Soviet power. Even while America still held a nuclear monopoly, Stalin's aggressiveness had been so dismaying that Winston Churchill had said, "If this they do in the green wood, then what will they do in the dry?" In 1950 Stalin too had the bomb, and probably had less compunction than Truman about using it.

What saved the situation was that the military commander in the field was able to stem the Chinese advance without recourse to nuclear weapons. MacArthur and his generals soon were able to convert their rout into an orderly retreat, forming a defensive line across the Korean penin-

---

* Was this an oblique indication of intent to bomb Peking?

sula. Eventually this line stabilized near the 38th parallel and became the new international boundary. Never again, however, not in the Korean War nor in any subsequent crisis, would the United States come so close to using the bomb. Even in the Cuban missile crisis of 1962, the issue was Soviet deployment of missiles in Cuba rather than their imminent use. President Kennedy never had to go further than to set up his naval blockade and to warn in a speech that "these actions may be only the beginning." It was a far cry from Truman's almost matter-of-fact attitude that the bomb was simply "one of our weapons."

It is easy—too easy—to read the current headlines and to conclude that tensions are rising, peace is fragile, war is not far away. But among powerful rival nations with conflicting interests, the natural state of affairs is just this state of continuing tension, this continued drum patter of minor quarrels and conflicts. The United States and the Soviets may be likened to two tectonic plates, vast continent-size masses of rock which fit together to form the crust of the earth, and which slowly slide past each other amid the rumbling of earthquakes where these plates meet along a fault line. The earthquakes are occasionally frightening, but they are not large, and they relieve stress along the fault. What is to be feared is that the plates would lock. Then the sliding would stop, tension and pressure would build to unprecedented levels and in time would bring an earthquake of truly catastrophic proportions.

On the diplomatic Richter scale, the crises and disagreements since 1950 have mostly been of magnitude 2 to 4, with a few medium-scale events of magnitude 6. But there has been no diplomatic lockage raising tensions so high that release could come only in that quake of magnitude 9 which would be a major war between the superpowers. Nor is there likely to be one.

Certainly the public mind will continue to be mesmerized by the fearsome sequence: World War I, World War II, World War III. But there is a stronger reason yet for disbelieving in the last of these, namely, the nature of the hopes and prospects that motivated the architects of the first two. After 1870, the world's military planners for decades were captivated by the prospect of wars that would be short, cheap, glorious, and profitable. These hopes were thoroughly cynical, but appeared entirely realistic, and their influence ran deep.

These plans called for war to be launched with a sudden invasion and to be fought on the enemy's soil, keeping the homeland safe from invasion or attack. It would be won quickly, at low cost to the invader. Then, at the peace conference, the victor would claim not only lands and territories, but would as well demand the payment of an indemnity, forcing the defeated foe to reimburse the victor's exchequer for its war expenses. The Franco-Prussian war of 1870–71 was fought precisely along those lines. So was the China-Japanese War of 1895. With these successful examples before them, it is little wonder that the general staffs of Germany and Japan for decades would be dazzled by the prospect of more of the same.

Today no such prospects exist, and this fact is known to all. No one is talking of a thermonuclear Schlieffen Plan, a latter-day counterpart of Germany's pre-1914 plan for the conquest of France in six short weeks, a war wherein the attacker will gain all the glory and the other side will have all the grief. Still, if we have indeed made it through the night, if World War III truly is one of those things that will never happen in our lifetimes, there nevertheless is the possibility of a nuclear war involving smaller powers. It is widely believed that Israel, for one, already has the bomb, or at least the capacity to produce it on very short notice. Certainly the leaders of such nations as Iraq and Libya have publicly proclaimed their hope of destroying Israel with nuclear weapons. However, Israel's leaders have never been content to sit still in the face of such threats. Her intelligence service, Mossad, is widely rated among the world's best. It would not at all be surprising if Israel has agents in the very households of Libya's Muammar Qaddafi or Iraq's Saddam Hussein, those nations' leaders. Her Air Force and air defenses are so strong that no Arab power has succeeded even in raiding an Israeli target, sending a few aircraft through Israeli defenses to drop explosives in what would be at least a propaganda raid.

What may well keep a Mideast nuclear war in the realm of threat and bluster is Israel's well-deserved reputation for hair-trigger fearsomeness. A few years ago some of her Air Force jets actually shot down an unarmed civilian airliner, simply because it had strayed over her nuclear facilities at Dimona. (Dead airliners take no photographs.) In 1981 Israel raided and destroyed an Iraqi nuclear reactor near Baghdad, which her leaders had reason to believe was to produce plutonium for an Iraqi bomb. They did this even before the Iraqis could load the reactor with nuclear fuel. Indeed, Israel has been so successful at defending herself that there is probably less cause for concern about Israel's security than there is for concern about the insecurity of her neighbors. It is considerably less likely that an Arab nuclear bomb will fall on Tel Aviv than that an Israeli preemptive strike will destroy the enemy's atomic facilities. If a cloud of radioactive debris is to rise someday over the Middle East, it will probably not be a mushroom cloud towering over a blasted Israeli city. Nor will it be the cloud from an Israeli bomb; the Israelis are far too skilled at using conventional weapons to require recourse to a nuclear blunderbuss. Rather, it will be the smoke and gas from an Israeli raid on some nuclear facility, which while conducted with conventional bombs nevertheless will blast a mass of radioactivity into the air. A raid on a center for producing or storing nuclear weapons would do just that. Had Israel waited for the Iraqis to fuel their reactor and let its interior become radioactive, a subsequent raid might have sent just such a plume of radiation drifting over Baghdad.

If nuclear war is unlikely between the superpowers, or between Israel and its neighbors, does that mean the atomic genie is safely in the bottle, that no one will be using the bomb anywhere? That is probably too strong a statement. There are nations like India and Pakistan, bitter enemies, who have the bomb now or are likely to get it soon. After India detonated a "peaceful" (sic) bomb in 1974, Pakistan's president declared his people "would eat grass" rather than permit India alone to have the bomb. The future will see other Indias and Pakistans. It is hard to say what is in the mind of some of the Third World's leaders or to know what temptations, what cautions, what sense of restraint or of new opportunity may present themselves to a nation's strongman when he decides his country should have the bomb. But it is very probable that people are forgetful, that long-familiar terrors tend to fade with time, and the lesson of Hiroshima may have to be relearned at intervals of fifty years or so.

Certainly, any leader who actually uses the bomb will face a wave of condemnation and ostracism such as would be reserved for the sponsor of a worldwide outbreak of bubonic plague. For weeks the news media would emphasize the carnage and destruction in the destroyed cities, the burying of the dead and the pitiful moans of the barely living. Other nations would rush to dissociate themselves from the act, speeding aid and assistance to the attacked country while imposing harsh sanctions and penalties against the attacker. The attacking leader, for his part, would find he had effectively violated an international taboo. For while nations might acquire the bomb and even build up nuclear arsenals and delivery systems, any use of such weapons would be universally seen as the most severe of threats to mankind itself, a violation of that thin bond of restraint by which humanity hopes to preserve itself.

Within a rather short time, however, it would be evident to anyone who cared to see that this isolated and restricted use of nuclear weapons would not have widened into a global nuclear exchange, would not have wiped out the human race. The names of the destroyed cities, their country, and the aggressor nation would stand as new and emotional symbols evoking the dread and horror of nuclear war. Within those cities, however, normal conditions would begin to return astonishingly soon. Most of the survivors would recover and heal, at least in their bodies; new people would come to the reviving cities from the surrounding countryside. Nuclear night would have fallen amid ashes and the stench of the unburied dead, but soon morning would come, amid construction cranes and the smell of wet mortar. The world would go on as before, free to repeat the follies and ambitions that had brought on the recent destruction. In its messy and uncertain way, history would continue to unfold; then for some countries, the ordinary developments of ongoing life might produce changes greater than would come from a major war.

Again, Israel will illustrate. Strong enough to defend itself against all attackers, powerful enough to deter or defeat any foe, Israel still may never overcome its essential weakness. It is a Western and European outpost which was introduced suddenly into the Middle East in 1948. It is surrounded by Arab powers, most of them hostile, and is coveted by the Palestinians whose land it formerly was. It may stand off any attack, repel any assault. Still, history is patient, and even without being defeated in war, Israel may one day be seen to have shared the fate of the other European enclaves established amid the Arab lands in earlier days. These included the Kingdom of Jerusalem, the counties of Tripoli and Edessa, and the Principality of Antioch.

These were the states of the Crusaders, set up in what is now Israel, Lebanon, and Syria, in the wake of the First Crusade about the year 1100. They were the focus of attention of Christian Europe, which sent crusade after crusade to defend them. In doing so, the Crusaders fought war after war with the local Arabs and other Moslems. But in the end, it was to no avail. In 1291, after two centuries of crusading, the last Christian stronghold (Acre) fell, and with it, the European presence in the Middle East. Not till the rise of the British and French empires in the nineteenth century would Europeans reappear there in force.

Why did these states prove weak, unable to sustain themselves? From the very beginning, early in the twelfth century, they proved unable to attract large numbers of European settlers. Nor were their leaders particularly skilled in diplomacy. It was the empire of the Turks with which the Crusaders had to contend, but time and again they spurned opportunities for peace or alliance with local Arabs, who were as opposed to the Turks as were the Crusaders themselves. Thus, for the most part, when threatened militarily the Crusader states could not rely on their own resources, but had to call on princes and popes to send an army.

The parallel with Israel is obvious. Here too there have been repeated wars, which increasingly have required the assistance of arms and other support from the United States, if Israel was not to lose the day. Here too a bellicose diplomacy has exacerbated tensions with the Arabs and left the Israelis increasingly isolated in the world's councils. In addition a stagnating economy featuring triple-digit inflation has encouraged people to emigrate, while immigration has fallen off considerably. Between 1969 and 1979, 510,528 Jewish citizens emigrated while only 384,000 came to Israel. Moreover, within Israel itself is a substantial Arab or Palestinian population, and the birthrate among these Arabic people has been far higher

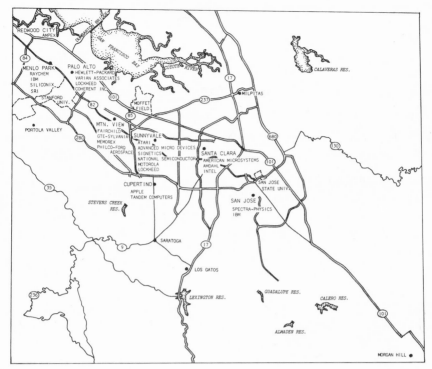

Fig. 33. *Silicon Valley, forty miles south of San Francisco.* (Map by Gayle Westrate.)

Fig. 34. *Palo Alto's Page Mill Road, in the heart of Silicon Valley.* (Photo by Don Dixon.)

Fig. 35. *The way it was: Jay W. Forrester, left, and associates fiddle with the racks of vacuum-tube circuitry in Lincoln Laboratory's Whirlwind, one of the most advanced computers of the early 1950s. (Courtesy MIT Lincoln Laboratory.)*

Fig. 36. *The way it is: today's micromainframes, on a single silicon chip smaller than a postage stamp, can do everything Whirlwind could do. The disk in her hand is a wafer of silicon, on which similar chips have been formed. (Courtesy Electronic Devices Division, Rockwell International Corp.)*

Fig. 37. *Electronic household.* (Art by Don Dixon.)

Fig. 38. *Automated bank teller.* (Photo by Don Dixon.)

Fig. 39. *Matsushita's robot television plant near Osaka, which builds Panasonic and Quasar TV sets. On the assembly line a total of five people are in view.* (Courtesy Matsushita Electric Industrial Co.)

Fig. 40. *Stanford University: where Silicon Valley got its start.* (Photo by Don Dixon.)

Fig. 41. *The DNA molecule: two parallel strands held together by base pairs. Here Dr. James Watson, co-discoverer of this structure, is lecturing on his discovery at the Cold Spring Harbor Symposium, June 1953.* (Courtesy Cold Spring Harbor Laboratory Research Library Archives.)

Fig. 42. *Ocean farm for oil bugs.* (Art by Don Dixon.)

Fig. 43. *December 1, 1950: the day the President had his finger on the button.* (Courtesy Millikan Library, California Institute of Technology.)

Fig. 44. *A warning for the future? Ruins of Crusader fortifications at Acre in Israel, abandoned in 1291.* (Courtesy Consulate-General of Israel.)

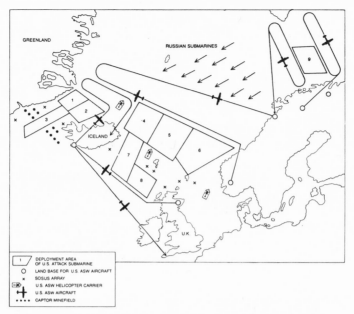

Fig. 45. How the United States and its allies would attempt to seal off the Greenland Straits against Soviet submarines. Note the optimism of the strategists in assuming our aircraft could operate freely out of northern Norway, close to the Soviets' heavily fortified Kola Peninsula. (Courtesy Scientific American.)

Fig. 46. High-rise buildings sprout in São Paulo, Brazil, in a cityscape typical of many in the Third World. (Courtesy American Association for the Advancement of Science.)

Fig. 47. *Village scene.* (Art by Don Dixon.)

Fig. 48. *What downtown Los Angeles will probably look like about the year 2050.* (Photo by Don Dixon.)

than among Israel's Europeans. Among the Jewish people in Israel the fertility rate was 3.2 births per woman in 1972; among that country's Arab population, 7.3. With 83.9 percent of the population in 1979 being Jewish (down from 85.6 percent in 1967), those trends predict that Israel's 400,000 Arabs of 1972 would grow to a majority in less than a century. What is more, there are well over a million Palestinians and Arabs in the Gaza Strip and the West Bank, lands the Israeli government has effectively annexed. It may prove quite difficult for Israel to maintain a separation or distinction between these people and the Arabs already living within Israel proper.

The future of Israel may well be determined less in Tel Aviv than in New York. Her future will rest on the question: is Israel still the natural homeland for Jews, or is it America? If the answer is the former, Israel will once again receive an invigorating influx of new immigrants. If the latter is the case, then Jewish people will still cherish and support that distant land, but they will not support it by going there to live. They will visit Israel for a vacation, a summer, perhaps a year or two as an experiment. But their lives will be so satisfying and free, lived in lands beyond Israel's seas, that they will not permanently stay. Within Israel other Jews, attracted by the magnets of America and perhaps of Europe, would themselves leave. The people who stay would be the Arabs. Of the Jews who stay, more and more would be Sephardi Jews, immigrants from Yemen, Morocco, and other Moslem lands, themselves often similar to the Arabs in culture and ways of living. Indeed, Israel's population today is already 50 percent Sephardi.

Should this state of affairs continue, Israel may be reduced to a hollow shell, militarily strong on the outside but weak on the inside, increasingly prone to suffer the fate of the medieval Crusader states. Preoccupied with their nation's physical survival, her people would lose the sharp edge of faith in Israel as a beacon of hope for the world's Jews. Within Israel, the Jewish religion would become tinged with rigidity and dogmatic fanaticism. Israel's leaders would cease to be Jewish leaders, inspired by Jewish ideals. Instead they would be captains of a garrison state, preoccupied with winning the next battle, and the next.

In time Israel would lose its European cast. It would become increasingly Asian, with Arab and Sephardic influences waxing strong. With European Jews a waning minority in the country that once was theirs, emigration to America and elsewhere would look increasingly attractive, and the Arabization of Israel would proceed apace. Eventually Sephardis, or even Arabs, would rise to positions of command in government and perhaps in the armed forces, and the transformation would be complete. The Middle East would have reclaimed its own. Israel in spite of itself would have become Palestine, as in the days of British rule prior to 1948:

an Arab state with a legally protected Jewish minority. This would not result from a dramatic battle in which the Muslim crescent would stand ascendant over the Star of David. It would instead have followed from gradual but inexorable processes of pressure and population trends. To the
Jews watching from Europe and America, it would be as when a desert
encroaches on and reclaims the land of an oasis.

While all this is happening in the lands west of the Jordan River, between the United States and the Soviet Union the future decades of
peace will see a continuation of business as usual. This does not mean the
friendly, close relations we enjoy with Japan or even with China, but
rather a continuation of the sharp rivalry and competition of interests we
have long known. It does not mean disarmament, or even serious negotia-
tions aimed at arms limitation, except as a means of steadying and chan-
neling future arms buildups. What it does mean is a continuation of that
institution of our times, the arms race. Pundits often write of the "mad
momentum" of the arms race and wonder in print how long it can go on
without leading to all-out war. Currently, anyone who reads the newspapers and newsmagazines has had a chance to read about this buildup,
with its aircraft carriers, MX missiles, and Stealth bombers, to match the
other side's SS-20 missiles and other modern weaponry. But it is not the
purpose of this book to dwell on topics treated in the weekly magazines.
Rather, what is appropriate here is to illustrate the nature of the arms
race, by drawing comparisons with an earlier arms buildup. The appropriate comparison is with the naval arms race between Britain and Germany in the years prior to World War I.

While Britannia had been ruling the waves for centuries, in the late
1800s Germany was still a land power, with little in the way of a powerful
Navy. In 1890, however, Captain Alfred T. Mahan, president of the U.S.
Naval War College, published *The Influence of Sea Power upon History*.
In its own field it would soon prove as influential as Darwin's *Origin of
Species* or Marx's *Das Kapital* in theirs. Its message, that the nation controlling the seas would be best able to control the future, fell on congenial
ears. As the Kaiser wrote in 1894, "I am just now not reading but devouring Captain Mahan's book. It is a first-class book and classical in all
points. It is on board all my ships and constantly quoted by my captains
and officers." With his characteristic impetuosity, he proclaimed that
"Germany's future lies on the water" and set out to build a fleet to rival
the British. In seeking this, he had certain advantages. The British fleet in
many ways was obsolescent, its leaders complacent. They lacked the cut-
ting edge of German assertiveness, and the fleet had not seen a general action since Trafalgar in 1805. Many of its warships were thinly armored
and slow; they even included a number of square-riggers that would have
offered few novelties to Captain Bligh of H.M.S. *Bounty*. Germany, by

contrast, could proceed from the start to build warships of the latest design, with heavy armor and armament.

The Kaiser, however, had not counted on having Admiral Sir John Fisher as England's First Sea Lord. On taking office in 1904, "Jacky" Fisher immediately set out to counter the German buildup with one of his own. He ordered a program of naval reform, aimed at scrapping over a hundred obsolete ships and replacing them in a massive construction program of fewer but larger ships, carrying the heaviest possible firepower. He was a man of florid style, and when he set forth the plans for his buildup, he wrote that "the country will acclaim it! the income-tax payer will worship it! the Navy will growl at first! (they always do growl at first!) BUT WE SHALL BE THIRTY PER CENT MORE FIT TO FIGHT AND WE SHALL BE READY FOR INSTANT WAR!"

Fisher's buildup brought quick results. At Queen Victoria's Diamond Jubilee naval review of 1897, many of the best ships in the fleet had resembled nothing so much as merchant vessels or passenger liners that just happened to have been fitted with 6-inch guns. When the fleet gathered for a similar review honoring the coronation of King George V in 1911, Portsmouth Harbor fairly bristled with massive low gray hulls mounting armor plate up to a foot thick, showing 12-inch guns in immense armored turrets, with great turbine engines below decks to speed at better than 20 knots. By 1914 the Royal Navy would have battle cruisers like the *Lion* of 26,350 tons, with eight 13.5-inch guns and a flank speed of 29 knots. Such dreadnoughts as the *Iron Duke* and the *Queen Elizabeth*, the latter mounting 15-inch guns, would foreshadow the size, speed, and armament of the battleships that would fight in the Pacific war thirty years later. The Germans, for their part, had built an entirely similar fleet; and the competition of the two naval powers had accentuated the international tensions which would erupt in the wake of Sarajevo.

Yet when the war broke out, the bellicose Germans showed a curious reluctance to seek naval battle. The historian Barbara Tuchman has written of that Navy, "As it grew in strength and efficiency, in numbers of trained men and officers, as German designers perfected its gunnery, the armor-piercing power of its shells, its optical devices and range finders, the resistant power of its armor plate, it became too precious to lose. Although ship for ship it approached a match with the British and in gunnery was superior, the Kaiser, who could hark back to no Drakes or Nelsons, could never really believe that German ships and sailors could beat the British. He could not bear to think of his 'darlings,' as [Chancellor] Bülow called his battleships, shattered by gunfire, smeared with blood, or at last, wounded and rudderless, sinking beneath the waves." Accordingly, he ordered his fleet to stay in port. Similarly, the British were in no hurry to seek a naval battle. Throughout the whole of the war, there was only one full-scale action between the British and German fleets: the Battle of Jutland, which pitted 99 German ships against 149 British. A total of 25 ships went down on both sides, yet not one was a battleship. During the war, while Allied and German armies were bleeding each other white along the Western front, one of the safest ways to see military service was with the British Grand Fleet or the German High Seas Fleet.

There are some noteworthy similarities to today's nuclear arms race. Here too we see two great competing powers spending vast sums as they vie with each other to build and deploy the most advanced and sophisticated of weapons. However, the main use of these weapons is for display in Moscow's Red Square parades and other military reviews. Both superpowers have shown considerable reluctance and restraint when it has come to doing more than boasting of their strength. Having invested so much treasure and effort in their strategic forces, both sides would tend to regard them as their "darlings," not to be hazarded in battle; but the mutual reluctance runs deeper than that. The "darlings" are not merely weapons systems or strategic arms; they are the nations' cities, industrial regions, and centers of national life. If the British and German commanders were unwilling to hazard a few dozen capital ships in World War I, how much more unwilling are the U.S. and Soviet leaders today to risk the very lives of their nations.† Had Britain, France, and Germany all been island nations, then rather than fight a war they might have carried forward a continuing naval buildup, as an institutionalized means of absorbing hostile energies. That is rather the situation of the superpowers today.

What this means is deterrence, a policy sometimes called Mutual Assured Destruction, or MAD. It amounts to the superpowers holding one another's populations hostage. This is hardly a comfortable position to be living in; yet it preoccupies few people and for the most part stays in the back of our minds. Even people in northern California, quick to demonstrate over offshore oil drilling or nuclear power plants, rarely are seen getting exercised over the fact that somewhere in the Soviet Union is a nuclear-tipped rocket targeted right on their hot tubs and Jacuzzis. Deterrence works. The mutual threat of mass destruction is not something one would seek in the best of all possible worlds. It does not describe the kind of relationship we would hope for with the Soviets, if we could come up with anything better. But it has the overwhelming advantage of allowing

† It should not be argued that the national leaders of 1914–18 sought to save their ships but were willing to spend the lives of their young men by the millions. In 1914 everyone expected a short war, and few anticipated its destructiveness. Sea battles were avoided through foresight and restraint, but in the land war, leaders and people alike were caught up in events.

the two nations to pursue their disagreements without blowing each other up. And in wondering if it will last, or for how long, it is worth remembering that deterrence worked even when the adversary was that most unlikely of practitioners of restraint, Adolf Hitler.

Prior to the invention of nuclear weapons, the nearest thing anyone had to weapons of mass destruction was poison gas. During World War I the Germans, followed by the Allies, made extensive and often devastating use of chlorine, phosgene, mustard gas, and hydrogen cyanide. Such gas warfare was prohibited after the war by the Geneva Protocol of 1925, signed by all the leading powers. As Nazi Germany rearmed in the 1930s, however, she built up large stocks of these chemical munitions. The Germans also pioneered in developing the far deadlier nerve gases, which unlike the earlier chemicals could kill by being absorbed through the skin, and which were lethal in quantities as small as a milligram. Even before the outbreak of World War II, the Germans had built production facilities for their nerve gases, had stockpiled quantities ready for use, and were proceeding with research which during the war would produce still deadlier forms of nerve gas.

When the war broke out in 1939, England, France, and Germany all exchanged assurances that they would abide by the Geneva Protocol. In 1943 President Roosevelt declared that gas warfare was "outlawed by the general opinion of civilized mankind" and "we shall under no circumstances resort to the use of such weapons unless they are first used by our enemies." The general opinion of civilized mankind would shortly make its weight felt in the war-crimes trials at Nuremberg, once it became known the Nazis were using the insecticide Zyklon B to exterminate millions of civilians in the concentration camps at Auschwitz and elsewhere. And Hitler had often proclaimed his scorn for the treaties and agreements of Geneva. But he did not use poison gas in the war itself. As he knew full well, the Allies also had ample stocks, quite sufficient to match his own, so he honored the Geneva Protocol. During World War II, in at least that limited sphere, deterrence worked.

As the arms race proceeds, the focus of attention will be the strategic arms, the missiles and nuclear warheads making up the superpowers' deterrent forces. But the arms race will not be a static process; it will be no mere matter of piling bomber upon bomber, submarine upon sub. There will be changes in the art of war, and these changes stand to transform quite completely many of our existing ideas about armed conflict.

The transformation in land warfare is already well under way. For decades the standard weapon has been the tank. For nearly as long a time it has been possible for an infantryman, armed with the bazooka or other antitank weapons, to stand his ground against them. But in this competition, the tank has been steadily losing ground. In the 1973 Middle East

War, as Israel's General Chaim Herzog later would write, Israeli military planners showed "a complete lack of appreciation of the new antitank capability within the Arab forces." On the Egyptian front, one Israeli tank brigade lost 77 tanks, three fourths of its complement, in sixteen hours of fighting. On Syria's Golan Heights the battlefield lay strewn with burned-out tanks and spiderwebs of missile-guiding wires; over 1,500 tanks were destroyed, many by antitank missiles. Egypt's General Sa'ad al-Shazli wrote "it was impossible to ensure the success of any attack, whether of tanks or of armored infantry, without destroying or silencing in advance the antitank missiles." Soviet Minister of Defense Andrei Grechko wrote that "combat actions in the Middle East . . . have put anew the question of the relationship of offense and defense . . . Tanks have become more vulnerable and the use of them on the battlefield more complicated."

The weapons which produced this carnage were Soviet-built Sagger missiles, which were wire-guided; they trailed a thin wire through which a gunner could send steering commands. They were a vast improvement over the unguided bazooka, but had the disadvantage of having to be steered by a manually operated joystick. The American antitank missile, the TOW (Tube-launched, Optically tracked, Wire-guided) needs no such joystick. The gunner need only keep the cross hairs of his sight on the target; smart electronics, based on a microprocessor, do the rest. Some thirty-three countries have bought more than 275,000 of these missiles, including 107,000 bought by the U.S. Army and Marines. The Soviets have been building new tanks at the rate of 2,000 a year. But the Army is procuring TOW missiles at a rate of 12,000 a year and is deploying them to be fired not only by infantrymen but also by helicopters. A TOW missile costs $7,000 and can wipe out a Soviet T-72 or T-80 tank costing a hundred times as much.

Even more advanced battlefield weapons are in use today. These are the "fire and forget" variety, which offer the prospect of "one shot, one kill." Typical is the Maverick air-to-surface missile, which comes with a choice of guidance systems. One version has a TV camera in its nose, transmitting a picture to a screen in the cockpit. The pilot picks out a target, locks the missile onto the target, and shoots; the missile's electronics do the rest. In the Marine Corps version, the missile homes in on a spot of laser light reflected from the target, the laser being pointed from the ground or from an aircraft. Such laser-guided "smart bombs" enabled the Air Force to strike successfully at heavily defended bridges and other targets in North Vietnam in 1972, and the Maverick is reported to have scored a direct hit in more than 88 percent of all its test firings. Another laser-guided missile, the Copperhead, is fired from a 155-millimeter howitzer. At a range of ten miles it can score a direct hit on the turret of an enemy tank.

TOW and Maverick are rather small missiles, respectively five and eight feet long. A larger and much more capable missile, currently under development, is the eighteen-foot Assault Breaker. It would rely on airborne radar to locate a tank formation deep within enemy territory. The target coordinates would then be fed to Assault Breaker, which would fly to the general area of the target carrying a warhead filled with homing submissiles, each about the size of a TOW and each having its own guidance system. The terminal guidance in the submissiles would home in on the infrared or millimeter-wave radio emissions which a tank naturally gives off from the heat of its diesel or turbine engine. If the target was a spread-out array of tanks, Assault Breaker would release its submissiles all at the same time, allowing them to home in. If the target was a column of tanks, the missile would use its own radar to guide it along the column, releasing submissiles one by one. Each submissile then would strike its tank at the top of the turret, where the armor is thinnest and the tank most vulnerable.

These developments mean the end of the blitzkrieg. No longer will heavily armed tank forces strike by surprise and seize great chunks of territory before the defenders can rally. As the defense analyst Paul F. Walker has written, "On a heavily armed front, such as central Europe or the Middle East, the side that strikes first, thereby giving away its position, will be the more vulnerable side. What can be seen, be it by eye on a clear day or by radar and homing guidance on a smoky and rainy field at night, will be hit and will probably be put out of action . . . Blitzkrieg attacks will become suicidal, and as a result major defensive positions will probably be stabilized throughout the world."

The Soviets, however, would not need to strike with massive tank forces into the heart of Europe to strike terror into the hearts of Europeans. A type of war that could well appeal to them would be a naval war. The world has not changed in this regard; what Captain Mahan wrote in the nineteenth century will certainly still be true in the twenty-first: the nation controlling the seas will control the future. Without sending a single infantryman across a national frontier, the Soviets nevertheless could strike for important strategic and political goals, in a way offering little risk of nuclear escalation, by lashing out at one of the visible and dramatic weaknesses of the Western powers: their oil supply.

Europe and Japan are heavily dependent, and the United States only somewhat less so, on an uninterrupted stream of tanker traffic from the Middle East. Seaborne commerce has often been the lifeline of nations, but as Noel Mostert wrote in *Supership*, "No other ships have been so universally important, none more political . . . there never was a time when the viability of life for millions in both hemispheres was inextricably linked to the daily unimpeded passage of any one class of merchantmen."

The supertankers often spend weeks in transit, lumbering slowly along routes many thousands of miles long, reaching across the Indian Ocean, the southwest Pacific, around the periphery of Africa. If set upon by attacking Soviet submarines, many would be caught days from land in the open ocean. Along the southern and western African coast there would be few harbors where they could seek shelter. Defenseless, far from safety, spread out along the sea-lanes where Allied naval power could not quickly be brought to bear, the burning tankers would send their smoke plumes skyward while the Soviet subs advanced, to kill and kill and kill.

In a matter of days such attacks could easily sink more than the 13 million tons of shipping lost to the Germans in World War II. Moreover, the Soviets might not have to sink many ships to accomplish their aims. After the Soviets had made their policy clear, backing it with only a few ship-sinkings, maritime insurance firms would invoke the war-risk clauses in their contracts, forcing shipping firms to pay unacceptably high wartime premiums, or else keep their ships in port. This would immediately precipitate a new oil crisis. The Allied navies would organize convoys, but their best hope would be to escort some ships through the Persian Gulf and Red Sea along the Suez Canal route to the Mediterranean. With over 300 attack submarines, many of them nuclear-powered, the Soviets might have little trouble in shutting down tanker commerce across the Indian and Atlantic oceans, and around Africa. The United States would then be largely isolated, and Japan would be wholly cut off.

By thus blockading Japan the Soviets would hope to force her to sever her ties with the United States and to follow policies made in Moscow. Success in this in turn would aid the Soviets in overawing China, compelling Peking to accept such concessions as a dismantling of the military forces facing the Soviets along their common frontier. The goals the Soviets would seek in Europe would be devastating: an imposed Soviet-oriented neutrality, the dissolution of NATO, and the granting of extremely favorable commercial and trade concessions on Soviet terms. America, for its part, would be left largely alone in the world. The world would declare that the Soviets had triumphed and would accept a *Pax Sovietica* backed by the same sort of unrivaled naval power that in its day brought the *Pax Britannia*.

The alternative would be for Allied navies to sweep the Soviet subs from the seas, blockading these subs from the sea-lanes. In this the Allies would be helped mightily by geography. Although it possesses the world's longest seacoast, the Soviet Union remains virtually a landlocked power. Its Pacific coast includes the Sea of Japan, the Bering Sea, and the Sea of Okhotsk; but the first two of these are screened from the open ocean by the island barriers of Japan and the U.S.-owned Aleutians. In the west there is the Black Sea, the Baltic, and the Barents Sea. The first two communicate with the Mediterranean or Atlantic through Allied-held straits and would be hopeless for the Soviets to use in wartime. Thus, everything would depend on how well the Allies could keep Soviet submarines bottled up in the Barents and Okhotsk seas.

Both these seas, in turn, connect with the oceans through straits that, while wide, are not so wide as the Soviets would wish. Across the opening to the Sea of Okhotsk stretch the Soviet-held Kuril Islands, but close by is Japan with its American bases. As for the Barents Sea, subs heading for the Atlantic would have to pass the stretch of water extending from Scotland to Iceland. The great sea battles of the war then would be fought for naval and air supremacy over these straits. In the Battle of the Kuril Islands, Allied power based in Japan and the Aleutians would contend with Soviet forces based in the Kurils and at Petropavlovsk on the Kamchatka Peninsula. It would be fought virtually in the Soviets' backyard and would be a severe test for our side. The strategically more important battle, however, would be fought under more favorable conditions for us. This, the Battle of the Norwegian Sea, would pit Allied strength based in England, Iceland, and Norway against Soviet forces from some distance away, operating out of the Kola Peninusla and perhaps Finland as well. If the Allies could win those battles, they would have a free hand in conducting antisubmarine warfare in those straits and in bottling up the Soviet subs.

To detect the subs, the United States would first rely on the Sound Surveillance System, SOSUS. This is a worldwide network of hydrophones emplaced on the seabed, which listen for the sounds of Soviet submarines. SOSUS arrays are permanently emplaced in strategic areas, including the Greenland-Iceland-Scotland ocean gaps, along the edge of the Barents Sea running north from Norway, and parallel to the Kurils, Kamchatka Peninsula, and the Aleutians. These arrays would be supplemented by two other hydrophone systems. SURTASS, Surveillance Towed Array System, would feature strings of listening devices towed behind surface ships. RDSS, Rapidly Deployable Surveillance System, would have fields of hydrophones sowed by aircraft or submarines, the hydrophones mooring themselves to the ocean bottom. Certainly we would need vigorous efforts to maintain the SOSUS and RDSS arrays, in the face of Soviet efforts to discover and destroy them. But while in place, these devices would communicate with central control stations via the Fleetsatcom satellite system. At these central facilities, powerful computers would sift through the data and search for signs of submarines. Any sub thus found would be fixed in position to within fifty nautical miles or less.

Antisubmarine-warfare aircraft or helicopters would then be sent to the area. The helicopters could lower sonar devices with a winch. These, descending a thousand feet or more below the surface, would set off small explosive charges as probes, producing sound waves the sub would reflect. Or the aircraft could drop sonobuoys, small floating buoys dangling a hydrophone from a wire. These would indicate the direction to the sub. The aircraft or helicopter then would follow along this line, dropping more sonobuoys. Near the sub, the hunters would employ a magnetometer to sense the presence of a large mass of submerged steel. Closing in for the kill, they would then launch an acoustic homing torpedo to detect the sub with its own sonar and head in the proper direction. An explosion from deep underwater would announce the death of the enemy.

In addition, the United States would deploy nuclear-powered attack subs, which the Navy considers the most effective antisub weapon in its arsenal. These include the powerful *Los Angeles*-class subs, the Navy's most advanced. At present we have some seventy-five such attack subs, barely a quarter as many as the Soviets. However, ours are generally acknowledged to be superior in the critical areas of quietness and sensor range. They carry sensitive sonar systems which can track a Soviet sub up to twenty miles away, while their own exceptional quietness makes them difficult to detect. They are designed to operate in waters near the Soviet Union and to perform their mission in the face of local Soviet air and sea superiority. These hunter-killer subs, also armed with acoustic homing torpedoes, would be deployed in a zone defense in the Scotland-Iceland gap. Even if the Soviets were to win the Battle of the Kurils, these subs still would act to screen Soviet sub movements in and out of the Sea of Okhotsk and of their base at Petropavlovsk. Strong antisub forces, supported by aircraft carriers, destroyers, and land-based aircraft, would also operate to the northeast of Japan and would add to the dangers faced by the enemy. As for the subs already at sea, their time would come when their supplies ran low and they had to return to Petropavlovsk.

Enemy subs, in turn, could seek to escape by diving deep and relying on their speed, or they could fight. Again their best weapon would be the acoustic homing torpedo, directed against attack subs or surface ships; but such ships with sonar could detect the sub well outside of torpedo range. The subs could launch antiaircraft missiles or antiship cruise missiles, but in doing so would give away their position. For instance, if the Soviets' antiship missile resembled our Navy's Tomahawk cruise missile, the sub could fire no more than two or four in a salvo, every half hour. (Loading the Tomahawk into a torpedo tube takes five minutes, but aligning the gyroscopes of its inertial guidance system then takes 25 minutes.) The missile has a rocket booster; its underwater ignition produces a loud sound detectable at great distances. Also the booster produces copious bubbles visible on the surface more than five minutes after launch, and as the rocket boosts the missile into flight, its exhaust can be seen fifty miles away. The sub thus might sink a destroyer or shoot down an aircraft, but in doing so would seal its own doom. Since nuclear attack subs are quite costly and valuable, we would say that trading a sub for a destroyer is a fair exchange. On the other hand, if the sub were to attack an aircraft carrier, it would find the carrier equipped with antimissiles, some of which in tests have shot down 4.5-inch naval shells in flight. Moreover, the carrier would be built to resist damage, with as many as a dozen torpedo hits being necessary to sink it. With all this, once our antisub forces were mobilized and deployed, any Soviet subs venturing toward the sea-lanes would likely lead a short, exciting life.

What is more, we could supplement these antisub forces with mines. Across the Greenland-Iceland-Scotland gap, and outside of the Sea of Okhotsk, we could sow fields of seabed-mounted acoustic torpedos, all pointing upward. They would be programmed to let surface ships pass, but would recognize and launch themselves against a submarine. They could be remotely activated and deactivated and could be laid by aircraft, surface ships, or submarines. For the British who would anchor such a minefield at its southeast end, Rudyard Kipling's lines, "Mines reported in the fairway, Warn all traffic and detain," would have new meaning.

The Soviets can be expected to improve their submarine forces. For example, some of their latest subs are built of titanium rather than steel to avoid detection with magnetometers. They also have made their subs quieter and thus harder to detect. Still, as former Defense Secretary Harold Brown has stated, "our Navy has maintained and, in some cases, even widened our technological lead." An important area of activity is our use of satellites in submarine tracking and detection. As Rear Admiral Donald P. Harvey, director of U.S. naval intelligence, remarked in 1977, "They [his superiors] expect me to tell them as soon as they [Soviet subs] cast off the last line" when leaving port.

In 1978 and 1979 the Navy conducted tests which for the first time showed a submerged sub under way generates a wake, producing waves that can be detected at the surface. These tests confirmed the presence and detectability of such waves and showed their characteristics could be measured. Satellites carrying appropriate sensors thus might serve to scan large ocean areas. Sensitive infrared scanners, also satellite-mounted, might detect plumes of heated water in a submarine's wake. Already it has been reported that spaceborne radar carried aboard the Seasat spacecraft can potentially detect the radar "signature" of a sub, and airborne submarine detection by a technique known as laser interferometry has already been successfully demonstrated. This method uses laser light to detect the slight disturbances produced at the ocean surface by a sub. All these developments raise the prospect that Soviet subs will be tracked from their pens to their operating areas. As on land, the defenses against sub attack might prove so strong that any such attack would quickly falter.

Certainly, we would hope it would. The methods of war will not be static, and it is hard to say there will ever be a day when submarines will be as readily seen and attacked as tanks on a battlefield. The depths of the oceans will continue to hold the promise of concealment, and Soviet military planners will continue to be attracted by the prospect of a naval war. The submarine as we know it today has existed in naval fleets for eighty years; in that time neither sub nor antisub forces have been able to gain the upper hand. Should this continue, for a long time there will be the threat of a Soviet war at sea, a war which would stand to win for them important strategic and political victories, yet which would count few casualties and would not lend itself to a quick escalation into nuclear conflict. The United States and its allies, particularly Europe and Japan, will not soon lose their status as island nations, utterly dependent upon seaborne commerce. Such nations are susceptible to military defeat without invasion. Advocates of air power have long noted that Japan capitulated in 1945 without being invaded, but that has been merely the latest chapter in a long story. Had Germany defeated Britain in World War I, it would have been by sea power, by using submarines to destroy Britain's commerce. Her cities would have stood unbombed, her land uninvaded, yet she would have surrendered. Over two thousand years earlier, Rome was able to defeat Carthage in the First Punic War, wresting valuable lands from the Carthaginian empire, by winning upon the sea. The opportunity for the Soviets to do likewise will not go away.

If there is to be a war with the Soviets, when may it come? There has been something of a cycle to the world's major wars, which can give us some idea of the answer. In the eighteenth century, Europe's powers fought a global land and naval war, the Seven Years War (1756–63), known in America as the French and Indian War. Following an uneasy truce of thirty years, this was followed by the far more widespread and destructive wars of the French Revolution and of Napoleon. These wars will bear comparison to the World Wars of our own century. They were not so costly in human life, but their political and historical consequences were just as sweeping, and in some ways they were even more far-flung. The Seven Years War was fought not only in Europe but in North America (including Canada) and India. It established Prussia as a first-class power and made Britain the predominant colonial power in both India and North America. The Napoleonic Wars, like World War II, pitted Europe against Russia. But whereas in 1941–45 the Japanese were unable to do much more against the United States than bomb Pearl Harbor, in 1814 the British burned the White House and Capitol in Washington, D.C.

After Waterloo in 1815, things in general were settled. For a century there was peace, or at least the absence of large-scale war. When 1914 came, the war would involve very different powers, alliances, issues, and modes of fighting. The victories of 1945 were similarly sweeping, and the ensuing decades have seen a reluctance to wage full-scale war quite like that of the nineteenth century. It would be far too facile to set forth a "99-year rule," that just as that duration separated Waterloo from the Battle of the Marne, a similar duration will elapse between the surrenders of Berlin and Tokyo and the Battles of the Kurils and the Norwegian Sea. But we still are living very much in the aftermath of 1945, and we have not had time for it to fade.

It will take time, many decades, for this fading to proceed. Local wars, as in the Falkland Islands in 1982, will demonstrate more conclusively the new techniques in war-fighting and illustrate by explicit combat experience their strengths and weaknesses. Somewhere, someone will use nuclear weapons. The world will still stand, but the memory of Hiroshima will be supplanted by another and far more recent horror. The evolving relationship between the United States, Japan, and China will continue to unfold. In time no one will be left alive who remembers 1945, or remains strongly influenced by its legacy. As in the nineteenth century, new powers will arise, new alliances, new issues, and methods of fighting. The day will come when the world again is ready for war.

If there is to be a World War III, almost certainly it will not be called by that name. Perhaps it will be called the Russo-American War. It will more likely be fought after, rather than before, the year 2050, and it will be a naval war fought for control of the seas and their commerce. This is the kind of war that will continue to offer the superpowers the chance to contend for global goals and for the determination of their relative degrees of strength and power. It is the only kind of war that could engage their energies without incinerating their cities, or leaving them with resentful conquered peoples to pacify. It thus will loom in the future as a continuing temptation, a threat, and for the Soviets, an opportunity.

10

POPULATIONS AND PEOPLE

Is another baby boom on the way?

It all depends on what you mean, but certainly manufacturers of rubber baby buggy bumpers can find reason to cheer in the statistics. In 1973 the number of births in the United States stood at a low of 3,137,000, the lowest number since 1945, when so many potential fathers were still overseas. In 1981 the number was 3,646,000, the highest since 1965; the number of births was climbing still further in 1982. The rise in baby-making also shows up in the birthrates. In 1975 and 1976 they bottomed out at 14.5 births per 1,000 of population, the lowest rate in U.S. history. By 1980 the rate was up to 15.8, and during 1981 was on its way past 15.9, the highest rate since 1971. Clearly, more women have been having babies, and what is more, they have been having more of them past age thirty. At Chicago's Northwestern Memorial Hospital, half the new mothers in 1980 were over age 33. To New York psychiatrist Donald A. Bloch, a family therapist, "there is a profound baby hunger around these days among women who have put off having children."

What does this mean then? Will we see a replay of the 1940s and '50s as motherhood waxes triumphant? Will Doris Day reemerge to make a film about the joys of large families? Will school districts stagger under the burdens of massive numbers of elementary-school children? Will bassinets and bottle warmers be the growth stocks of the 1980s? Don't bet on it. What is happening is no mass rush to the maternity wards, but rather an upward easing of the birthrate from the abnormally low to the merely low. It is not a baby boom, not even a baby boomlet, and would scarcely be noticeable at all except the statistics are so complete. A few additional babies each year, even a few hundred thousand, will not change the basic pattern. America, like the rest of the developed world, is committed to low levels of fertility and is well on the way to achieving a population which grows only slowly, if at all.

To understand what is happening, it will be worthwhile to present a short history of the time-honored trade of making feet for children's stockings. In 1820, the Census Bureau for the first time recorded the birthrate: 55 per 1,000 of population. Those were the days when motherhood was Motherhood. In the words of Parson Weems, inventor of the legend of George Washington and the cherry tree,

> My friends, 'tis population, 'tis population alone, that can save our
>   bacon.
> List then, ye Bachelors and ye Maidens fair, if truly ye do love your
>   dear,
> O list with rapture to the decree
> Which thus in Genesis you may see;
> Marry, and raise up soldiers, might and main,
> Then laugh ye may, at England, France, and Spain.

Even then, however, not everybody was listening with rapture. By 1860 the birthrate was down to 44 per 1,000. Half a century later, in 1910, it stood at 30; but even that meant 2,777,000 births in 1910, over three fourths the 1980 figure, in a population two fifths the size. Between 1920 and 1930 the birthrate declined further, from 27.7 to 21.3. During the Depression it continued to slide, hitting a low of 18 in 1936. The birthrate then began to recover, rising to 19.4 in 1940 and to 23 in 1943 as wartime prosperity took hold, but easing to 20.4 in 1945.

Then the Baby Boom hit. In 1947 the boys were home from war, and the birthrate leaped to 27. It then stayed close to this level, not far from the birthrates of the early 1920s, for over a decade: 24.1 in 1950, 25 in 1955, finally peaking at 25.4 in 1957. Then it began to ease, but slowly: 23.7 in 1960, 19.4 in 1965. After 1965, birthrates dropped further, to a Depression-level 18 or so in the late 1960s. After over thirty years, the rate was finally back where it started. And then it went down further. Between 1970 and 1973 there was almost the reverse of the Baby Boom, as the rate fell dramatically from 18 to 14.6. It stayed there for a few years, then in 1977 began slowly nudging upward toward a level of about 16 in the early 1980s.

With these statistics, it appears possible to make a fearless forecast. By 1984 or 1985, the birthrate may actually get to 16.5. What would be even more amazing, it may actually reach 17 before beginning again to decline. But then even at the peak of the present rise, the birthrate would still be lower than at its nadir in the depths of the Depression, which for decades was regarded as an impossibly low level that could never be sustained. Things have certainly changed, then, if the unsustainable low of the 1930s is higher than what will likely be the unsustainable high of the 1980s. With birthrates showing a long-term slide dating back a century and a half, even a rate of 17 in 1985 may someday look almost as surprising as the rates at the peak of the Baby Boom.

These birthrates can be translated into fertility rates, which help in giving a clearer picture. The fertility rate in a particular year is the average number of children a woman would have, if throughout her childbearing years she would have children in accordance with the birthrates for the different age groups, for that year. During the Depression, 1936–39, this was about 2.15; that is, a random sample of 100 women would have been expected to have about 215 children in all. During the Baby Boom the fertility rate leaped above 3.5, with a peak of 3.724 in 1957. By the late 1960s it was down below 2.5, and in 1970–73 fell from 2.48 to a remarkable level of 1.90. It bottomed out at 1.768 in 1976 and since has climbed, but only marginally, to a bit under 1.9. It is hardly too much to say that from 1957 to 1973, fertility rates fell from nearly 4 children per woman to fewer than 2.

Particularly influential here are women's expectations as to the number of children they will have. The Gallup Poll over the years has asked the question, "What do you think is the ideal number of children for a family to have?" The number answering "Four or more" was 49 percent in 1945, 45 percent in 1960, and still 40 percent in 1967. By 1971 it was 23 percent and by 1974, 19 percent. An even more dramatic shift is seen in answers to a question asked of married women by the Census Bureau: "How many babies do you expect to have in your lifetime?" In 1967, among wives aged 30 to 34, who had largely completed their childbearing, 55.3 percent answered "Four or more"; only 18.8 percent replied, "Two." In 1975, among wives in this same age group, only 19.8 percent expected four or more. As for younger women, in 1979, among married women aged 18 to 24, only 7.1 percent expected four or more while 54.9 percent anticipated only two. Among all women aged 18 to 34, in 1980 the average of all expectations was 2.059 births. For the younger group, aged 18 to 24, it was 2.023.

The big change, which took place in the early 1970s, coincided neatly with the rise of the feminist movement, and it is tempting to say the two developments were related. The widely publicized ideas that women should be persons in their own right, should be able to hold jobs or continue their education, should choose the time and occasion for having their children—all these doubtless were influential. Or, if you will, the idea that women should cease to be passive receptacles for their men and should reject the attitude that anatomy is destiny—these also may have influenced the demographic changes of the early 1970s. But it is not necessary to weight these changes with heavy ideological baggage. What we are talking about, after all, is the sum total of tens of millions of decisions made quite naturally by women and their husbands or sweethearts, in the course of their everyday lives. Ben Wattenberg, a former aide to President Lyndon Johnson, has illustrated how a woman might decide that two babies are enough:

> Suppose [she] doesn't get married until, say, age twenty-four, and then continues working for, say, three years. The couple think they might well like to have a large family—three or four kids. But, by age twenty-seven she's likely got a fairly well-paying job, and possibly an interesting one. She and her husband have been able to sock away a little money on their two-earner income, they've bought a nice sofa and some nice chairs, they've taken a vacation in the Scandinavian countries and have gone out in the evenings to see the good movies.
>
> They decide that now is the time to have a family. She gets pregnant. At age twenty-eight, she has a baby. A year later she is pregnant again. She has another baby. She's now thirty. The oldest baby is chewing on the leg of the beautiful couch. There are two sets of diapers now. The couple hasn't seen a movie in half a year. The last vacation was a weekend drive to the beach, with diaper bags, car seats, high chairs and bottle warmers. There is only one paycheck.
>
> Another year passes. She's thirty-one. The one-year-old is chewing the leg of the sofa. The three-year-old has found a dead bird in the yard. The bird is now on the living-room rug. Does she have another baby? Another pregnancy? More diapers? No paycheck? The decision is made: no. Two is enough.

So the Baby Boom is over, and it's over for good. The current modest rise in the birthrate is little more than a delayed echo of that earlier boom. In fact, it would be surprising if there were no such rise. Birthrates are given per thousand of population, but young women have the babies, and there are quite a lot of them nowadays. Because birthrates stood close to 25 per 1,000 from 1946 to 1960, it was a certainty that come 1980, more than one seventh of the U.S. population would be women aged 18 to 34. But they are having fewer children than their mothers did. Come the year 2000, there will be about 28 million women in that age group. We know they will be here; they have already been born. But they will then be little more than one tenth of a rather larger U.S. population. A change from one seventh to one tenth is a rather significant drop and reinforces all the more strongly the prediction that the birthrate by then may be plumbing new lows.

All this means a lot for the future. To begin, the '60s and '70s are over. Whether you liked them or not, they are gone with that Baby Boom of which to a large degree they were the product. Again, with birthrates touching 25 per 1,000 in 1946–60, it was evident that come 1970 and throughout the '70s, more than one fifth of the entire population would be young people aged 14 to 24. True, there had been such inundations of youth before. Back in the days of Parson Weems, the high birthrate meant the United States was largely a nation of children and teenagers. (The median age in 1820 was 16.5 years! In 1980 it was a more mature 30.0.) But earlier generations of young people had vanished quietly into its farms and factories. Between 1960 and 1975, however, the number of students enrolled in college leaped from 3,570,000 to 9,697,000. For several years in the 1960s the nation was opening new colleges, mostly community colleges, at nearly one a week. And there is a vast difference between being a college student and being a young wife with babies or a new worker enrolled in an apprenticeship program, which is more nearly what young people had done in previous decades.

With so many people under 25, with so much affluence and free time on their hands, it is little wonder that many of them felt free to indulge what the essayist Lance Morrow has called "a swelling sense of their own inevitability, their unique moral rightness. Everything they did was done in the incandescent certainty, the grand optical illusion, that it had never, ever been tried or felt before." Although few of them realized it, they were hardly the first to believe that the history of the world had begun, or at least begun anew, with them. As the Italian writer Giovanni Papani wrote of his generation after World War I, "For the 20-year-old man, every old man is the enemy; every idea is suspect; every great man is there to be put on trial; past history seems a long night broken only by lamps, a gray and impatient waiting, an eternal dawn of that morning that emerges today finally with us."

Well, not much is left today of hippies and flower children, of campus protest demonstrations, of students canvassing door-to-door for Senator Eugene McCarthy, or of the days when the University of California at Berkeley was widely hailed (or feared) as a model for the nation's universities. Some things have persisted; for instance, marijuana seems here to stay. But certainly things are quieter. Part of what has happened has been a national weariness with political passions, coupled with a waning of the widespread 1960s attitude that anything wished for could be had simply by demanding it loudly. But also influential has been the natural passage of time. Again, to Lance Morrow,

Just now the baby boomers, in their early-to-mid 30s, are grappling for the first time with life's serious, mundane and (in many cases) long postponed business: trying to discover living arrangements more permanent than mere roommating, finding ways to raise children, shelter them, nurture them, serve as models for them and otherwise turn them into the next generation—a hopeful and sometimes painful drudgery that is inevitably hard on narcissists. The aging baby boomers are now daddies and mommies with careers to build and all kinds of adult banalities to face: failures and divorces and alcoholisms and, yes, now deaths.

If the 1960s and '70s were years of peace marches, the '80s and '90s stand to be years of peace and quiet. They also will be years of prosperity. With so many wives working, with children so often postponed or served up in small batches, and with something like half of the baby boomers having had at least some college education, they will not find it difficult to keep on living the good life. One way that this becomes evident is in the restaurant industry. When people are poor, they eat bread and cereals. When they enter the middle class, they eat meat. And when they become affluent, they start going to restaurants. In such prosperous areas of the country as Orange County, California, it is literally impossible for anyone to sample dinner in all the restaurants. Every day a new one opens up.

Another place where affluence is visible is on Broadway. One thing people are glad to do, when they have the chance, is to take vacations and visit places like Manhattan, and to see the current plays. (The prices of hotel rooms in Manhattan have risen in response to this demand.) On Broadway, the number of ticket buyers has jumped from 5.4 million during the 1972–73 season to 11 million in 1980–81. This has meant longer runs for the hits. For decades the longest-running Broadway plays were *Life with Father* with 3,224 performances and *Tobacco Road* with 3,182. Then in 1972 *Fiddler on the Roof* broke the record, running to 3,242 performances. In 1980 this was surpassed by the musical *Grease* with 3,388, and any day now the palm for longest run may go to *Oh! Calcutta*, which has been running irregularly since 1969. It is not that these plays have intrinsically greater merit than such beloved old standards as *Oklahoma!* or *My Fair Lady*. It is just that there are more people who want to see the shows, but there are only so many seats in the theaters. The inevitable consequence is longer runs.

With the baby-boom generation now working and earning salaries, they are also helping to swell the coffers of the U.S. Treasury. As children and teenagers in the 1950s and '60s, their growing numbers brought a widespread demand for more schools and colleges. Today they are repaying that debt. As Ben Wattenberg has written, "The child grows up and he . . . can produce whatever public needs the society decides it wants . . . The society does that through an interesting device called taxes. Adults pay taxes. Children don't. But children become adults. When they do they pay taxes." Still, their taxes include Social Security; and a very real question is what will happen to that system as they approach retirement.

When the Social Security system was first set up, it featured huge numbers of workers paying tiny sums to underwrite modest benefits for a relatively few retirees. From 1937 to 1949 the most anyone had to pay in Social Security taxes was $30 a year, and in 1950 there were 16.5 payroll taxpayers to each beneficiary. By 1965 there were only 4 such taxpayers; in 1980 only 3.2, and this number is likely to fall to 3.0 by 2000, 2.3 by 2020. To cope with this, Social Security taxes have been rising. In 1970 the rate was 4.8 percent on the first $7,800 of income, an equal amount being paid by the employer. In 1981 the rate was 6.13 percent on the first $29,700, and by 1990 will go to 7.65 percent on the first $66,900. Hardly anyone will avoid paying that 7.65 percent on their full income. Even so that will not be enough. Long-term projections show a rate of 21.5 percent by 2055, again with an equal share being paid by employers. Unlike the income tax, the Social Security tax features no deductions and no exemptions. Even today more than half of all American families pay more in payroll taxes than they do in income taxes, and this situation can only get worse.

The Social Security system, like the rest of the federal government, has been running at a deficit, but the deficits have not been large. In 1975 the system paid out $69.2 billion in benefits and the deficit was $1.6 billion. In 1980 the figures were $124.5 billion and $3.6 billion, and in 1984 the projection is for $203.3 billion and $6.2 billion. In 1985 this problem will ease as the government applies the good old-fashioned method of hiking the tax rate, with a scheduled increase to 7.05 percent of the first $43,500 of income. Since any interruption in benefit payments would be politically unthinkable, Congress will meet any actual shortfall by appropriating money out of general Treasury funds. But as we look beyond the next few years, such appropriations would be no more than a financial Band-Aid. Within the Social Security system as a whole, surgery will be necessary.

The coming reforms will balance the political power of the nation's wage earners and of its retirees. Today any proposal to trim or defer benefits is treated as if it were a prelude to abandoning the system, but with vigorous presidential leadership this need not always be so. The day will come when a President will pose the alternatives: "We can maintain the present program of steadily increasing benefits. But then your Social Security tax rates will have to double, during the careers of people now working. Or we can keep the tax rates steady. But then we will be forced to impose massive cuts in benefits, if the system is not to go broke. Our plan will cut the rise in benefits. But we will maintain a basic package of payments to retirees, which is the pledge of the American government. Our plan will raise taxes. But the raises will not be large, and will be phased in slowly, over several years." The American people are not fools, nor do they expect pie in the sky; the issue will come down to not whether, but by how much, the system must be trimmed while taxes rise. If the issue is seen as saving the system from bankruptcy, there will be give on all sides, and the system will be successfully reformed.

First to go will be marginal programs, such as student payments available to children of beneficiaries. Benefits will also be cut for workers eligible for more than one pension, as will the minimum payment of $122 per month. Those who would have received this minimum instead will get $60, $80, or whatever they are entitled to, considering their history of earnings, in accord with the regular formula for payments. Some lump-sum death benefits also may go. The tax law too may be revised, so benefits remain untouched for those retirees to whom Social Security is a mainstay, but will be taxed when a retiree has substantial additional income. There will be changes in the way benefits are indexed or revised upward to meet inflation. The present law provides for benefits to rise rather faster than the cost of living as measured by the Consumer Price Index, which in turn overestimates the inflation that the retired population as a whole actually experiences. For example, this index includes a component reflecting rising prices for new homes; but few retirees buy such houses, and many own theirs outright. A special Retirees' Price Index may be set up to track the actual changes in their consumer prices, with benefit levels keyed to that.

In addition the retirement age will have to go up from 65, and benefits to early retirees will be trimmed noticeably. When the system was set up in 1935, relatively few Americans lived much past 65; today a majority do. Since 1960 there has been a trend to early retirement and a trend for people to go to college or otherwise delay entry into the work force. These trends could conjure up a vision of people staying in college to age 30 and retiring at 50, then living in retirement to 90; but that will not happen. By raising the retirement age, workers will pay taxes longer and draw benefits for fewer years. Nor need the increase be large or sudden. Economists who have studied this problem typically have recommended raising the age for receiving full benefits by something like two months each year, till after 18 years the retirement age would reach 68.

The Social Security tax will have to go up still more and will probably be broadened to include all earned income, with no earnings cutoff. But it may be possible to halt the rise at around 10 percent. Eventually the future of Social Security will be a prime political issue, but not an exercise in demagoguery whereby Democrats attack Republicans for wanting to tear up the old folks' Social Security cards. Both parties will offer serious, well-considered reform plans, and the debate will be lively. But neither party can afford to hike tax rates to levels likely to provoke a tax revolt. Nor can they allow the budget to run twelve-digit deficits to bail out the system year by year. So there will be change and probably well before the year 2000.

A nation with a bulge of young adults building their careers, a nation of prosperity and affluence, but one in which an eventually aging population will strain the retirement system—that is America in the 1980s. That is also Europe, Japan, and much of the Soviet Union. In the world at large, however, the picture of populations and their prospects is rather different. The United Nations publishes surveys of world population trends, by country and by region. Not all the data are as complete or accurate as those found by the U.S. Census, but the available data do give a useful picture of the trends. The table on page 200, "World Population Trends, 1950–2000," reflects the best available information on regional populations and growth rates and also gives educated projections for the next two decades, in eight principal areas of the world.

This table shows the sharp division between the advanced nations of North America and Eurasia and the rest of the world. The Europeans are in the vanguard of trends to zero population growth. They are growing at less than half a percent per year (0.39 percent between 1975 and 1980), and between 1980 and 2000 their population stands to increase by only 7.4 percent. Such a steady population means a slowly aging population, and the European nations may well be the laboratories which will first wrestle with the problem of social security and pension reform. These countries have a strong tradition of social welfare, and their governments have maintained extensive packages of tax-supported benefits for their people. Certainly those governments do not operate under the same constraints as ours; many have been able to continue to raise taxes, and they have found it quite possible to cut military expenses. Still, the issue of social security reform may come to the forefront there before it does here, and we will learn from their decisions and their mistakes. There is a precedent for this. The European governments in the decades before World War I long preceded our own in introducing retirement programs and other social benefits, in the first place.

Japan is lumped together with China and other nearby countries as "East Asia," so her remarkable advances deserve special mention. Her birthrate in 1978 was 15.1 per 1,000, lower than that in the United States and lower even than the level of 15.7 established that year for Europe and the Soviet Union. Japan is a very crowded and thickly populated land, in which birth control has been a key government priority, but if they had to, no doubt the Japanese could go a lot lower. In 1966 birthrates plummeted nearly 25 percent. That was the Year of the Fiery Horse, which comes every sixty years. In Oriental astrology, girls born that year may

WORLD POPULATION TRENDS, 1950–2000

YEAR	WORLD	UNITED STATES AND CANADA	EUROPE	SOVIET UNION	EAST ASIA	SOUTH ASIA	AFRICA	LATIN AMERICA	AUSTRALIA, OCEANIA
			Population, millions						
1950	2,513	166	392	180	673	706	219	164	13
1955	2,745	182	408	196	738	775	244	187	14
1960	3,027	199	425	214	816	867	275	215	16
1965	3,344	214	445	231	899	979	311	247	18
1970	3,678	226	460	244	981	1,111	354	283	19
1975	4,033	236	474	254	1,063	1,255	406	323	21
1980	4,415	246	484	267	1,136	1,422	469	368	23
1985	4,830	258	492	280	1,204	1,606	545	421	24
1990	5,275	270	501	292	1,274	1,803	630	478	26
1995	5,733	281	510	302	1,340	2,005	726	541	28
2000	6,199	290	520	312	1,406	2,205	828	608	30
			Average annual rate of increase, percent						
1950–55	1.77	1.80	0.79	1.71	1.85	1.86	2.16	2.72	2.25
1955–60	1.95	1.78	0.84	1.77	1.99	2.24	2.36	2.78	2.18
1960–65	1.99	1.50	0.90	1.49	1.94	2.44	2.49	2.77	2.09
1965–70	1.90	1.11	0.66	1.09	1.75	2.52	2.61	2.67	1.96
1970–75	1.84	0.87	0.61	0.84	1.62	2.45	2.71	2.64	1.82
1975–80	1.81	0.83	0.39	0.94	1.32	2.49	2.91	2.66	1.47
1980–85	1.80	0.96	0.36	0.94	1.16	2.44	2.97	2.65	1.41
1985–90	1.76	0.91	0.35	0.85	1.14	2.31	2.93	2.58	1.37
1990–95	1.66	0.76	0.37	0.70	1.01	2.13	2.81	2.46	1.30
1995–2000	1.56	0.61	0.38	0.64	0.95	1.91	2.64	2.34	1.19

Source: United Nations, *World Population Trends and Prospects by Country, 1950–2000. Summary Report of the 1978 Assessment* (United Nations, New York, 1979)

murder their husbands. Evidently many Japanese parents decided to avoid the risk of having a daughter who would prove a poor marriage prospect.

Neighboring China is also turning out to be rather a remarkable success. Well into this century, China was subject to drought and famine. In such times poor villagers would actually sell their children for food, though in a bad crop year a teenage girl would fetch barely a hundred pounds of grain. However, as Ding Chen, secretary general of the Shanghai Federation of Industry and Commerce, has written, "Today 977 million Chinese, nearly a fourth of the world's population, are secure against famine, flood and epidemic disease." The birthrate in that country was 34 per 1,000 in 1965, a very high rate typical of poor nations. In 1978 their government announced a birthrate of 18.34. If true, this would be a remarkable achievement, a reduction to the level of the slowly growing Soviet Union (18.2 in 1978) and nearly to the level of the advanced nations. Good statistics from China are hard to come by, and some experts have proposed that the birthrate actually was 22, with the population growing at 1.4 percent per year. Even so, this still would be a remarkable decline from the growth rates of 3 percent or more that prevailed briefly in the 1950s, rates which would have doubled the population in less than 25 years. The Chinese government has announced the goal of reducing the population growth rate to 0.5 percent in 1985 and to zero by 2000. They probably won't make it, at least not that soon. Nevertheless, the United Nations estimates in the table indicate that by the turn of the century, East Asia should show a population growth rate below 1 percent a year, a level heretofore associated with the wealthy nations of the West. This hopeful population trend will surely speed the emergence of East Asia as the world's next major power center, a region whose economic and political influence may rival that of Europe and North America.

Across the Himalayas, South Asia stands as a very extensive region encompassing the Arab countries of southwest Asia as well as Iran, Pakistan, India, Bangladesh, Indonesia, the Philippines, and the lands of southeast Asia. In this vast area India will serve as an illustration. It is by far the largest single country (population 694 million in 1980) and seems to be everybody's favorite candidate for widespread famine. In their 1967 book *Famine—1975!*, Paul and William Paddock listed India among the nations which "can't be saved," whose people should be left to starve in order to conserve shipments of scarce food for other nations which, though desperate, are not so hopelessly far gone. Today India is self-sufficient in grain production. She has built up substantial reserves of food, and among government officials in New Delhi there has even been talk of India becoming a grain exporter. India's birthrate appears to have fallen from 45 per 1,000 in 1965 to 36 in 1975. During the 1960s her population grew at 2.24 percent per year; this appears to have been slowed to 1.97 percent

in 1976, and even pessimistic official projections for 1981–1991 stand at
1.7 percent. Meanwhile, from 1949 to 1977 the growth rate in agricultural
production was 2.6 percent. The economist Raj Krishna, a leading advisor
to his government, has written:

> There are many reasons to think the agricultural growth rate can be kept
> at 2.5 to 2.6 percent per year over the next two decades. First, the
> amount of land in India that can be brought under irrigation can still be
> doubled, from [128 million acres to 277] . . . Second, the yields that have
> been realized on farmers' fields under a National Demonstration Program
> are many times the actual average yields. Even in Punjab, the Indian state
> where agriculture is the most advanced, the yield of wheat can be dou-
> bled. In other states it can be raised three to seven times . . . India now
> has an autonomous and well-organized agricultural research and extension
> system . . . Only in the event of two or three successive droughts that ex-
> haust the nation's grain reserve would India have to draw on world
> reserves, and then only to a small extent.

If even India shows such prospects, can the rest of the world's poor
nations be far behind? The statistics appear to suggest that in the area of
population growth and food supply, as in so many other troublesome
areas, the world is managing to muddle through. The world population
growth rate appears to have peaked just below 2.0 percent in the early
1980s, with a falloff to about 1.5 percent projected for the year 2000. That
would suffice to see the population explosion reduced to merely a baby
boom. With further reductions in birthrates and population growth rates,
world population might level out around ten billion. That is a very large
number, but provided it does not come about too quickly, it should be
possible to provide table settings for all. According to Roger Revelle, director of Harvard's Center for Population Studies, if the world as a whole
were to use the agricultural technology of present-day Iowa corn farming,
then "a diet based on 4,000 to 5,000 kilocalories of edible plant material,"
some of which would go for meat, egg, and milk production to give a diet
such as Europeans and Americans enjoy, "could be provided for between
38 and 48 billion people."

During the years 1948–1980, the world's population growth rate was
passing through its peak. In those years a United Nations index of per
capita food production, published by the Food and Agriculture Organization, rose from 100 to 129. On the average, the 4.415 billion people of
1980 each had 29 percent more food available than did the 2.513 billion
of 1950. Moreover, the amount of "arable and permanent cropland" in
these surveys increased from 3.467 billion acres to 3.724 billion in the
years 1961–74, with the spread of irrigation and as new lands were
brought under the plow. Even in the bad year of 1971–72, which was a
time of widespread drought, the UN food index dropped no farther than

from 125 to 120. (It rebounded to 126 in 1973.) At the peak of that drought, people on average were still 20 percent better off than they had been twenty years earlier.

During those times of drought there were widespread reports of starvation in the Sahel region south of the Sahara, and the world became aware that Africa was not sharing in these prospects. The table shows Africa with the world's highest population growth rate, which only now appears to be peaking just short of 3 percent per year. Africa's population stands to nearly quadruple between 1950 and 2000, and her food production has not been keeping pace. During the high population-growth years of 1953 to 1971, per capita food production nearly kept up but declined a total of 1.1 percent. In 1971–80 her per capita food production fell another 9 percent. Africa was the only major world region to show such a decline.

Africa certainly will be lagging behind the rest of the world in its approach to development. Even in 2000, the UN estimates show that continent with a population growth rate exceeding the peak in South Asia. The Sahel drought has been widely held up as a warning of the future to come, and it may indeed be; still it is worthwhile understanding to a degree just what happened. In his message to the UN Desertification Conference in 1977, Secretary General Kurt Waldheim stated, "Who can forget the horror of millions of men, women and children starving, with more than 100,000 dying, because of an ecological calamity that turned grazing land and farms into bleak desert?"

The statement was widely quoted, but it raised the question: in an area of the world where statistics are meager and often of poor quality, how could these numbers be vouched for? Julian L. Simon, an economist at the University of Illinois, succeeded in tracking the basis for Waldheim's statement to a one-page memo by Helen Ware, an Australian expert on African demography, written in 1975 while she was a fellow at the University of Ibadan, Nigeria. From calculations of the normal death rate for the area, together with "the highest death rate in any group of nomads" during the drought, she estimated "an absolute, and most improbable, upper limit [of] a hundred thousand . . . Even a maximum [this estimate] represents an unreal limit." In a later letter to Simon, she wrote: "The problem with deaths in the Sahel is precisely that there was so little evidence of them—rather like the photograph of the dead cow which kept turning up in illustration to every newspaper story." Another demographer, John Caldwell, an expert familiar with the area and who spent 1973 there, has gone even further: "One cannot certainly identify the existence of the drought in the vital statistics . . . nutritional levels, although poor, were similar to those found before the drought in other parts of Africa. The only possible exception was that of very young children."

Against this backdrop, it is appropriate to review the notion of Limits to Growth. This certainly is a matter no survey of the future can ignore, for in recent years Limits to Growth has dominated much thought about the future. Indeed, it has influenced public policy, encouraging what many have seen as a too easy assumption that hope for economic growth must be abandoned.

Limits to Growth echoes a theme of economic hopelessness and inevitable decline that is recurrent in history and goes back very far. In his *Ad Demetrium* of 250 A.D., Cyprian, Bishop of Carthage, wrote:

> You must know that the world has grown old, and does not remain in its former vigor. It bears witness to its own decline. The rainfall and the sun's warmth are both diminishing; the metals are nearly exhausted; the husbandman is failing in the fields.

In 1798 Thomas Malthus published his enormously influential *An Essay on the Principle of Population*, setting forth his argument of exponential population growth contrasted with no better than an arithmetical increase of resources. With no effective refutation for over a quarter-century, Malthus concluded it would be useless to improve the lot of the poor. He argued if their wages were increased or they were granted poor relief, they would merely have more children and breed themselves back into poverty.

A similarly despairing view of the future inspired H. G. Wells' *The Time Machine*, which foresaw mankind evolving into the childlike Eloi, kept alive for their pleasure by the cannibalistic Morlocks. Needless to say, this view of the distant future was controversial. When Wells met President Theodore Roosevelt, master builder of the age, by Wells' account: "He became gesticulatory, and his straining voice a note higher in denying the pessimism of that book . . . 'Suppose after all that should prove to be right, and it all ends in your butterflies and Morlocks. *That doesn't matter now*. The effort's real. It's worth going on with. It's worth it—even then.'"

Without being gesticulatory, it is quite possible to offer a similar response to the specific books and studies that launched the widespread awareness of Limits to Growth: *World Dynamics* by Jay W. Forrester (1971) and *The Limits to Growth* by Donella and Dennis Meadows, Jorgen Randers, and William Behrens (1972). The second of these was essentially a popular version of the first. The point of departure for their work was the obvious fact that the world and its resources are in some sense finite, so exponential growth of population and industry must eventually bump against physical limits. The problem lay in their interpretation of the phrase, "in some sense." The sense they chose was to lump all the world's resources together into a single one, and to assert that this super resource could only be depleted or used up, not replaced, substituted for, or added to. At the same time, they used assumptions which biased their analysis against the reduction of birthrates, the improvement of methods in agriculture, and the control of pollution. The resulting work might be compared to modeling the world as an enclosed box to which nothing is to be added and within which rabbits multiply at will. It would not be surprising if the rabbits soon would get into trouble, and Forrester's work indeed showed human prospects collapsing within a century or less, amid resource exhaustion and a massive dying of peoples.

Economists were quick to offer criticisms. The fact that so many different things had been aggregated, or lumped together, was one sore point. Typical were the remarks of Allen Kneese of Resources for the Future: "I really don't see what value this model has for the real world. How can you define meaningful relationships with such a high degree of aggregation?" Other economists noted that describing a resource as "finite" can be downright misleading, since to be useful such a description must tell us how to calculate its availability in the future. But the future quantities of a resource such as oil cannot be calculated even in principle, because of newly discovered deposits, new methods of extraction, and variations in the grades of oil-bearing formations; because oil products can be made from coal; and because of the vagueness of the boundaries within which oil might be found—including the seabed as well as deposits of tar sands and oil shales heretofore unexploited. Even less possible is a reasonable calculation of the future services we are now accustomed to getting from oil, because of substitution of other energy sources such as the various possible forms of solar and nuclear energy, including fusion.

One of the most telling criticisms of Forrester's work came from Robert Boyd of the University of California, in 1972. He started with the basic Forrester world model, with its five variables: population, pollution, food production, industrialization, and resource consumption. But to these he added a sixth, technology. He wrote equations representing assumptions that the growth of technology could serve to hold steady the supply of available resources and could reduce pollution, increase the output of agriculture, or improve the quality of life. Boyd also relaxed Forrester's assumptions regarding population growth, which were biased in favor of rapid growth rates.

The results given by Boyd's world model were much more hopeful, predicting a leveling off of population amid improving nutrition and quality of life, as well as rising living standards. In 1973 and again in 1975, two important reviews of economic history appeared in *Science*; both strongly supported Boyd's approach. Chauncey Starr and Richard Rudman, in 1973, described "technology's historical exponential growth," and wrote, "This growth pattern would be substantially altered only if we assume that knowledge is bounded or if society makes a conscious decision to stop the flow of resources into the production of new technological options." Glenn Hueckel in 1975 stated that "the history of technological advance suggests an optimistic outlook for future economic growth."

Faced with these and other criticisms, Forrester himself retreated. Later in 1975 he was quoted as believing that debate about the physical limits to growth is counterproductive, since it "invites the rejoinder that technology can circumvent such limits," and further that since "only nations have effective political processes," problems of growth must be addressed on a national and not a global basis.

The issue of Limits to Growth then tells us little about the economics of the future, but by studying it we can nevertheless learn much. How is it that so poor an economic study, a study so redolent with faulty assumptions and bad approaches, could nevertheless prove so influential? Why could it have been released in a blaze of publicity and thereafter retain its hold on thought and policy, even in the face of its subsequent refutation and partial retraction?

To answer these questions, we must look not to economics but to the spirit of the times. The early 1970s were no time of calm, reasoned thought. They were a time when even very reputable opinion makers felt free to indulge in wild exaggerations and predictions of doom. A few quotes from some intellectuals of the day will illustrate the spirit of those times:

Susan Sontag, 1967: "The white race is the cancer of human history. It is the white race and it alone—its ideologies and inventions—which eradicates autonomous civilizations wherever it spreads, which has upset the ecological balance of the planet, which now threatens the very existence of life itself."

Gore Vidal, 1973: "I fear the United States has always been a nation of ongoing hustlers from the prisons and disaster areas of old Europe . . . I do not think that the American system in its present state of decadence is worth preserving. The initial success of the United States was largely accidental. A rich almost empty continent was . . . exploited by rapacious Europeans who made slaves of Africans and corpses of Indians in the process."

Senator Edmund Muskie, 1971: "The Attica tragedy is more stark proof that something is terribly wrong in America. We have reached the point where men would rather die than live another day in America."

James McGregor Burns, 1972: "The nation is essentially evil and the evil can be exorcised only by turning the system upside down."

John Fischer, *Harper's* Magazine, 1970: "In these past months I have come to understand that a zooming Gross National Product leads not to salvation but to suicide."

Norman Mailer, New York *Times*, 1968: "I think American society has become progressively insane because it has become progressively a technological society."

Wayne H. Davis, 1970: "Blessed be the starving blacks of Mississippi with their outdoor privies, for they are ecologically sound, and they shall inherit a nation."

Stewart Udall, 1973: "We've become a gluttonous, greedy nation acquiring wealth we do not need, and as we do this, we extinguish the hopes of other, poorer nations."

Paul Ehrlich, 1969: "Most of the people who are going to die in the greatest cataclysm in the history of man have already been born."

Senator Walter Mondale, 1971: "The sickening truth is that this country is rapidly coming to resemble South Africa . . . And our apartheid is all the more disgusting for being insidious and unproclaimed."

Supreme Court Justice William O. Douglas, 1971: "We must realize that today's establishment is the new George III. Whether it will continue to adhere to its tactics we do not know. If it does, the redress, honored in tradition, is also revolution."

Democratic Party platform, 1972: "We are not sure if the values we have lived by for generations have any meaning left."

The response to Forrester's work was quite similar. In 1972 the British journal *Ecologist* used that report as a basis for calling for an end to economic growth in Britain and for a reduction in British population to 30 million, compared to the 1968 value of 55 million. Earlier in 1972, according to the columnist Anthony Lewis in the New York *Times:* "The conclusion of the scientists [is that] . . . there is only one way to avoid the pattern of boom crashing into earthly limits. That is to moderate all the interconnected factors: population, pollution, industrial production. The essential is to stop economic growth." In their textbook *Economic Growth vs. the Environment,* authors Hardesty, Clement, and Jencks used Forrester's work to support their contentions that "all developed countries, capitalist or socialist, must give up their unquestioning allegiance to the credo that 'more is always better,'" and "Blind faith in continuous future technological developments is foolish when action can be taken now to change man and his institutions."

Significantly, these statements and attitudes were not a response to the 1973–74 oil embargo and its dramatic energy crisis. They were published years in advance of that challenge. The spirit of those times, and the statements of the cited intellectuals and of many, many others, thus must be seen as nothing less than a moral collapse. In the absence of challenge or crisis, these writers had already abdicated their responsibility for clear and critical thought, a responsibility that went with their status as makers of opinion and influential recommenders of policy. They had given themselves over to an inbred and supercilious rejection of America

and of Western civilization. They had embraced with glee an environ-
mentalist pettifoggery that would bode our nation ill. When America indeed would face new and difficult challenges in its domestic economy and
foreign relations, these opinion leaders would stand morally bankrupt, unable to lead, unequipped to offer America the hope for the future our people have always prized.

If Limits to Growth can be dismissed as little more than an intellectual fad of the 1970s, what is it reasonable to say about the prospects
for the world's poor people and nations? To begin, for all the hopeful
words about a revolution of rising expectations, throughout the world the
predominant pattern will be one of people coping with adversity. However, most people are rather good at this when they have to be and often
make out better than one would expect. A bit of history from Merrie England will illustrate.

In the early seventeenth century, London with a population of
300,000 was probably scarcely more pleasant for its ordinary citizens than
is Nairobi or Calcutta today. The city was subject to visitations by smallpox, which today no longer exists anywhere on earth, and by plague,
which today is well controlled. There was nothing resembling public
health or sanitation; the Thames, as the city's common sewer, was
polluted and smelly. Infant mortality was appalling; life expectancy was
short. The Statute of Apprentices (1563) had effectively frozen laborers'
wages while prices rose, thereby making poverty compulsory. According to
J. E. Rogers' *Six Centuries of Work and Wages*, real wages of carpenters
stood at less than one third their level in 1480. London filled up with
poor people dwelling in slums and frequently unemployable. At the fu-
neral of the Earl of Shrewsbury in 1591, some 20,000 beggars applied for a
dole.

Certainly London in those days offered conditions at least as
wretched as any in the world today. In 1662 John Graunt published a review of vital statistics for London during 1603–1624, when 229,250 deaths
were reported to the parish clerks in the churches. Commenting on the
causes of death, Graunt wrote, "My first Observation is, That few are
starved. This appears, for that of the 229,250 which have died, we find
not above fifty-one to have been starved . . . The Observation, which I
shall add hereunto is, That the vast numbers of Beggars, swarming up and
down this City, do all live, and seem to be most of them healthy and
strong . . ." In London in those days, as in the world today, people
coped.

It is also worth remembering that the world's poor are not baby machines mindlessly breeding. They may be illiterate, but they generally
have a shrewd sense of their interests and of what actions will advance or
detract from their livelihoods. If they have large families, it is because in

their experience children are an economic asset. It is true that the minute a calf is born, per capita income and wealth go up; the minute a child is born, per capita income and wealth go down. But at a surprisingly early age, children can begin to help in farm work. Significantly, while high population-growth rates have slowed development, they have generally not reversed its course. Having babies may be a poor way to raise living standards, but if babies actually and immediately detracted from these standards, the world's peasants would show much keener interest in birth control.

This raises the point that people's expectations generally are set by their own experiences and by the common experiences of their societies. Ordinary Americans, working to pay their bills, do not measure themselves against international jet-setters who live lives of languor upon their yachts. Ordinary peasants in India or Latin America do not measure themselves against the world's prosperous peoples, of whom in any case they often hear little more than vague rumors. Everywhere people will count themselves fortunate if they are a little better off than three years ago, if their lives have improved in the last five. In comparing the world's prosperous and poor, it is worth recalling the French tale about the wealthy man who lost a fortune on the stock exchange, found himself down to his last $20,000, and died of misery. When his poor nephew inherited the $20,000, he died of happiness.

It is important to appreciate that there is no such thing as Spaceship Earth. Certainly there is a worldwide trade in food and other commodities, and a drought in the Punjab may jiggle grain prices on the Chicago Board of Trade. But our world is one of sovereign nations, the governments of which—and not we ourselves—are responsible for their peoples. Some of these leaders are interested in promoting domestic economic growth. Others seek to build up their military forces and acquire advanced weapons. Still others pursue ideological designs or agitate for international redistribution of income. But it is their own people who in the main must put up with such enthusiasms, suffer the excesses of a murderous Idi Amin in Uganda or a fanatical Ayatollah Khomeini in Iran. The rest of the world has a way of going on as before. When a nation falls behind in its development or fails to meet its Five-Year Plan, it is not the victim of neocolonialism or of some plot hatched amid the rich nations. The fault lies in its own capital city.

Nevertheless, famine is rare, and likely will be rarer. D. Gale Johnson, an economist who has extensively studied the history of famines, has estimated that only a tenth as many people died of famine between 1950 and 1975 as between 1875 and 1900, despite today's much greater population. The reason is that today there are not only larger food reserves, but also much better methods of transportation, so food can be rushed where it is needed. Indeed, one of the principal problems that has inhibited the sending of aid to such hunger-lands as the Sahel, Ethiopia, Bangladesh, and Cambodia is their lack of modern road systems. By contrast, India has an extensive road system. China since 1950 has built some 500,000 miles of paved roads and 20,000 miles of railroads, part of which has been electrified.

It is often said that the rich nations get richer while the gap between rich and poor countries grows wider. To this one may answer, what else is new? The important concern is that the poor not get poorer. So long as they advance, even if slowly, they will be on the right road. While their progress may be slow, it is definite and measurable. Many people have written of the cities among the poor nations where districts of high-rise office buildings and apartments stand mere blocks away from scabrous slums. But at least the high-rises and commercial districts are there; thirty years ago they often were not. In 1961, in his inaugural address, President Kennedy spoke of the "peoples in the huts and villages of half the globe struggling to break the bonds of mass misery." It is a measure of the improving world situation that by the turn of the century these lands of mass misery may cover not a half but only a quarter of the globe. By the year 2000, the principal remaining regions of too-rapid population growth stand to be Africa and Latin America, which will number only 23 percent of the world's people. Any food shortages there will be only local and temporary, and the rest of the world will be well able to help this remaining impoverished one fourth.

In the long run, the world at large may share the common present-day experience of the advanced nations, an experience that stands Malthus' doctrine on its head. In North America and Europe, it is population that increases arithmetically, when it increases at all, with gains being eked out slowly and haltingly. It is agricultural production that increases exponentially, as capital investment and technology flow in, the process feeding on itself. When nations develop, increasingly powerful economic and psychological pressures limit most women to no more than two children, a fertility rate that merely replaces the population from one generation to the next. But let us remember Roger Revelle's vision of a world employing modern agricultural techniques to grow enough varied and high-quality food to feed forty billion well where today four billion manage only modestly.

A common fantasy pictures Limits to (population) Growth being achieved when there is standing room only, everyone being crowded with no place to sit down. But an alternate vision pictures Limits to Growth being achieved when all the world's arable land is planted and cultivated using seeds and techniques which have raised food yields to the limits achieved at agricultural research centers and model farms. And why should this vision not someday be accomplished? It is already close to reality in the Midwest. In the world the land is there, as is the technology. The capital investment is potentially there, as is the energy, particularly when we consider solar or fusion energy, or breeder reactors. The approach to this ideal may take a century or more. But a century ago there were no tractors in Illinois, nor was there hybrid corn. Indeed, when world population levels off well below Revelle's food-production limit, then the world may actually have to limit its crops, to avoid a glut that would drive down prices to the farmer. The phrase "Limits to Growth" then would take on still another meaning.

All this lies far in the future, but for now there is a village scene that exists in an increasing number of lands. Today it could be in China, Egypt, or Brazil. Tomorrow it may be in India, Indonesia, even in Zaire.

It is early evening amid a cluster of small homes lining a road. Crickets chirp nearby; a hen cackles. The road is thinly paved with asphalt, broken in spots and with a few potholes, but it is paved; trucks use it occasionally. The homes are weatherbeaten cottages, often needing paint, but with good roofs. Their owners evidently are proud of them, even though they have outhouses nearby. Though drab they are neat; many of them have gardens planted with vegetables, some with flowers. A couple of motorbikes are parked here and there. Many more people have them in the cities, but apparently some are even out here.

What is good to see, though, is that along the road is a row of poles carrying an electric cable. From the cable, lines lead to connections on most of the houses. This means many of the familes have a radio, a washing machine, perhaps a refrigerator or a sewing machine. No doubt still others are saving up to buy such items. As night falls, the lights through the windows come not from kerosene lamps, but from electric. And through one or two of the homes' windows, there is the familiar pale blue glow of what can only be—a television set.

11

LIVING IN THE FUTURE

In 1905 the photographer Edward Steichen was an early specialist in the recondite art of color photography. On a misty late afternoon in winter, he took his equipment down to Broadway and 23rd Street in Manhattan and amid the gathering dusk captured one of the famous early color photos. His "Flatiron Building—Evening," shot through leafless branches overhanging from a nearby park, portrayed that famous early skyscraper, a landmark in the development of high-rise buildings. Across the street through the mist was 170 Fifth Avenue, topped with a curious small dome like that of a Mexican church. In the foreground, dimly outlined, were two or three horse-drawn carriages parked against the curb, their horses appearing as mere suggestions against the wet pavement, their drivers bundled in coats and scarves, wearing top hats. Along the street were paired circular lights resembling auto headlights but which in fact were street lamps. Three quarters of a century later, Steichen's scene appears peaceful and serene, giving little hint of what the city would become.

Go down today to that same spot. The Flatiron Building still is very much there, quite unchanged in its appearance. The same park is there, still with young trees planted along the curb. Even 170 Fifth Avenue is there, still with its curious dome. There have been changes, of course; the ground floor of that building now is a pizza restaurant, and the hansom cabs are long gone. The traffic today is the usual New York madhouse of buses and taxis interspersed with brave automobiles, and nobody today would wear a top hat. But what is surprising is how little fundamental change there has been in the neighborhood. The buildings are more brightly lit, but remain the same buildings. The traffic features horse-power rather than horses, but it travels along the same streets, and with not much more speed than in 1905. The district remains one of business offices full of secretaries, file cabinets, telephones, and typewriters. It has not been converted to high-rise condominiums, nor has it given way to parkland interspersed with electronics plants or low, campuslike medical research facilities.

A city of high-rises is a rather imperturbable place. For example, the Pan Am Building is one of the few major additions to a midtown Man-hattan skyline that has changed little since 1940. Buildings up to twenty or so stories may be knocked down and cleared to make way for new con-struction, but above that level most high-rises will be nearly as permanent as the cities themselves. If the cities last for hundreds of years into the fu-ture, so will their tall buildings. Europe's great cathedrals have shown pre-cisely this sort of longevity, and a steel-frame high-rise is more strongly constructed, less susceptible to rot and ruin. After all, churches and castles were usually built with heavy timbers for structural frames. In an age be-fore creosote such timbers would decay, lose strength, send the roof crashing in, and convert a fine medieval building into a dilapidated wreck. Steel girders are made of sterner stuff.

So if you want an idea of what your city will be like in the middle of the next century, hop on the freeway and take a drive through the high-rise district. Come the year 2050 the high-rises will still be there, as will the freeways. Rapidly growing cities like Denver and Houston will look very different then; downtown business districts will have grown; small groupings of tall banks and hotels or office buildings will have spread to become larger ones. But today's individual high-rises, such as Houston's 75-story Texas Commerce Tower, will blend in with the new construction. Anyone who wishes can emulate Edward Steichen and go take a color photo of whatever group of tall buildings that looks interesting. Come the middle of the next century, that city block and its buildings will be quite recognizable; little may have changed beyond fashions in dress and styles in auto and bus design. If a skeleton of brown steel girders is rising high above the nearby streets, the completed tower, decorously clothing the stark steel nakedness, will be there to greet your grandchildren. If a dra-matic new design has recently gone up, an architectural fancy in prestressed concrete and tinted glass, it too will be there to tell the next century about our tastes and ideas of novelty.

Among the most pleasant forms of new construction in cities has been the development of downtown malls and centers for pedestrians, particularly when they have been built alongside a river or harbor. Many cities have had old, run-down harbor districts full of abandoned wharves, railroad yards seldom used, factory buildings long fallen into disrepair. Such districts have often stood not far from highways, however, and have commanded potentially pleasant and attractive waterfront views. The old factory buildings also have often remained structurally sound, and even when this has not been so, it has been easy to knock them down for their land. Thus, cities have been turning parts of these harbor districts into marinas, shopping malls, convention centers, and other islands of pleasantness.

In San Francisco, just such a transformation has given rise to Fisherman's Wharf. Only a few blocks from working piers serving a functioning city harbor, tall-masted private boats and yachts are docked gunwale to gunwale beside an array of fine shops, restaurants, and cafes. Here the pedestrian is king; no vehicular traffic intrudes. One of the most intriguing places is The Cannery, formerly a red brick factory for canning fish, its interior now completely remodeled to make room for bookstores, boutiques, and gift shops. Similarly, Ghirardelli Square is nearby, with gaslights in a park fronting on what was once the Ghirardelli chocolate factory. In the evening, emerging from a seafood dinner, you may see the lighted masts and spars of the *Balclutha*, a turn-of-the-century windjammer saved from the Cape Horn gales and now on permanent display.

Baltimore's city planners have done even more. In what was once a downtown industrial wasteland-on-the-waterfront, they began by clearing out thirty-three acres to build Charles Center, a high-rise office development. Next came the 28-story World Trade Center designed by I. M. Pei, as well as a convention center and the Maryland Science Center. The performing arts were given their share in the harborside Outdoor Music Pavilion, which has featured performers ranging from the Baltimore Symphony Orchestra to Judy Collins. In 1980 came Harborplace, a quarter-mile of charm and chic where pedestrians promenade alongside the water, next to two-story pavilions savory with the aroma of Maryland seafood and filled with all manner of attractive shopping. Here too is a ship, the U.S.S. *Constellation* built in 1797, its masts and rigging outlined in strings of lights. Topping it all off is the most recent addition, the National Aquarium, with a whale skeleton hung in midair within its open central space.

There are many other places like Fisherman's Wharf and Harborplace, places where people can stop to buy a chocolate éclair or cotton candy, fresh-baked bread and cheese, or an unusual book. Or enjoy lunch in an outdoor cafe. Or lounge on a bench and watch the boats, the clouds, and the people go by. Or dangle feet in the water. Or drop into a gift shop with a name like Remembering You and buy a lava-light, its ever-changing globulous shapes rising and circulating within its vase of blue water. Boston has Faneuil Hall Marketplace, built around the restored Faneuil Hall with its red brick and leaded windows, built in 1713. In Detroit is Renaissance Center, a hotel and convention complex built along the riverfront, with Canada across the water. Manhattan is building South Street Seaport, just a short walk from Wall Street near the southern tip; it will feature a renovation and expansion of the famous old Fulton Fish Market. Seattle has its Science Center and Space Needle, originally built for the 1962 World's Fair. In Los Angeles Marina del Rey is north of the airport, where palm trees are interspersed amid fine restaurants and shopping along Admiralty Way, with luxury apartments and a marina full of boats quite close by. Chicago, St. Louis, San Diego—here too are similar oases, full of delightful diversity.

In all this, one common theme is the cherishing and reuse of old buildings. In many Eastern cities there are districts of once elegant brick or brownstone row houses, long fallen into decay but still structurally sound and salvageable. People are buying them, sometimes for as little as one dollar each, and refurbishing them. Often they rip out virtually the entire interior, then rebuild with all new plumbing and baths, wallboard, wiring, and ceilings as well as new doors, windows, and plaster work. The new owners get more than a fine place to live, more than a sense of having helped to preserve part of the city's history. They often get twelve-foot ceilings and hand-carved stone for exterior trim, or thick hand-hewn ceiling beams and pegged hardwood floors—things which today can't otherwise be had for love or money.

In Baltimore, dozens of blocks of fine old town houses have been salvaged this way. In New York's Brooklyn Heights and Washington's Georgetown, this trend is well advanced. In Chicago, old warehouses and factories are being salvaged, with windows being punched in the solid red-brick walls, and are being converted to loft apartments. A loft apartment is a large room with modern plumbing, having 1,000 square feet or more of undivided floor space. Renters or purchasers use room dividers, furniture, and portable closets to mark off bedrooms, dining rooms, and the like. Some of these buildings feature restored marble lobbies, brass caged elevators, and elaborate wrought-iron railings. Others were originally so run down that no one had paid taxes on them for years; no city records indicated who owned them; and when finally bought they sold for something like fifty cents per square foot. Refurbished as loft apartments, they now cost $80 per square foot.

To be sure, only a small minority of Americans will ever live in these lovely new-old buildings. Loft apartments and rebuilt brownstones will account for 1 percent or less of cities' populations. In Chicago, for instance, they can provide space for about twelve thousand people in a city of three

million. Their significance is not in providing mass housing for our urban
tens of millions, but rather in their demonstration that run-down city districts can be made valuable and joyous. For the sake of the vast majority
of Americans then, it is appropriate to take a look at a few other places in
the country, to see how people will be living in the future. A good place
to start is in a most unfashionable but wholly typical place: the working-
class districts of Brooklyn, New York.

This is lower-middle-class America, the home of recent immigrants
and of the children of earlier newcomers, strong in their ethnic ties. Here
is a city of medium-rise apartment buildings, from whose windows
mothers lean out to call to their children in the streets below. Along these
streets are small mom-and-pop stores and markets, opening early and closing late. Subway trains rattle and clatter, never far away; radios blare, garbage cans stand near the sidewalks. The people often are scarcely less drab
than their buildings. Most of the cars are several years old and in need of
a good mechanic. Overall, Brooklyn has not so much thrived as endured.

Has Brooklyn changed much in the last fifty years? The auto styles
have changed, and now there are TV antennas on the roofs along with
the pigeon coops. Yet today, as half a century ago, there are the same subways and tenements, the same whey-faced and harried people, the same
sense of making do through hard effort. The rainy days are just as cold
and damp as they were, the winters as bitter, the summers as sweltering.
We could say that one of the most important changes in Brooklyn occurred much earlier in this century, when the subway replaced the old
elevated railways, whose smoke and soot had been the despair of many
neighborhoods. We could say equally that an important recent development has been the invention of the spray-paint can, which has greatly
eased the work of painters of graffiti.

What will the next fifty years bring to Brooklyn? Probably nothing
much. Cuban or Mexican immigrants may appear in force, replacing the
earlier Puerto Ricans, Italians, and Jews. Still Brooklyn is hardly likely to
cease being a working-class city of ethnic newcomers. While it will not
likely drown in flood or blaze in a nuclear torch, neither will Brooklyn
give way to open greensward interspersed with crystal towers. As long as
America remains a place of hope for people in the world seeking better
lives, there will be places like Brooklyn, providing the intitial footholds
for new arrivals and for their children. Through times to come, Brooklyn
may resemble some of the long-established cities in England or Europe
that have preserved their character for centuries.

We leave these vistas and inquire further about the future in a
different place: Ann Arbor, Michigan. Here is a college community, home
of the University of Michigan; far more than most, this town and university have stood at the vortex of swirling winds of change. Its repertory the-

ater has often featured the avant-garde. It was a major center for the anti-war movement in the late 1960s and was home to founders of Students for a Democratic Society. Early in the 1970s it was a center for the nascent environmental activists and stood at the focus of challenges by black students seeking better educational opportunities. Outside of the university area recent decades have seen suburban subdivisions grow amid farmland, with freeways and shopping malls building apace. The city today is home to such advanced enterprises as KMS Fusion, Bendix Aerospace, University Microfilms, and the auto-emissions testing laboratory of the Environmental Protection Agency. In addition it stands only a few miles from the large airfreight terminal, Willow Run Airport, and boasts one of the nation's most extensive medical centers.

If any city is engaged in building the future, surely it is Ann Arbor. Yet if you visit the downtown area, apart from a few high-rises and a new city hall, you will find a town eminently familiar to those who have lived there since before World War II. There are the blocks and blocks of white-painted frame houses along tree-lined streets, so typical of long-established towns in the Midwest. The suburban tracts are like suburbs everywhere. The university has grown extensively, but its central quadrangle, flanked by a college-town business district, remains recognizable to returning alumni. As for the upheavals of the Vietnam years, in 1972 the student precincts actually succeeded in electing two city council members on their homegrown Human Rights Party ticket. But their reelection bid failed, and in recent elections, as in those prior to 1970, Ann Arbor has generally gone Republican.

Fifty years ago, Ann Arbor was a small, quiet university town. Today it is a large, quiet university town. Only for a few years were the students involved in activism and demonstrations; as in previous decades they now appear mainly interested in their courses, their dates, and the football team. There are few natural limits to Ann Arbor's physical growth, and fifty years from now no doubt it will be larger still. It may by then be a Midwestern Palo Alto, the center of its own Silicon Valley. Yet it will still be Ann Arbor. It will not lose its collegiate character, certainly not while the university is there, not to mention its stadium which fills up on autumn Saturdays with a hundred thousand vigorously partisan Michiganders.

Is there anywhere we may find change pure and simple from the last half-century? We may look to southern California and particularly to the coastal area southeast of Los Angeles. Well past the 1940s, the area was given over largely to the orange groves that gave it its name, Orange County. Also there was other fruit farming and an occasional small town. Today it is a suburban sprawl from the sea to the mountains inland. Land that once grew citrus trees now sprouts a luxurious crop of $200,000 homes. The fifteen years following 1965 saw the University of California at Irvine grow from nothing to achieve an international reputation. Nearby are the headquarters of the Fluor Corporation, famous for its synthetic-fuels plants in South Africa. Aerospace and electronics firms are well established, as are high-fashion shopping malls, close to the myriad sailing craft docked at Newport Beach. For those preferring other entertainment, a few miles away is Anaheim Stadium, home to the California Angels and Los Angeles Rams; in that same city is Disneyland. Since 1950 Orange County has consistently stood among the nation's fastest-growing areas.

And yet, and yet. The story of Orange County has really been a continuation of the story of California. Even before its admission to the Union in 1850, this state had justified the hopes of the sixteenth-century Spanish explorers who gave this land the name of a fictional utopia in a novel of the period. Spared from the Civil War, insulated from the harshness of early industry, endowed with oil, fertile farmland and a most excellent climate, California has seen recurrent booms, nearly continual growth and prosperity. The people who came to Orange County after 1960 were not so different from the transplanted Midwesterners who made Los Angeles grow in the 1920s. Even the important north-south freeways, Highway 101 and Interstate 5, follow for long stretches the route of the old Spanish road, the Camino Real, which linked the coastal missions.

From this standpoint, the next fifty years may well see Orange County echoing, with a delay of perhaps several decades, the patterns of growth of its precursor, Los Angeles proper. During the 1970s, while the population of Orange County grew by 36 percent to nearly two million, that of Los Angeles County grew by less than 6 percent and stayed near seven million. Thus, notwithstanding its reputation for rapid growth, by 1970 L.A. indeed had grown nearly to its available limits. In the next decades, similarly, Orange County will fill up and its growth will level off.

In the nation as a whole, what can we say about some of the trends that will be influencing people's lives? Some futurists see people abandoning the suburbs, returning to the downtown areas of cities; they anticipate a boom in apartments or town houses. Certainly much has been written about the delights of urban life as contrasted with the garish plastic and ugly ticky-tacky of suburban sprawl, and perhaps someday America indeed will follow these fond hopes. The U.S. Census data, however, show no indication that America as a whole is doing anything of the kind. As it has been for decades, the trend continues to be out of the cities and into the suburbs—crabgrass, ticky-tacky, and all. Between 1970 and 1980, the total population of our cities, that is, of the incorporated metropolitan areas, went from 67.8 million to 67.9 million, a decline from 33.3 percent to 30.0

percent of the U.S. population. The fraction of city dwellers has been declining for decades—it stood at 35.5 percent in 1950—but the 1970s saw something new: a leveling off in their numbers. Meanwhile the growth of suburbia continued apace, going from 85.6 million people and 42.1 percent of all Americans to 101.5 million and 44.8 percent between 1970 and 1980. In addition, millions of people today live in suburban tracts that twenty or thirty years ago lay just outside the city limits but which since have been annexed, incorporated into the central city. These people's lives have not changed; they still live in two-story houses on Elmwood Street and spend Saturday afternoons mowing the lawn. But for census purposes they are listed among the city dwellers. If they were listed as the suburbanites they still feel themselves to be, America would be close to having a suburban majority.

Urban planners like James Rouse, builder of Baltimore's Harborplace, have declared that the suburbs "sucked the blood out of the central cities and left behind some of the urban basket cases we see today." The growth of the suburbs has indeed been facilitated by federal policies, which built the freeways, provided low-interest FHA mortgages, and permitted income-tax deductions for mortgage interest. However, there is a word for such policies: democracy. The suburbs are what the people wanted; they are what they got; and the forces which could reverse this trend are nowhere in sight. Increasingly the suburbs are attracting jobs and industries, which establish their new plants where the people are. There still are few symphony orchestras outside the cities, but such traditionally urban amenities as the major medical center, the quality shopping district with its fine department stores, the excellent university—all may be found in the suburbs. Parks and beaches are there too and are usually less crowded. The cities retain their business districts, but branch offices now bring their services within easy reach. And in professional sports, some of the great pro teams are from Hackensack, New Jersey; Irving, Texas; Bloomington, Minnesota; and Foxboro, Massachusetts. Sports fans know them as the New York Giants, Dallas Cowboys, Minnesota Vikings, and New England Patriots.

With all this, and with only three tenths of America's people now living in incorporated central cities of over 50,000 population, the sort of city love advocated by many political leaders may appear almost as nostalgic and irrelevant as the hankering of old-time Republicans for their small towns and for life on the farm. Unquestionably, the future for much of American life lies in its suburbs. Gertrude Stein once said of Oakland, across the bay from San Francisco, that "there's no *there* there," and far worse things have been written of the suburbs. But it is worth recalling Ben Wattenberg's statement that "Suburban Sprawl is a pejorative phrase that describes perhaps the most comfortable mass residential living conditions in history."

A common prediction is that rising prices for homes will drive people to smaller houses or to apartments. Again the Census data show the opposite trend. Among all occupied housing units, the fraction occupied by their owners went from 62.9 percent in 1970 to 65.6 percent in 1980. What's more, people have been getting more living space. The median number of rooms, in houses and apartments, rose from 5.0 in 1970 to 5.1 in 1980; that is, half the homes had more than this number, half fewer. This was not a large change, but in that same time the median number of persons per home fell from 3.0 to 2.6 in owner-occupied houses, from 2.3 to 2.0 in rented apartments. Evidently many people were leaving home to set up their own households. In addition, new houses have been getting larger. In 1970, 36 percent of newly completed homes were under 1,200 square feet in area; 21 percent were over 2,000 square feet. In 1980 only 21 percent were under 1,200 and 28 percent were over 2,000. People's homes have also been getting cozier. In 1960 only 1.9 percent of all homes and apartments had central air conditioning. In 1970 this was 10.8 percent; in 1980, 26.2 percent, and that year 63 percent of all new houses were being built with central air conditioning. This is the same sort of trend past decades have shown for telephones, washers and dryers, automobiles, television, and other amenities; no doubt succeeding decades will show a similar trend for home computer systems. And while people's homes have been adding these things, the number of truly substandard homes and apartments has been rapidly dwindling. The percentage of "units lacking some or all plumbing facilities" was 13.2 percent in 1960, 6.5 percent in 1970, 2.7 percent in 1980.

To pay for these nice new homes, people have been earning more money. Statistics on earnings call for some interpretation. Thus, to say that median family income went from $3,319 a year in 1950 to $21,020 in 1980 looks impressive until we remember the cost of living jumped 342 percent in that same time. Even so, measured in constant 1979 dollars, this income went from $10,008 in 1950 to $19,684 in 1973. Then, with inflation outstripping wage gains, median family income in 1979 dollars stayed at a plateau, fluctuating around $19,000. Yet on a per capita basis, real income has continued to increase. In 1979 dollars, "disposable personal income" (personal income less taxes) went from $7,048 per capita in 1973 to $7,183 in 1976 and to $7,721 in 1980—a 10 percent increase in a time when real family income was barely holding its own. What has brought this about has been more wage earners per capita. Thus, between 1970 and 1980 the percentage of women in the work force rose from 43 percent to 51 percent. At the same time, the size of U.S. households—

which include single or divorced people living alone, retired couples, and young couples without children—fell from 3.14 people to 2.75. Clearly, the demand for continued improvements in people's standards of living is quite strong and will put continued pressure on women to seek work outside the home, to put off having children, and then to have very few.

Americans have also been getting more education. In 1950 the "median years of school completed" was only 9.3; we could say the typical American that year was no more than a junior high school graduate. In 1960 the median was 10.6, and the typical citizen had advanced to the status of a high school dropout. By 1970 the median had jumped to 12.2 and America had earned its high school diploma. Since then it has risen even further, to 12.5 years of education completed as of 1980. Among young people aged 25 to 29, the median was even higher, 12.9 years, which puts the typical young American nearly at the level of a beginning college sophomore. As with money income, however, there has been a sort of educational inflation. These added years of school may not be quite as significant as they would have appeared back in 1940, say, when only one out of twenty young people received the bachelor's degree. The widespread stories of illiterate college students make it easy to believe that to a large extent we are merely relying on our colleges and junior colleges to do the work that previously was done by our high schools.

As an example of this educational inflation, here is a letter written by a young woman to her sweetheart back in 1852, when education was on the gold standard:

> My Dear Sir,
> Time flies, so I can no longer delay informing you that our better years are passing rapidly away and we enjoy not the blessings within our reach —I refer to the joys of wedded life. You are aware that this is Leap Year & we maidens are inclined to improve our privileges. For this purpose, I now address you. Why will you longer hesitate to declare? Surely did you but realize the half of my affection for you, you would hasten the day that should see us united in happy wedlock. I would sign my name but dare not. We will meet during the Fair & dance at the ball.
> I long to sign myself "Mrs. William Ragan."

Her ploy worked; two months later they were married. At the time she wrote this letter, she was fourteen years old.

We can set aside for the moment what we would now call her social and sexual precocity in anticipating marriage at an age when many young girls today are hardly prepared to go steady, though in those days such youthful marriages were common. What is certainly surprising, though, is her flowing command of the English language, at an age when she could not have gone further than the eighth grade—a command which would do credit to plenty of today's college seniors.

Nevertheless, just as real incomes have risen in the face of inflation, it is evident that people have been learning more despite educational inflation. At least they have been qualifying for the better jobs, in increasing numbers. A short summary of the principal Census Bureau job categories will tell the tale:

CIVILIAN EMPLOYMENT CATEGORIES, 1960–1980

	TOTAL EMPLOYED (× 1000)			PERCENT		
	1960	1970	1980	1960	1970	1980
TOTAL	65,778	78,627	97,270	100.0	100.0	100.0
White-collar workers	28,522	37,997	50,809	43.4	48.3	52.2
Professional and technical	7,469	11,140	15,613	11.4	14.2	16.1
Managers and administrators	7,067	8,289	10,919	10.7	10.5	11.2
Salesworkers	4,224	4,854	6,172	6.4	6.2	6.3
Clerical workers	9,762	13,714	18,105	14.8	17.4	18.2
Blue-collar workers	24,057	27,791	30,800	36.6	35.3	31.7
Craft and kindred workers	8,554	10,158	12,529	13.0	12.9	12.9
Operatives, except transport	11,950	13,909	10,346	18.2	17.7	10.6
Transport equip. operatives			3,468			3.6
Nonfarm laborers	3,553	3,724	4,456	5.4	4.7	4.6
Service workers	8,023	9,712	12,958	12.2	12.4	13.3
Farmworkers	5,176	3,126	2,704	7.9	4.0	2.8

In 1978 the United States passed a significant milestone: it became a majority white-collar nation, with more than half the work force so listed. Within this category the most desirable occupational groups, "Professional and technical" and "Managers and administrators," have risen to over 27 percent of the work force. At the same time, there has been a modest decline in the proportion of blue-collar jobs to 32 percent. In many categories, however, their actual numbers have risen, and in 1980 there were 19 percent more carpenters, construction workers, machinists, and mechanics than in 1972. There also were 23 percent more plumbers, raising the prospect that it will become easier to find someone to unclog the drain when it gets stuffed up. Service workers have shown a modest gain, a category that includes cleaning service workers, restaurant and hospital workers, barbers and hairdressers, as well as policemen, firemen, and guards. As for the farmers, they have continued their long-term decline. If only a couple of million more leave the field (and the fields), there will hardly be any left.

What of new life-styles and social forms for the American people? This is an area in which many futurists love to fly away on flights of fancy; but the title of this book is *The Real Future*, and we need not get too excited over these supposedly new social forms. Certainly it is all too easy to seize upon some widely publicized novelty and proclaim it the wave of the future. For example, few living arrangements have caused more ink to flow than that of a man and woman setting up housekeeping without taking the trouble to get married. This sort of arrangement indeed is delightful. It has shown a remarkable upsurge in recent years, rising from 327,000 such households in 1970 to 1,136,000 in 1980, and increasing by over 150,000 per year recently. But this still is barely 1 percent of all the people in the United States. The real story, as always, has been one of people getting married and living in families. In 1980 there were 58,426,000 family households, numbering a total of 191,600,000 people. Of these households, 48,180,000 featured a husband and wife living in their own home. Most of the rest consisted of a divorced woman and her children. Not much of the avant-garde there.

As for new social forms there simply aren't all that many, if one seeks something truly and definitely unique in the human experience. This has been the bane of social utopians, whose attempts to design perfect societies have repeatedly foundered on the fact that societies are made up of people, and people aren't perfect. Nor do they change very fast. One reason so many alternative life-styles remain "alternative" is that they don't work too well, at least not for very long. In 1978, as talk show pundits were pontificating on the death of the nuclear family, the anthropologists Mary Leakey and Paul Abell were finding unmistakable fossil evidence for the existence of the nuclear family some 3.5 million years ago.

But hasn't there been the sexual revolution? Hasn't that been an important and dramatic change? A popular view is that the invention of the Pill spawned this presumed revolution in morality. This recent upsurge in sexual freedom, or at least in awareness of sexual freedom, has spawned an enormous amount of cant. We have heard repeatedly that by their sexual liberation, young people today have become freer, more open, more naturally human, in contrast to the cramped, narrow, restricted morals of their elders. The truth is somewhat more interesting.

For a background, it is useful to present something of a short history of sex among young people. In the Middle Ages, according to the historians Will and Ariel Durant,

> Boys reached the age of work at twelve, and legal maturity at sixteen . . .
> by the age of sixteen the medieval youth had probably sampled a variety

Fig. 49. *Edward Steichen's "Flatiron Building."* (Courtesy Metropolitan Museum of Art, New York.)

Fig. 50. *Same scene today.* (Photo by Aryeh Brodsky.)

Fig. 51. *Baltimore's Harborplace.* (Courtesy The Rouse Co.)

Fig. 52. *Loft apartment in Fulton House, a Chicago industrial warehouse remodeled into a collection of chic apartments.* (Photo by Photogroup courtesy Harry Weese and Associates.)

Fig. 53. *Street scene in Brooklyn's Crown Heights section.* (Photo by Aryeh Brodsky.)

Fig. 54. *Ann Arbor, Michigan.* (Courtesy Bentley Historical Library, University of Michigan.)

Fig. 55. *Orange County, California.* (Photo by Don Dixon.)

Fig. 56. *"The Marriage of Giovanni Arnolfini and Giovanna Cenami" by Jan van Eyck, 1434. Far from being a love match, this union more than likely was arranged by their parents to seal a business or political alliance of their families.* (Courtesy National Gallery, London.)

Fig. 57. *In lovely old residence halls like this one, college women up to the late 1960s lived lives amid restrictions like those of a minimum-security prison.* (Courtesy Michigan State University.)

Fig. 58. *Hogarth's "Gin Lane."* (Courtesy Metropolitan Museum of Art, New York.)

Fig. 59. *Big Brother is watching: a wall-size poster of Mussolini in 1934.* (Courtesy National Archives, Washington, D.C.)

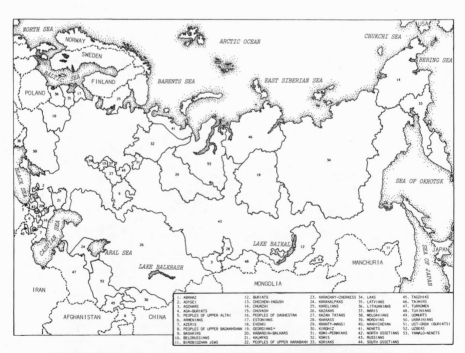

Fig. 60. *The world's last great empire: peoples and nationalities of the Soviet Union. Borders shown are territorial and administrative boundaries.* (Map by Gayle Westrate.)

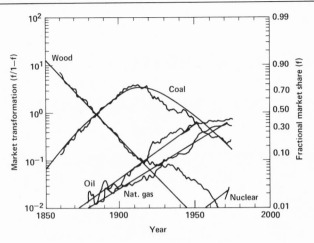

Fig. 61. *Trends in energy substitution in the United States.*
(Courtesy Michael J. Monsler, Lawrence Livermore Laboratory.)

Fig. 62. *Pumping water over the Tehachapi Mountains. The aqueduct terminates in a pool feeding two pumping stations, which force the water into pipes of twelve and a half feet in diameter.* (Photo by Don Dixon.)

Fig. 63. *The proposed North American Water and Power Alliance.* (Courtesy *Scientific American.*)

Fig. 64. *The Neolithic Revolution and the Industrial Revolution. Top, 10,000 and 8000 B.C. Bottom, 1750 and 2000 A.D.* (Art by Don Dixon.)

of sexual experiences. Premarital and extramarital relations were apparently as widespread as at any time . . . Knights who served highborn dames for a kiss might console themselves with the lady's maids; some ladies could not sleep with a good conscience until they had arranged this courtesy. The Knight of La Tour-Landry mourned the prevalence of fornication . . . [and wrote that] some men of his class fornicated in church, nay, "on the altar" . . . Marriage came early. A child of seven could consent to a betrothal . . . The normal age of consummation in the girl was presumed to be twelve, in the boy fourteen . . . State and Church alike accepted as a valid marriage a consummated union accompanied by the exchange of a verbal pledge between the participants, without other ceremony legal or ecclesiastical.

About 1230, Jacques de Vitry described students at the University of Paris:

[They were] more dissolute than the people. They counted fornication no sin. Prostitutes dragged passing clerics to brothels almost by force, and openly through the streets; if the clerics refused to enter, the whores called them [gay] . . . It was held a sign of honor if a man kept one or more concubines. In one and the same house there were classrooms above and a brothel beneath; upstairs masters lectured, downstairs courtesans carried on their base services; in the same house the debates of philosophers could be heard with the quarrels of courtesans and pimps.

In Italy of the Renaissance, things were not much different. Again quoting the Durants,

There must have been considerable premarital adventure; otherwise it would be difficult to account for the extraordinary number of bastards to be found in any city . . . To have [bastards] was no serious disgrace . . . To be a bastard was no great disability; the social stigma involved was almost negligible . . . Robert, Bishop of Aquino . . . described the morals of the young men . . . they explained to him, he tells us, that fornication was no sin, that chastity was an old-fashioned taboo, and that virginity was on the wane . . . There were 6,800 registered prostitutes in Rome in 1490, not counting clandestine practitioners, in a population of some 90,000. In Venice the census of 1509 reported 11,654 prostitutes in a population of some 300,000. An enterprising printer published a "Catalogue of all the principal and most honored courtesans of Venice, their names, addresses, and fees" . . . A daughter unmarried at fifteen was a family disgrace.

In those days, marriage was an affair of property. Family councils assigned young people their marriage partners to seal family unions of economic or political interest. In sexual matters, men had full opportunity to be as free as they pleased and could only be lured into matrimony by a bride with a bribe; that is, a large dowry. Amid such marriages of convenience, few

husbands expected or cared to be faithful, and most felt justified in taking
a mistress or engaging in other diversions. They felt perfectly justified in
seeking Venus in the flesh and not in statues.

That was in Italy. What of the pious lands of northern Europe, fertile ground for the Protestant Reformation? We learn that "the girls who
on Sunday bowed demurely before statues of the Virgin rouged their
cheeks hopefully during the week, and many of them got themselves se-
duced." Martin Luther in 1544 wrote of matters at the University of Wit-
tenberg:

> We have a great horde of young men here from all countries, and the race
> of girls is getting bold, and run after the fellows into their rooms and
> chambers and wherever they can, and offer them their free love; and I
> hear that many parents have ordered their sons home . . . saying that we
> hang wives around their necks.

It would be easy to assemble similar quotes concerning subsequent
centuries. Even in the France of Voltaire, when "nearly all women were
virgins at marriage," we find that "in the great majority of cases, even in
the peasantry, marriages were still arranged by the parents . . . The legal
age of marriage was fourteen for boys, thirteen for girls . . . Parents married off their daughters as soon as practicable to avoid untimely deflower-
ing . . . Girls in the middle and upper classes were kept in convents until
their mates had been chosen; then they were hurried from nunnery to
matrimony, and had to be well guarded along the way . . . Adultery was
accepted as a pleasant substitute for divorce."

In his *Persian Letters*, Montesquieu wrote that in Paris, "a husband
who would wish to have sole possession of his wife would be regarded as a
disturber of public happiness, and as a fool who should wish to enjoy the
light of the sun to the exclusion of other men." In these matters, England
was not far different. Courtesans, mistresses, and bordellos of all classes
flourished openly. Captain Frances Grose's 1811 "Dictionary of Buckish
Slang, University Wit, and Pickpocket Eloquence" is full of such defini-
tions as, "Balum Rancum, a hop or ball, wherein the women are all prostitutes. N.B. The company dance in their birthday suits." Not till 1929
was the legal age for marriage in England raised from the medieval ages
of twelve for the girl, fourteen for the boy; common-law marriage, of a
man and woman living together, was legally recognized.

With such a rich background of venery, whatever could have possessed the Victorians to invent Victorian morality? There were several
reasons. To begin, there were the romantic novelists. Television today has
often been accused of influencing (or corrupting) the morals of the
young. A hundred and fifty years ago the comparable influence was the
fictional literature of the day, which exalted love and painted a glowing

picture of marriage for love. In an era when demands for political liberty were waxing strong, young people increasingly gained the right to marry partners of their choice. The decline of the parent-dictated marriage eliminated much of the need for the mistresses and courtesans of earlier times and permitted a climate wherein their activities could be made illegal.

When marriages became love matches, the partners had much greater interest in giving and demanding fidelity. The Victorian code thus worked, when it worked, to hedge a marriage with restrictions tending to protect a wife against another man's attentions, at which in earlier times a husband might have winked. There were economic reasons also. A career of wenching and mistress-keeping has always required much time, effort, and money. In Victorian times these resources could be invested in some enterprise that would advance one's prosperity. In an age of economic growth many people prospered, particularly in the middle and upper classes; those of a moral turn of mind could naturally credit their successes to their virtue, thrift, and sobriety. The fact that those who cheated in business or marriage also prospered was one which could be conveniently overlooked.

Finally, there was that elusive concept, the spirit of the times. The Victorians were tremendously impressed by the achievements of their age, in government, commerce, engineering, and the building of empires. They could easily incline to a smug belief that they were living in a new era, much advanced over and having few ties to the past. (It is an attitude quite familiar today.) They thus could be attracted to a new morality which exalted love and fidelity and appeared loftier, purer, and higher in tone than the rude morals of an earlier day. To this was added, in America, a strong streak of the self-righteous piety which has recurred frequently in our history, from Jonathan Edwards' "Great Awakening" of the 1730s to Jimmy Carter in the late '70s and the Moral Majority today.

Victorian morals worked a real hardship when marriages turned loveless, since divorce was legally difficult and socially unacceptable. Withal, the Victorian code lasted over a hundred years and evidently it satisfied many people, since it was kept voluntarily. The work of Freud, with his emphasis on the importance of sex, did not weaken the code since it was quite consistent with the connection of sex with marriage. In any case, men seeking sexual variety could usually find it if the price was right. Consenting adults, particularly in the upper classes, could do pretty much as they pleased provided they were discreet.

What brought an open challenge to the code was not the Pill, but the post-World War II advent of mass college education. American universities had always had restrictive rules, particularly for their women students, but such rules reflected general social mores and usually were accepted by students as part of the process of gaining that rare and valuable boon, a college degree. In the postwar years, however, the universities were flooded with millions of young people. Many of them regarded college simply as a part of life's expectations, an extension and natural continuation of their earlier education. Often they had grown up amid liberal parental rules, driving cars at an early age, coming and going much as they wished.

However, prior to 1972 the legal age of maturity was usually twenty-one; universities often required their students to live in residence halls on campus or in similarly supervised fraternities or sororities. Apartments off campus were often few and expensive. The university had the legal right to act *in loco parentis*, "in the place of a parent," but in practice the resulting restrictions were nothing at all like the liberal family rules the students had known. It was not hard to understand why. Universities were dependent largely on donations from alumni and appropriations from state legislatures, and university presidents were loathe to touch so politically explosive a topic as the relaxation of rules designed to protect the virginity of their students.

If you were a coed of twenty (but not a male student), as late as 1967 your dorm life likely was not so much that of a family as of a minimum-security prison. Your dress was regulated; woe to the woman who wore slacks, let alone a miniskirt. If you were found with a man in your room you would likely be expelled from college. This by itself would not have been too severe a restriction, for inventive lovers could find opportunities in friends' off-campus apartments, in parked cars, even in the woods in warm weather. But in addition, your comings and goings were thoroughly supervised and regulated. If you left the dorm at night, even to go to the library, you signed out in a book; on returning, you signed in. You were subject to a curfew, and if late you received demerits. With enough demerits (the number was never large), you might be subjected to a modified form of house arrest. You would have had to stay in your room for an entire weekend, except for brief trips to eat or to visit the bathroom.

Escape from all this was very difficult. At the witching hour the housemother would lock the main door, which your key could not open. If late, you could get in only by ringing a bell, whereby your shame (and demerits) would be obvious to all. If you turned away to stay elsewhere, you would be found out during bed check, when the dormitory supervisors would open each door to certify that everyone was in her room. You might arrange for a night or a weekend off campus, but you would have to declare your destination and receive official permission. If your true destination was your boyfriend's apartment, and you were found out, then without due process or habeas corpus you would receive summary punishment, which often meant expulsion.

Such arrangements of course were insupportable and in the late 1960s were broadly relaxed and later removed entirely. This was widely seen as part of the sexual revolution, and from the perspective of the students involved we can understand their view of matters as a Manichaean conflict between the light of sexual freedom and the darkness of an outmoded Victorianism. But in a larger perspective, what happened was more the sort of political adjustment that has been frequent in the American system: the students demanded and won their rights. From an even broader perspective, it is not the recent upsurge in acceptance of youthful sexuality that is the unusual event, except that it was delayed so long. What was really unusual was Victorian morality itself, a time of prudery preceded and followed by an almost unbroken history of young people's sexuality being openly acknowledged and provided for, even given free rein.

The New Morality then, if such it may be called, will certainly remain popular, since it offers an amorous liveliness rivaling that of Renaissance Italy, without requiring early marriage or widespread illegitimacy. As for the future, we must be cautious in predicting new or unusual social forms; the communes of the 1960s stirred much appreciative comment but proved little more than a passing phase. For a general prediction, it is best to be content with the comment that for young people to want each other is only slightly more unusual than the sun rising in the morning.

Sex among young people, of course, is easy to sensationalize, particularly when it involves teenagers. In 1980, a cover story in *Newsweek* emphasized the social pressure faced by some teenagers, leading them into sex long before they were emotionally ready. But this sort of excess tends to be self-correcting; it is a safe bet that if enough young teenagers have unpleasant or hurtful sexual experiences, they will quickly rediscover the word "No." On the other hand, some of these teenage girls do get pregnant; in fact a lot of them do. Each year some one million teenagers get pregnant. This is a nice round number which attracts attention, and much has been made of this supposed epidemic of teenage pregnancy. The popular imagination pictures near-children of 13 and 14 getting badly burned from biological dynamite, but the birthrate for young girls of ages 10 to 14 has stayed nearly level since 1970, at 1.2 per 1,000. In this age category, the number of births has declined from 10,600 in 1974 to 9,500 in 1979. Overall, of the million teenage pregnancies each year, about half end in abortion. Of the births, more than half are to married teenagers, most of whom are age 18 or 19. The real problem is with births to unmarried girls age 15 to 19. These have increased somewhat, from 210,800 in 1974 to 253,200 in 1979, and the birthrate among these unmarrieds has gone from 22.9 to 25.4 per 1,000 during those same years. That certainly is not a trend in the right direction, though it is hardly an epidemic. Still, this means that in the country as a whole, only about one person per 1,000 is an unmarried teenage girl who will have a baby this year.

What are the most significant changes we can anticipate, as we look forward to the millennium which the year 2000 will usher in? To begin, things will cost a lot more. It is something of a fool's errand to project inflation rates only one year ahead, let alone twenty, but let us be optimistic and imagine inflation is brought under control. This will not mean a return to the halcyon days of the early 1960s, when the Consumer Price Index was going up about 1.2 percent per year (not per month). Still, if the inflation rate could be kept in single digits, even kept below 5 percent, financiers would dance in the streets. Let us imagine, hoping it is not just a fantasy, a 7 percent average inflation rate between 1980 and 2000. That would quadruple prices during those years.

Gasoline would be cheap at $5.00 per gallon. A week's shopping for a family will run to $400 without much trouble. Airlines will announce discount fares of $999 for a one-way flight coast to coast. An automobile of only modest attainments will run to $20,000, more than the cost of a hand-built Lamborghini or Aston Martin not too many years ago. You will be able to go to the bank and take out a mortgage for a cool quarter-million, but it will no longer buy a home on the beach at Malibu; quite the contrary. In many parts of the country it will buy only an ordinary house in a neighborhood that is nothing special. A home selling for a quarter-million may be snapped up as a bargain. The penny postcard may have given way to the dollar postcard, and it will probably cost more to mail out the checks to pay your bills than it would have cost to pay the bills themselves, only a few decades ago.

How will people cope with such prices? As always, they will get used to them, for incomes will keep pace. A century ago, only the president of a railroad could command a $10,000 salary. A generation ago it was a bright goal to which the most able and ambitious might strive. For 1981 the federal poverty line was set at $9,287 for a city family of four, up from $8,414 for 1980; hence today $10,000 represents nearly poverty-level wages. In 2000 the median family income may be topping the fabulous level of $100,000. If people are fortunate, that will represent something like $25,000 in 1980 dollars—a modest but welcome advance over the actual figure for that year, which was $21,020.

Autos will be smaller; that will be one of the most visible changes. Today's small car by then will probably be tomorrow's standard, and in 2000 a movie showing highway traffic in the 1950s will perhaps look like a parade of limousines. Other quite visible changes will put a lot more electronics in people's everyday lives. What with cars and appliances having microprocessors, cable and satellite TV channels, video games and bank tellers, and all manner of video terminals and displays used routinely at work, people will expect as a matter of course to have the information, entertainment, and services they want, electronically.

People will have still more education. Perhaps by 2000 the United States will become the first nation to get a college degree. By the rarefied standards of academia it will not be much of a degree—Associate of Arts, signifying graduation from a two-year community college. But it will mean a median educational level of 14 years or more, compared with 12.9 for young people aged 25–29 in 1980, and graduation from at least a community college will have become the sort of routine expectation with which people now regard a high school diploma. And there will be still more white-collar workers, more professional and technical jobs, more managers and clerical workers; which is to say, more bureaucrats.

Overall, the year 2000 will be amazingly like 1980. People will still live in houses and apartments, not techno-cubes or Buckminster Fuller-type "machines for living." These homes will be remarkably like the ones we are familiar with today (in fifty million or so cases they will *be* the same ones), except that a majority will have central air conditioning. With their high electric bills, people will go easy on using their central air, but on particularly hot days they will appreciate it. In 2000 some people will work at computer consoles in their homes and will telecommute, but at least as many, probably more, will be self-employed professionals and craftsmen who have always worked out of their homes. Most people will commute to work, which at least for more people will be nearer by in the suburbs, and they will drive cars, not use car pools or mass transit. Their cars in turn will burn gasoline, not alcohol or extract of sunbeams. People will still be cheering (or booing) the New York Yankees in the World Series and the Pittsburgh Steelers in the Super Bowl, though office managers will warn employees not to bring their desktop portable TV sets to work during the week of the former. People will say that taxes are too steep, especially Social Security taxes, and prices too high. They will worry about relations with the Soviet Union and with their in-laws, as well as whether their kids are getting a good education. Many but by no means most will be going to church on Sunday, and on November 7, 2000, many, again not most, will be turning out to vote.

And the year 2020? Prices will be higher still; a million-dollar house may be a rarity, and will probably have so low a price only because of being in a run-down neighborhood. Perhaps by then people will think of dollars as the Italians today do of lira, where the exchange rate is something like 1,500 to the dollar. Perhaps there will be a currency reform, the government lopping two zeroes off prices by declaring that the new currency unit is the eagle, worth $100. (Early in this century, "eagle" was the name given to a $10 gold piece.) Certainly educational levels will be even higher, with a bachelor's degree perhaps becoming the standard. Some things will still be predictable, however. Taxes will be too high. August weather will be too hot. Election Day will be November 3. And if the ghost of Stephen Vincent Benét's Daniel Webster should return to ask, "Neighbor, how stands the Union?" he will get his reply, "She stands as she stood, rock bottomed and copper sheathed, one and indivisible."

Though predictable in view of America's vast strengths, this last comment deserves more than a simple assertion. This chapter has dealt principally with averages and median values, but everyone knows this can be no more than a neat device to ignore large deviations from the mean. In particular, America's black people have historically lagged far behind the white majority. In 1968 the Kerner Commission made headlines with its assertion that we are "moving towards two societies separate and unequal." The following year the Urban Coalition reported that "A year later we are closer to being two societies, one white and one black, separate and unequal." It thus is important to rephrase Daniel Webster's question and ask, "Neighbor, how stand the black people within the Union?"

The good news is that in several key respects, black people in the 1970s were much more like white people in the 1970s than like black people in the 1950s. In 1960 the median education level for blacks was 8.0 years of school completed; we would say the average black then was a junior high dropout. This contrasted with a median of 10.9 years for whites, a difference of nearly three years of school. In 1980 this gap had shrunk to six months: 12.0 for the blacks, 12.5 for the whites. Among young people aged 25 to 29, the gap was almost nonexistent: 12.6 years for the blacks, 12.9 for the whites. As recently as 1950, the median for all blacks was only 6.8 years. We would say that in only one generation the typical black person has advanced from being an elementary school dropout to being a high school graduate, while the typical young black, like the typical young white, is going on to begin college.

It should not be surprising, then, that black people have been getting better jobs. In fact, in 1979 an important milestone passed quite unnoticed. Among blacks and other nonwhite minorities, for the first time more were in white-collar jobs than in blue-collar. In 1980 it was 39.2 percent white-collar, 35.8 percent blue. This is not the same milestone that white people passed back in 1968, when they gained not a plurality but an absolute majority of white-collar jobs. Still, for the blacks this is a vast improvement over 1960. In that year only 16.1 percent of the nonwhites wore white collars while 40.1 percent wore blue. The advance has been even more dramatic in the Census Bureau job categories of "Professional and technical" and "Managers and administrators." In 1960 hardly one nonwhite person in fourteen held one of these most desirable jobs. In 1980 it was better than one in six.

With all this, the stereotype of the poor and hopeless black, mired in his poverty, should be gone with the wind. A useful measure of poverty is the federal poverty line, which is a group of income levels, set by family size, and raised each year to take account of inflation. In 1959, 18.1 percent of white people were living below the poverty line; for black people the number was an astonishing 56.2 percent. By 1980 these levels had been cut nearly in half, to 10.2 percent for the whites, 32.5 percent for blacks. But there is a catch. These numbers refer only to people's money incomes. In 1959 that was virtually all people had. For instance, in that year less than three fifths of the nation's workers were covered by unemployment insurance, and the average benefit was only $21 a week. By the mid-70s such traditional benefits as unemployment insurance and Social Security had been greatly expanded, while a plethora of new benefit programs had sprung up. By 1976 the available programs included Social Security, railroad retirement, government pensions, unemployment insurance, workmen's and veterans' compensation, veterans' pensions, Supplemental Security Income, Aid to Families with Dependent Children, food stamps, child nutrition, housing assistance, Medicare, and Medicaid. In that year a study for the Congressional Budget Office showed that when these benefits were included with people's incomes, the poverty rate fell to only 2.9 percent for white people and 11.5 percent for black. Not much poverty any more; certainly very little of the hollow-eyed, thin-limbed, hookworm-and-tuberculosis, tarpaper-shack variety that in some areas of the country used to be so common, even among whites.

That is the good news. The bad news, simply stated, is that black incomes have continued to lag well behind those of whites, with little prospect for a quick catch-up. In 1950, median family income for blacks was $1,869 a year (yes, that's right), 54 percent that for whites. By 1970 the gap had narrowed considerably, to $6,516 vs. $10,236, or 64 percent. However, since then the gap has actually widened. In 1980 the median family incomes were $12,674 and $21,904, for a ratio of only 58 percent.

Is this an indictment of American society, a confirmation that as the black author James Baldwin has written, "All of the civil rights legislation is absolutely meaningless, and it was meant to be meaningless"? Not so fast. To begin, these statistics are for families. But to a disproportionate degree, black families are headed by women: 41.7 percent in 1980, compared with 11.9 percent for the whites. (In 1950 the corresponding figures were 18 percent and 9 percent.) Not only do women as a rule earn less than men, they often can only work part-time. Other matters are also influential. Blacks continue to live disproportionately in the South, where everybody's income tends to be lower. Blacks are younger; their median age in 1980 was 24.9 years, while that of whites was 31.3. This is significant because people's incomes increase as they get older and advance in their careers. In fact, families headed by people in their mid-40s to mid-50s earn income greater than the income of families headed by people 25 and under—by a greater amount than white incomes exceed black incomes. Lastly, it is mostly the younger blacks who have had the new opportunities and who have been able to benefit from them. There still are plenty of older blacks who really were held back in their earlier years and who would find it hard to catch up with their sons and daughters who have been forging ahead.

This means we should expect to find the core of black poverty in families headed by women with no man at home. As we peel away the factors which hold back black incomes, we should find these incomes approaching closer and closer to parity with whites. That is exactly what we find. Among households headed by black women, the median 1980 income was only $7,425. This was below the poverty line, $8,414 that year for a family of four, showing that over half such families required public assistance to stay out of poverty. Here is truly the hard core, and James Baldwin is right in the sense that civil rights laws cannot easily touch these people. But then it is hard to think of any law that can keep husbands and wives together when one of them wants to leave, or prevent an unmarried woman from having a baby. (In 1979, 55 percent of black births were to unmarried mothers, a rate nearly six times higher than among whites.) Or force the men to pay child support when they want to make themselves scarce. Or make a mother work full-time when she really thinks she should work only part-time so as to take better care of the kids. In these poor black families there is considerable human tragedy, but it does not flow from racism or discrimination. Rather, it results from people making bad decisions when they are in a position to understand what they are doing, and then having to live with the consequences. It is one thing to keep whites from hurting blacks. It is more difficult to keep blacks from hurting themselves.

However, among families whose householder, man or woman, was a year-round full-time worker, the median incomes in 1980 were $26,865 for whites and $20,037 for blacks. Black income thus was 74.6 percent that of white—a big improvement from 58 percent.

Among all husband-wife families, regardless of age or location, the median incomes were $23,501 for whites, $18,593 for blacks—a ratio of 79.1 percent.

When husband and wife both worked, the ratio was 83.7 percent. Where the householder was under 35 years old, the ratio was 85.3 percent.

In husband-wife families of all ages outside the South, both spouses working, the mean income for white families was $28,673 and for black, $25,722. The ratio was 89.7 percent. Among such families, including the

South but with the householder under 35, the medians were $21,448 and $19,588. The ratio was 91.3 percent.

Now, here's the real zinger. As early as 1971, among such married couples outside the South with both spouses working, if he was under 35 then the black couples earned 105 percent of the white couples' income— a median of 5 percent *more*.

The American Dream lives. This does not mean there will nevermore be discrimination or unfair treatment for black people. It means that to the degree bias exists, it will not take the blatant and overt form of forcing a Satchel Paige to spend a brilliant career as a baseball pitcher in the minor leagues or preventing a black college professor from voting. There will always be discrimination. But increasingly it will be the common, garden-variety form that everyone experiences, in the sense that we all know what it is to hear the word "No" when we feel entitled to hear "Yes." That, however, is no matter for the civil rights laws. If everyone went through life hearing only "Yes," we all could retire at age forty and spend the rest of our days surf-casting in the Bahamas.

The American Dream lives. Particularly among families that avoid having children without benefit of a husband present, the present generation of blacks may well be the last to show significant differences in earnings and job categories from those of white people. Already the education gap has all but vanished, and as for income, there is another significant straw in the wind. In the United States are not only native blacks, but also immigrant blacks from the West Indies. In their physical appearance and usually in their accent, they are indistinguishable from those born in the United States. But while blacks in general have an income only 62 percent that of whites, West Indian blacks have an income ratio of 94 percent. Perhaps within another generation, then, there will be larger statistical differences between subgroups of whites, such as between those of Jewish and Polish descent, than between intact black and white families overall.

The American Dream lives. But still there are the rotten rat-infested slums, where unemployable black people nod in an alcoholic stupor or wait to buy heroin; where gangs of teenage dropouts stare vacantly into a future that has no use for them; where felonies are an everyday matter, and graffiti proclaims the general despair. What can we say about that?

We can recall an etching by William Hogarth, "Gin Lane," in 1751. With only modest exaggeration, he portrayed the filth and destitution of those consumed by drink. On the steps in the foreground sits a dirty and slatternly woman, oblivious to the world; as she reaches to sniff some narcotic, her baby falls head first to the pavement below. A few steps below her, an emaciated drunk with a gin glass lolls in his last extremity. Behind her a gimlet-eyed pawnbroker scowls at the meager coats and kitchen pots pawned by a threadbare couple on their way to the gin mill below, which sports a sign, "Drunk for a Penny, Dead drunk for two pence." In the background, a tumbledown house wall discloses a man hanged from a roof rafter. On the street outside, a baby wails piteously as a man and woman lower the half-naked corpse of the mother into a waiting coffin.

Significantly, all the people in the scene are white.

12

BEYOND THE PRESENT ERA

There is a kind of utopian thinking of which many futurists are all too fond. It consists of asserting that we are on the verge of, perhaps indeed already experiencing, a sudden and dramatic change that will usher in a new era. Such sudden changes are far from unknown. There really have been years in which the world on Christmas Eve faced very different prospects and had undergone very significant changes since the previous New Year's Day. However, these have not been the sort of years most of us would care to live through; they have included 1945, 1917, 1914, 1789 . . . Such years of upheaval have generally been rather unpleasant, and our reaction to any prediction of a new addition to the list can only be, "I hope not."

Fortunately, that is not the only way to begin a new era. The changes of the nineteenth century offer another way, that of slow and gradual change over a term of decades. Following Waterloo in 1815, European history was free to unfold at its own pace during the subsequent century. In 1848–49 there were widespread upheavals all over that continent, as people rose in revolt against their autocratic rulers, demanding liberty and the establishment of republics. Those years do not stand out as a later 1789, for the very good reason that the upheavals were crushed or co-opted and proved to have little immediate or direct effect. No new constitutions were promulgated, at least none which lasted long; no king lost his head. The real changes in Europe during that century, in politics, in industry and technology, in the shifting powers of states, the growth of empires, and the lives of the people, were generally much more gradual. Yet they were as real as any the revolutionaries might have sought. Without pinpointing the change as focused in any one year, certainly 1890 represented a very different era than 1820.

This is the sort of change we can anticipate, an unfolding of events at their own pace, a continuation and extension of familiar themes. There will be plenty of dramatic headlines along the way, that much is sure. There will be a full measure of unpleasant surprises, like the OPEC oil-price hikes of 1973 and 1979, or the Soviet invasion of Afghanistan. But we should not anticipate anything so dramatic as a global war, or a revolution of such consequence as to make some future year a new 1917. Slowly, imperceptibly, at a pace scarcely discernible amid the headlines of the day, the world nevertheless will ease into a new era. Again without being able to pinpoint the change as centered in any one year, the world of 2050 indeed will represent a very different era than did 1980.

One of the most welcome changes would be a waning of the revolutionary spirit. The word "revolution" has been applied to everything from a new hair spray to the rise of an important new industry like microelectronics, to the advent of a new scientific theory, to a riot in the streets, to a change of government by coup d'etat, and on to a major and thoroughgoing overturning of a society and its institutions. Advocacy of revolution can be thought of as a sort of secular religion, not far removed from belief in the Messiah or the Second Coming. All these belief systems share the fond hope that by some sudden and dramatic epiphany, this turbulent and troublesome world will suddenly be put to right. It is too much to hope that even by 2050, people will have grown up enough to reject such notions. But we may hope that by then the revolutionary spirit will have largely worn itself out.

The great enthusiasms have had a way of doing just that. Between Luther's Ninety-Five Theses in 1517 and the Peace of Westphalia in 1648, Europe was rent by religious upheaval and conflict, the bloody quarrels of Catholic and Protestant being quite as consuming as those between any of our modern political factions. These conflicts culminated in the Thirty Years War, the deadliest and most destructive war Europe was to know until our own century. By the time it ended in 1648, Europe was physically and emotionally spent, and at last was amenable to a live-and-let-live attitude toward different nations' religions. A century of political upheavals and conflicts, in our own time, may leave the participant nations and societies similarly spent, similarly inclined to live and let live. In this connection it is worth noting that the revolution disease is not particularly contagious. Nations that have experienced major revolutions, such as France, Russia, China, Iran, have been widely hailed or feared in their time, as examples which would be copied. In the wake of its revolution the Soviet Union indeed was able to export its system to Eastern Europe and other neighboring regions, but that was less a matter of domestic emulation than of conquest, an age-old game of the powerful. The demands of African and Asian nations for independence from colonial status furnished other opportunities for revolutionaries, in Indochina and elsewhere, while countries like Cuba and Libya have been notorious for their open sponsorship of insurrection and violence. But unless conquered or subverted by main force, nations bordering a revolutionary state have surprisingly often been able to avoid infection with the revolution disease. Nor have revolutions recurred in nations that have experienced them; it is as though the experience has immunized them. Seen up close, instant utopia has rarely proved particularly appealing.

We can say there is little to choose between such revolutionary parties as the Jacobins, the Bolsheviks, the Hezbollah (Iran's "party of God"), but there is more to revolution than that. The great revolutionary totalitarianisms of our time have included Communism, Maoism, Nazism, Fascism. We think of them mostly for their cruelties, but in fact every one of them was in its time an exercise in applied futurism. All drew their strength from the totalitarian mystique, which traces its roots to Rousseau's assertion that society is the work of men and hence can be shaped to fit our desires. All were based on this mystique, with its belief that technology and methods of social manipulation had given men the means to shape the future and to give it a character that would be transcending, exciting, heroic. All denied the right of ordinary people to seek the common human happiness that is there to be found. To the contrary, their leaders emphasized that such everyday concerns as people's workaday lives and wishes were somehow ignoble. No, the nation was to mobilize for the sake of a transcendent vision. And for those possessed of the vision, what was to be done but kill the unbelievers who stood in the way of its realization? The resulting piles of corpses have repeatedly disfigured our world.

When the revolutionary dictators reached power, if they lasted for any length of time they often found themselves in a war, as their transcendent visions rubbed up against the wishes of other peoples simply to be left alone. Or else, if not destroyed by war, they found their techniques not so powerful, their citizenry not so malleable, the nature of man not so subject to being molded, and their tools for molding society not quite so sharp as they had wished. Time is a great leveler, and it has repeatedly worn away at the dreams of those who spoke of a new Soviet man or of some other kind of new human being. After all, the vast majority of people do not want to be mobilized for the sake of somebody else's dream. They have their own dreams, which include preeminently the hope of being left to live their lives in peace. A change of government may improve their circumstances or their situations. It cannot so easily improve them. The pseudosciences such as Marxism-Leninism that have buttressed our century's revolutions have sprung from the peculiar conceit that people are manipulable, that by appropriate laws or practices, educational systems, or Gulag Archipelagoes, they can be molded into a heroic form. They can't be and won't.

With the waning of revolution fervor, a waning spurred by the harsh experience of failed hopes, there may come a new set of commonly accepted realizations. That prescribing revolution as a cure for society's problems is on a par with prescribing heroin for the problems of everyday life. That civilization is fragile. That what we have is easy to destroy and hard to build back up. That a modest but real change which helps people is much to be preferred over a sweeping condemnation of the times. That no good can come of mass movements bearing a hammer and sickle, a swastika, or some other hot symbol. That when revolutionaries call for the destruction of everything old, what they really mean is the destruction of everything.

Revolutionaries and other futurists often invoke the vision of a struggle for the heights, proclaiming we must either reach for the stars or sink into the dust. The preeminent expression of this attitude has been Nazi Germany. They were to rule Eurasia, build a Thousand-Year Reich, establish themselves as the master race—that, or perish and be invaded by what Hitler in his last testament called "the stronger Eastern nation." But we do not have to reach for the stars, sink into the dust, or present the two as alternatives. There is another way. To live. To seek what happiness can be found. To try to make life better for people, knowing our successes are apt to be few and hard-won, and rarely will there be any final victories. To appreciate that a few people some of the time will be heroes, but most of the people most of the time will simply be concerned with their private lives and hopes, and to realize that in this there is nothing at all worthy of scorn. Our century has been blighted repeatedly by the excesses and enthusiasms of applied futurism, and one can only hope that the real future will offer a different set of prospects. If we cannot achieve President Woodrow Wilson's goal and make the world safe for democracy, perhaps it can be made safe for simple decency.

One of the outstanding literary expressions of this conflict, between ordinary decency and revolutionary militancy, is Boris Pasternak's Nobel Prize-winning novel *Doctor Zhivago*. Most Americans know it best through director David Lean's movie. It is the story of two lovers, Lara Antipova and the poet Yuri Zhivago, caught up in the war and revolution that engulfed their Russia. A brief scene in the movie version summarizes this central conflict. Yuri, having unexpectedly met Lara at a library, is

walking with her toward her apartment past a gray-brick wall. On the wall
is painted a revolutionary slogan, and those who can read Russian would
recognize it as SVOBODA, "freedom." They will not have the freedom
to be together long, Lara and Yuri. Soon the architects of that slogan will
press-gang Yuri into military service with the Red Guards and will allow
the corrupt and scheming Komarovsky to take Lara away. But in the next
century, amid a waning of revolutions, perhaps the world's Laras and
Yuris will be able to find the modest happiness that can be theirs, if only
they can keep from being interrupted.

The decline of ideology, the waning of revolutionary fervor, will have
important consequences for the Soviet Union. Increasingly it will be
evident—it is already much more true than in the 1950s—that the Soviet
challenge does not stem from its possession of a galvanizing, widely appealing plan for the future, held by successors to Lenin. It stems from the
status of that country as a major and powerful nation, ruled by successors
to the Czar. It has always been possible and often easy for powerful states
to meddle in the affairs of weak ones, and those who have acquiesced to
the meddling have often been motivated by no vision grander or more
hopeful than the desire to be on the winning side. No one has ever accused the Germany of Kaiser Wilhelm, early in this century, of standing
in the vanguard of the world's progressive political forces or of holding
high a beacon of hope for the oppressed people of the world. Yet because
they appeared to be winning, the Kaiser and his government were able to
negotiate an alliance with the Turkish Empire giving Germany effective
control over the Dardanelles (a goal that has continually eluded Moscow). They also furnished the bankrolls and ran the guns to various private armies struggling for control of Mexico, including that of Pancho
Villa, and even offered Mexico an alliance for war against the United
States. The activities of the Soviet government have been somewhat similar.

The Soviet state has shown considerable skill both at rolling with the
punches and at delivering quite a few of her own; predictions of change
must be made with care. One prediction has gained wide currency in such
works as Andrei Amalrik's *Will The Soviet Union Survive Until 1984?*
and General Sir John Hackett's *The Third World War: August 1985.*
This is that the Soviet Union will break up, dissolving into its constituent
national states. Such predictions follow from the fact that the Soviet
Union is indeed an empire. More than a hundred nations and nationalities lie within its borders, their people speaking more than a hundred
languages, and with all that distinguishes them as peoples: history, race,
traditions, religions. In addition to such well-known groups as the Ukrain-
ians, Belorussians, Armenians, and Kazakhs there are little-known peoples
such as the Birobidzhan Jews, Komis, Buriats, and Evenki; these last three

hold lands in Asia roughly the size of France. This century has not been kind to such empires, and it is tempting to suggest that the Soviet one will go the way of the others.

The problem with such a prediction is that it has already happened. The Soviet empire may be the only one left over from the last century, but it is also the only one to have broken up and subsequently reformed. Following the 1917 revolution and the German victory over Russia, its empire collapsed. In Poland and the Ukraine, in western Russia and the Caucasus, in Finland and along the Baltic coast as well as in central Asia and the Far East, peoples who had been subject to the Czar broke away and began to establish their own nations and institutions. For some time Lenin did not challenge their actions, convinced that his revolution would soon ignite others in Europe. But the European tinder proved damp. At the same time, however, it was becoming increasingly apparent that many of these formerly czarist lands lacked the strength or the outside support that would maintain them as viable nations. Thus, in 1920 and 1921, Lenin's government, the Russian Soviet Federated Socialist Republic (RSFSR), negotiated bilateral treaties with these neighboring governments, under which they were to maintain their independence but live within a Russian sphere of influence. Lenin then proceeded to reduce their independence, hastening the process by sending the Red Army to invade the lands of recalcitrants. Aided by his Commissar of Nationalities, who happened to be Stalin, Lenin then was able to prepare a plan for the union of most of the states of the former czarist empire, under the name "Union of Soviet Socialist Republics," to be dominated by the RSFSR. With the Red Army readily available as an enforcer, the revival of the Russian empire proceeded to completion.

It is not easy to see how any future breakup of the Soviet Union would end otherwise. Mexico's president once said "Poor Mexico, so far from God and so near the United States." Similarly, any breakaway Soviet state would find itself too close to the Russians and too far from outside help. That is why Hungary, which revolted in 1956, could be crushed when the United States was at the height of its power. That is why the rebels in Afghanistan have found it hard to make headway. Still in Eastern Europe, at least, there may be change. For centuries these lands have been prey to the rival ambitions of Prussia, Russia, Austria, Germany; it was to provide for their development under German guidance that the Kaiser forced upon Lenin the Peace of Brest Litovsk. It thus is little wonder that with the Soviets more powerful than any European state, or even any possible coalition of Europeans, they would impose their system throughout the region. However, the aborted rise of Poland's Solidarity labor union may yet stand as a milestone in converting the states of Eastern Europe from Soviet satellites into nations resembling Finland.

Finland, formerly ruled by the czars, stands today as the only former czarist province to have avoided reconquest. She has her own domestic institutions, including a parliament and a liberal constitution; her people share in the prosperity of Scandinavia and participate in free elections. But all political parties agree on the need to maintain good relations with the Soviets and to avoid any action that would challenge their interests. Finland's diplomatic neutrality is much more Soviet-oriented than that of neighboring Sweden, which has a well-developed domestic arms industry and a good Air Force. Indeed, should anyone escape over the Soviet border to Finland, he still will not be safe. Finnish police will arrest him, if they find him, and deport him back.

Still for both Finland and Scandinavia, this state of affairs is a vast improvement over one in which Finland would be ruled from Moscow. A revival of Solidarity in time may offer a similar prospect for Poland. Poland may enjoy a far greater degree of genuine independence in the future, its people free to acknowledge their Catholic religion, its workers seeing their interests genuinely represented and spoken for, its press and television granting access to the voices of Solidarity and of the Church. But no party will challenge the Soviets directly. Provided Poland maintains at least a nominal adherence to communism and a membership in the Warsaw Pact, and provided she can receive trade and credit from the West, like Finland she may succeed to a considerable degree in maintaining her own institutions.

Should Poland prove an example for Eastern Europe, or even for the Soviet Union itself, that would not mean totalitarian communist states would be evolving into democracies. But it could mean they would be developing into authoritarian states. The distinction is important. Totalitarians will crush the rise of any independent labor union. Authoritarians will permit it to exist, provided it does not challenge the rule of the dominant clique. Totalitarians will work vigorously to destroy religion in all its forms. Authoritarians will often encourage religion, provided it does not serve as a cover for political opposition. Totalitarians ruthlessly suppress all that is not in keeping with the party line. By contrast, authoritarians carefully censor the press and media, but often are indifferent to material that does not directly threaten their security.

A totalitarian government in the Soviet Union, or in Poland or other states, will proclaim itself the heir of Lenin and Marx, entitled to control every aspect of people's lives for the ostensible purpose of building the workers' paradise. An authoritarian leader, by contrast, may pay lip service to Lenin but will recognize that he owns no true legitimacy, being where he is merely by being strong enough to get there and stay there. An authoritarian regime can and will recognize important independent centers of power, in a labor union, a religion, or among national minorities.

Rather than crushing them it will work to maintain their allegiance or at least their acquiescence. Such a regime will also encourage higher living standards, as a way of building popular support. That is why an authoritarian Poland or Soviet Union could represent an important advance in human liberty and dignity over its totalitarian counterpart. Looking even further ahead, it is noteworthy that while no totalitarian state has evolved into a parliamentary democracy, some authoritarian states have done so, including Spain, Portugal, and Greece.

The most dramatic changes, however, may be not in Poland but in Cuba. Fidel Castro is an anomaly in Latin America. For other countries, even powerful ones like Brazil, the name of the president is rarely much known in the outside world, but everyone knows the name of Castro. Those who have met him have described him as exceptionally strong in leadership and personal magnetism, a virtual Napoleon ruling no more than a small Elba. As the ruler of a minor island nation who has claimed with considerable seriousness to be a world leader, Castro would certainly appear to be that. But he is not the 25-year-old commander in baggy green fatigues who launched his revolution on July 26, 1953 with an attack on the Moncada Barracks. His beard, cigar, and fatigues are still there, as is the mystique of his guerrilla war in the mountains of the Sierra Maestra, but now he is in his mid-50s. In another quarter-century he will be gone.

His successor will probably be some Moscow-trained army commander, but that man can hardly have Castro's magnetism and personal charisma, his mantle of the Sierra Maestra. Nor may he share Castro's overriding passion for revolution, the more so since a gray and aging Castro in his declining years will be a most improbable revolutionary. Instead, this new leader may be open to approaches from the United States, as we offer such blandishments as trade, economic assistance, even a renegotiation of the status of Guantanamo Bay, the only U.S. military base in a communist country. Little but Castro's own relentless pro-Sovietism has prevented Cuba from regaining her traditional ties with the United States. As Francoism in Spain died with Franco, as Maoism did not long outlive Chairman Mao, so may the death of Castro return Cuba to the status of merely another army-ruled Latin American dictatorship.

In the world as a whole, population trends will be among the most influential forces of change. Today the advanced nations are approaching what will be their peak populations. Their birthrates are low, and their societies are full of people in the productive middle years. The nations of Asia, Africa, and Latin America, by contrast, are largely nations of children and teenagers. Their populations have been growing so rapidly that half or more of their people are under age eighteen. In the middle of the next century, however, this situation may be almost reversed. Falling birthrates will mean that in today's developing nations, the populations will be growing only slowly. No longer will population growth come close to outstripping gains in economic growth; decades of more moderate birthrates will have given these nations' economies large and real gains. Within those countries will be relatively far fewer children. Also there will still be relatively few of the elderly; the majority, as in Europe and America today, will be in their working years and at their most productive. But what of the United States and other lands of slow population growth today? With birthrates continuing to fall, their populations will peak out, begin to decline, and slowly age. By 2050 these countries will face the problems of an enormous increase in the numbers of the elderly and retired. When the United States has to cope with a flood of retirees, the problems may be surprisingly similar to those of today's India with its flood of children. But in 2050 India may be one of many Asian, Latin, even African nations with the right population mix and may be among the countries advancing most rapidly.

A variety of scenes in present-day China point to this future. On the streets of Peking and other cities are tens of thousands of bright-colored motorbikes, in green, yellow, blue, or fire-engine red. They zip along through the crowds, dodging the traffic, giving their owners the characteristically Western experience of being able to hop on a machine and in a moment be off to wherever they like. Motorbikes such as the domestically built Jialing cost a year's pay for a typical worker but are in hot demand. They even are advertised on Chinese TV:

> Jialing, Jialing, Jialing gets you home twice as fast,
> Jialing, Jialing, everyone wants to ride on Jialing.

They are only one of the products of a government policy to emphasize consumer goods. Much of China's defense industry has been shifted to producing such items, as a way to keeping its production capacity busy in a time of peace. Chinese aerospace plants are building bicycles and refrigerators; tank arsenals are turning out sewing machines, washing machines, and stoves; electronics plants are switching from military radios to civilian TV sets, tape recorders, pocket calculators, and digital watches. Official reports list over 5,000 items coming from these industries, including motorbikes, detergent, cameras, hunting rifles, fishing nets, record players, clocks, fans, air conditioners, electric motors, clothing, shoes, furniture, fertilizer, and construction materials. As for automobiles, China has been building some 130,000 a year.

Another factory manufactures computers. As early as 1971 its people built one based on integrated circuits on silicon chips and assembled with locally produced components. By Western standards this factory will hardly be competing with Hewlett-Packard. It has built only about ten such systems per year, and a U.S. computer costing a hundredth as much can do the same calculations at the same speed. Before it started building computers, however, that factory produced—doorknobs and door handles.

In reporting a 1979 tour of China, William D. Carey, publisher of *Science*, told of visiting a middle school, with classes up to tenth grade:

> We stayed for English and algebra sessions, and the performance was superb. Despite a dreadful street racket of honking trucks and traffic, to say nothing of 20 staring Americans, neither teachers nor students missed a beat. Later, we met with the faculty and two students, a boy and girl. When we asked the girl what her ambition was (she was a science major), the answer came back, "to fulfill the four modernizations" (industry, agriculture, defense, and science and technology). Asked what they knew about America, for a moment they were stumped. "Very little," said the girl. "Nothing," said the boy. But after some fast thinking, the girl said, "Americans are friends of China, and they are very clever."

Next door to China is Japan, rich in technology, advanced industry, and investment capital. The Chinese may well prove their apt pupils. Two centuries ago the Industrial Revolution began in England and spread from there to the rest of Europe. Japan may play a similar role for China. Until well into this century Europe was the great center of human advance, the place where the most important ideas and techniques were born. Following World War II this center crossed the Atlantic to the United States. By the middle of the next century it may again cross an ocean, this time the Pacific. Quite possibly by then China and Japan will stand as the leading center of human dynamism, the place where the most influential new ideas and approaches to improved living standards are coming forth. These may be the nations standing as examples, the models for a world that aspires to prosperity.

The prospects for prosperity, of course, will depend heavily on the world energy supply. The best picture we have today of the world's long-range energy development comes from studies at the International Institute for Applied Systems Analysis (IIASA), near Vienna. Their work has not looked so far ahead as 2050, but they have prepared scenarios detailing projections through 2030. Depending on the worldwide rates of economic growth, their "high" scenario projects energy use increasing 4.4-fold over 1975, their "low" scenario projecting only a 2.7-fold increase. Their corresponding rates of economic growth are 3.4 percent per year in the high scenario, 2.4 percent in the low. Both scenarios call for world population to double by 2030, to eight billion people. Even so, however, and even in the low scenario, there is considerable worldwide improvement in living standards.

The IIASA projections measure energy in terawatt-years. A terawatt (trillion watts) is a billion kilowatts; hence a terawatt-year is 8.76×10^{12}

kilowatt-hours. This is a convenient measuring unit since today world energy use is about 10 terawatt-years (8.21 in 1975), and IIASA projects a rise to 22.4 in the low scenario, 35.7 in the high. More particularly, they have divided the world into seven regions, and they project the following regional changes:

PROJECTED ENERGY USE BY REGION (terawatt-years)

REGION	BASE YEAR 1975	HIGH SCENARIO 2000	HIGH SCENARIO 2030	LOW SCENARIO 2000	LOW SCENARIO 2030
North America	2.65	3.89	6.02	3.31	4.37
USSR, Eastern Europe	1.84	3.69	7.33	3.31	5.00
Western Europe, Japan, South Africa, Australia	2.26	4.29	7.14	3.39	4.54
Latin America	0.34	1.34	3.68	0.97	2.31
Central Africa, South Asia, Southeast Asia	0.33	1.43	4.65	1.07	2.66
Middle East, North Africa	0.13	0.77	2.38	0.56	1.23
China, Indochina, Korea	0.46	1.44	4.45	0.98	2.29
World total*	8.21	16.84	35.65	13.59	22.39

* Includes 0.21 terawatt-year from bunker-fuel oil

Since energy use is closely associated with a high standard of living, these projections point to some dramatic improvements. For instance, in the high scenario, energy use in Latin America, Africa, India, and China stand to increase tenfold by 2030. Their populations will somewhat more than double, leaving a fourfold real gain. This means that these regions will see an enormous growth in industry and hence in the efficiency of their agriculture; their people will be eating better.

Where will the energy come from? The IIASA work has also addressed that:

PROJECTED ENERGY SUPPLIES BY SOURCE (terawatt-years)

PRIMARY SOURCE	BASE YEAR 1975	HIGH SCENARIO 2000	HIGH SCENARIO 2030	LOW SCENARIO 2000	LOW SCENARIO 2030
Oil	3.62	5.89	6.83	4.75	5.02
Gas	1.51	3.11	5.97	2.53	3.47

PRIMARY SOURCE	BASE YEAR 1975	HIGH SCENARIO 2000	2030	LOW SCENARIO 2000	2030
Coal	2.26	4.94	11.98	3.92	6.45
Light-water nuclear reactor	0.12	1.70	3.21	1.27	1.89
Fast-breeder reactor	–0–	0.04	4.88	0.02	3.28
Hydroelectricity	0.50	0.83	1.46	0.83	1.46
Solar	–0–	0.10	0.49	0.09	0.30
Other†	0.21	0.22	0.81	0.17	0.52
Total	8.21	16.84	35.65	13.59	22.39

† Includes bunker fuel oil, wood, geothermal, methane from compost

The surprising figures are the modest role for solar energy (mostly rooftop collectors), the vastly expanded need for nuclear power, and the increasing production of oil and gas. By 2030, in many world regions oil from conventional wells will be nearly exhausted. The main sources will include tar sands, oil shales, heavy crudes, and enhanced recovery techniques to get more oil from existing fields. These techniques include injecting carbon dioxide to make the oil thinner and easier to pump, injecting detergent to wash the oil from its confining rock, and mining the oil-bearing formation. In addition a considerable synfuels industry will be producing liquids from coal. On top of this, OPEC will still be producing, and its reserves will still allow the oil to flow at some 33 million barrels per day.

In looking ahead to the coming energy transitions, there are historical data on the rise and fall of various energy sources. These data show quite consistently that in the United States it has always taken about sixty years for a new energy source to take over 50 percent of the market. For the world as a whole the figure is a hundred years. That is why, even by 2050, we cannot expect some miraculous cure to have put the world's energy problems permanently to rest. Even then we still will be muddling through, using a combination of all available sources. But the world will face two large-scale energy transitions between now and then.

The first is a shift from conventional and cheap fossil fuels to unconventional and expensive ones. Today there are limited supplies but no genuine shortages of any fuel, not even conventionally pumped petroleum, other than temporary ones resulting from OPEC's political manipulations. Tomorrow there will still be plenty of fossil fuels, but at a price. This shift is already under way, with the growing use of coal, the start of the tar-sands and oil-shade industries, the incipient beginning of synfuels.

None of these sources are so cheap or convenient as oil from wells, but at least they are available. By 2030 they will dominate the fossil-fuel picture.

The second and longer-range transition is from fossil to nonfossil and particularly to renewable sources, including solar, hydroelectricity, the breeder, and controlled fusion. The next half-century is far too brief for this transition, though by 2030 perhaps one third of the world's energy will come from such sources. But after 2030 there will be real limits and constraints on all fossil fuels, similar to those existing today for domestic U.S. petroleum, and the nonfossil sources will have to come to the fore. Among these will be fusion. Today it is a laboratory experiment, an exercise whereby physicists struggle to harness the thermonuclear processes that produce the energy of the sun. By 2000 a few demonstration fusion reactors may be operating. By 2030 a true fusion industry will exist, just as today there is a nuclear industry; by 2050 fusion will be coming into its own. As George Keyworth, President Reagan's science advisor, stated in 1981, "There is no doubt in my mind that controlled fusion will work and will be the ultimate power source in the future . . . I believe that some time in the twenty-first century fusion will replace most of the commercial sources of energy."

In the United States, water supplies may then be rivaling energy supplies as a matter of national concern, particularly in the West. The growing population of the western states, the demands of their agriculture, the depletion of supplies of groundwater stored in their aquifers, and the needs of the synfuels and oil-shale industries all will put massive pressure on the available water sources. Much of the West relies on water from rivers like the Colorado and Rio Grande, which are among the nation's longest but which do not even qualify for the U.S. Geological Survey list of thirty-three rivers with the highest discharge. The Colorado, for one, is a mainstay of the entire Southwest, including much of California. Its waters are conserved and reused so extensively that barely 10 percent of its flow is allocated to Mexico. Yet its annual flow is only 13.5 million acre-feet, about the same as the Hudson River in New York. As James E. Smith of the Environmental Protection Agency has observed, "The western states never accepted the wisdom of God when he didn't put all the water there."

The traditional Western response has been to say that water flows uphill to money. Since early in this century its people have interlaced much of the West with dams, reservoirs, aqueducts, and irrigation canals. Such dramatic projects as Hoover Dam, Glen Canyon Dam, and Grand Coulee are only some of the most visible instances of their determination. In addition to conserving and managing existing flows, they have also rerouted water from where it is in surplus to where it is needed. As early as 1913, William Mulholland and other engineers built an aqueduct to supply semiarid Los Angeles, tapping the Owens River, 230 miles away, north of the city and on the eastern side of the Sierra Nevada's highest peaks. In the 1960s came the California Water Project, featuring an aqueduct down the length of the state's central valley. It featured pipes and pumping stations to lift its water over the Tehachapi Mountains north of Los Angeles.

These may be only a foretaste of what the Southwest and Rocky Mountain West will build, for just as southern California has long cast covetous eyes on the well-watered lands of the state's north and center, so the Southwest may look to the abundant rivers of the Pacific Northwest. The consulting engineer William G. Dunn has promoted a plan to transfer 2.4 million acre-feet of water a year from Idaho's Snake River to a point below Hoover Dam on the Colorado. Another experienced consultant, Frank Z. Pirkey, has proposed a plan to transfer 15 million acre-feet annually, more than the entire flow of the Colorado, from the Columbia River to Lake Mead, behind Hoover Dam. The water would be pumped 4,900 feet over the mountains to Goose Lake on the Oregon-California state line and then to Shasta Lake behind Shasta Dam in northern California, before finally flowing to Lake Mead. Such projects pale, however, next to the grandest and most sweeping of them all: the North American Water and Power Alliance.

This audacious project was proposed in 1964 by the Ralph M. Parsons Company, a leading contractor on Hoover Dam. It would reach for water, not in northern California or the Pacific Northwest, but in Alaska and western Canada. It would begin by collecting water at the headwaters of the Yukon River. To this would be added water from Alaska's nearby Tanana River, as well as from rivers extending south to the Peace River in western Alberta. All this water would flow into the Rocky Mountain Trench, a 500-mile-long reservoir extending along the backbone of British Columbia, made by damming the upper reaches of the Kootenay, Fraser, and Columbia rivers. Still more water would enter the system from the Snake, Clark, and Salmon rivers in the U.S. Northwest, to bring the total flow to 160 million acre-feet a year, or nearly a dozen Colorados.

Along the way, hydroelectric plants would generate all the power needed for pumping and in addition provide a surplus of at least 70,000 megawatts, one fourth of the present average U.S. electric production. A far-reaching system of canals and aqueducts then would channel the water into the thirsty Southwest along the valleys of the Rockies to the Colorado River to serve Utah, Arizona, Nevada, and California. Then eastward through New Mexico to the Pecos and Rio Grande rivers bringing water to western Texas and northern Mexico. A branch would carry water through eastern Colorado to supply that state as well as Kansas, Nebraska, and Oklahoma. There would even be surplus water to send east. A diversion from the Rocky Mountain Trench would send water across Canada's Great Plains, as far eastward as the Great Lakes and the upper Mississippi. The project would take thirty years to build. The cost was estimated in the original 1964 report as $200 billion; it would be much higher now.

Such a concept calls to mind the Mars envisioned by those astronomers, a century ago, who thought they could see canals on its surface. They envisioned a dry planet inhabited by superengineers, who had interlaced its surface with a network of channels to carry water from the polar cap to the temperate regions. Perhaps in another century parts of Earth will be transformed in this manner; to a traveler in space, the Rocky Mountain Trench might appear as one of the canals of Earth. Many observers would dismiss such a scheme today as preposterous, but by 2050, this may be the scale of the works we will be prepared to undertake. If Louisiana, southern Florida, Holland, and other lowlands are to be saved from inundation, preserved from the rising sea levels as the West Antarctic Ice Sheet breaks up, they will need massive seawalls and dikes, and those works too may be of this magnitude.

Certainly there will continue to be vast opportunity for human ingenuity. Today there are mature technical industries such as automobiles and aviation, which change relatively little from year to year. There are rapidly growing and advancing ones, like electronics, computers, and communications, whose importance rivals the older industries. There are promising newcomers such as recombinant DNA, robots, synfuels and oil shale. Other industries today are gestating, not yet born; these would include controlled fusion, seabed mining, fast-breeder reactors, space manufacturing. Still other ventures are physically feasible but today exist only as speculations. These include using deuterium or heavy hydrogen from the sea as a limitless source of clean fusion energy, communicating with extraterrestrial civilizations, curing cancer, and incorporating genes into food crops to allow them to produce their own nitrogen fertilizer or to grow in seawater. From the most mature of our industries and activities to the most speculative, there is a hierarchy or pipeline of coming change, and we would say the pipeline is full. With promising new ideas being fed in at one end, with established industries growing at the other, there is nothing now to lead us to predict a slackening in the pace of change or advance. Moreover, most of the predictable new inventions or opportunities stem from past advances in the physical sciences. We have yet to fully appreciate the implications of a statement by the British scientist J. B. S. Haldane, that in the wake of the 1953 discovery by Watson and

Crick of the structure of the DNA molecule, biology and not physics would be the most important spur to future advance.

Even so, from the perspective not of 2050 but of 2500, the next century may stand as no more than a part of a longer and more sweeping era, but one with a beginning and end. This would be the years 1800–2100, more or less, representing the most rapid and thoroughgoing changes the human race has ever seen and likely will see. These centuries may compare in importance to the much earlier Neolithic Revolution, which took people out of their ancestral caves and hunting grounds to give them agriculture as well as life in towns and villages. The Industrial Revolution, as a term applied broadly to the changes of 1800–2100, would be described as giving people industries, taking them out of their towns and farms to live in cities and suburbs. Yet in important respects our current pace of change cannot continue for centuries to come. We cannot sustain a 2 percent annual worldwide growth in population; in only 350 years such a growth rate would take our numbers from a billion to a trillion. As was true before 1800, after 2100 the world's population will be approximately steady, rising or falling rather little from century to century. Also we cannot sustain our use of fossil fuels and other depletable energy sources. As was true before 1800, after 2100 we will have to rely predominantly on renewable and inexhaustible sources.

What of inventions and technical advances? Today their pipeline is full, but all industries mature and their capacity to lead the pace of change passes to newer ones. For many decades aviation symbolized this century's achievements. We may think of the dramatic advance from twin-engine airliners to the four-engine Constellation and DC-7 which opened up transcontinental and transoceanic flights, to the first generation of jets like the Boeing 707, and on to the wide-body jetliners. But the current generation of new jets offers no improvement or change as striking as the progression from the 707 and DC-8 to the 747 and DC-10. The Boeing 767 and other new jets are quieter and more fuel-efficient, but they look like any other wide-bodies and offer the same services.

Similarly, the chemical industry today counts some seven million known compounds and substances, but its innovations tend to fall into well-understood and familiar patterns; it no longer stands among the principal catalysts for change. The day may come when this will be true much more widely. Perhaps all of electronics and computers will resemble the television and telephone industries as we have known them until very recently, which for decades offered a standard service based on a standard set of equipment. Perhaps all the industries which will grow out of recombinant DNA will come to resemble today's activities in pharmaceuticals, which have not been coming up with much that is really new. Perhaps when fusion power plants are built, people driving past will give them no more thought than people today give to the big coal-fired plants with their tall smokestacks.

Should these things happen, we still would not have learned everything, made all the inventions. But, we would have at least partially exhausted the capacity to come up with new themes or patterns of change. Innovations increasingly would offer little element of surprise but rather would fall into familiar, well-defined patterns. It would be as though the fashion designers were running short of really new ideas, so that the big news in 2050 would be the return of the '30s look—that is, the 2030s. Even so, the pace of change might appear slow only by comparison. In his *New Atlantis* of 1624, Francis Bacon pointed to the compass, the printing press, and gunpowder as evidence that civilization could rise beyond the heights reached in antiquity. All these great inventions were widespread in Europe before 1500. It would not take many inventions of similar import, in the centuries after 2100 or 2500, to leave people believing that their times too would be times of progress and advance. The will to believe that one's times are special has always been strong. The followers of Peter Abelard, at the University of Paris in the early twelfth century, called themselves *moderni;* in 1287 the bishop of Exeter spoke of his century as *moderni tempores,* "modern times." As the historian Will Durant has noted, "our age of coal and oil and sooty slums may some day be accounted medieval by an era of cleaner power and more gracious life."

In this century, harsh experience has made many people unwilling to face the future with hope or in a sanguine mood. To many, it is much more reasonable to look ahead with dread, with premonition of disaster, or alternately with euphoria—anything other than to say that people will still be living and working and muddling through, as they always have. The excesses of this century have given the future something of a bad name. Still, the world got along quite nicely before the twentieth century began and will little miss it when it has ended. Perhaps in time it will even be possible to forgive its great totalitarians, as driven by an exuberant sense of the new powers and possibilities which the times had provided, without having yet gained a sense of their limitations. But the future should not be seen as one of people made slaves by technology or oppressive governments, blasted by famine or nuclear war, or, alternately, living lives of leisure in crystalline spheres.

The future is in America, Europe, and Japan, for these are the models for the world. These countries point to a future where people will not merely survive, they will thrive. Here people enjoy the abundance and prosperity, the opportunities, the sense of security and ease that the rest of the world aspires to share and that it too will gain in time. Here are the liberal, democratic, pluralistic laws and governments that in a post-revolutionary age must surely stand once again as examples to engage and inspire anew the world's seekers after change. And within these favored lands the way the majority of people live today, with their cars, homes, small families, and office jobs, is, with allowance for relatively modest and acceptable change and novelty, the way they will be living in the future.

REFERENCES
AND BIBLIOGRAPHY

CHAPTER 1. SOME ROOTS OF THE FUTURE

Allen, Frederick Lewis. *The Big Change*. New York: Harper & Row, 1969.
Ardrey, Robert. *African Genesis*. New York: Dell, 1967.
Becker, Carl L. *Modern History*. Morristown, N.J.: Silver Burdett, 1958.
Bernhardi, Friedrich von. *Britain as Germany's Vassal*. London: W. Dawson and Sons, 1914.
Boskin, Michael J. "Prisoners of Bad Statistics." *Newsweek*, Jan. 26, 1981, p. 15.
Canfield, Leon H. and Anderson, Howard R. *The Making of Modern America*. New York: Houghton Mifflin, 1964.
Chaloner, W. H. "The Age of Steam," in *Age of Optimism* (Alan Palmer, ed.). New York: Newsweek Books, 1970, pp. 55–61.
Durant, Will and Ariel. *The Age of Louis XVI*. New York: Simon & Schuster, 1963.
———. *The Age of Voltaire*. New York: Simon & Schuster, 1965.
———. *Rousseau and Revolution*. New York: Simon & Schuster, 1967.
Erickson, Charlotte. "The Birth of Big Business," in *Age of Optimism* (Alan Palmer, ed.). New York: Newsweek Books, 1970, pp. 132–37.
Harris, John R. "The Rise of Coal Technology," *Scientific American*, August 1974, pp. 92–97.
Howarth, David A. *The Dreadnoughts*. Alexandria, Va.: Time-Life Books, 1979.
Manchester, William H. *The Arms of Krupp*. New York: Bantam, 1970.

McAlpern, David. "How the Poor Will Be Hurt," *Newsweek*, March 23, 1981, pp. 23–24.

Nef, John U. "An Early Energy Crisis and Its Consequences," *Scientific American*, November 1977, pp. 140–51.

Rousseau, Jean Jacques. *The Social Contract and Discourses*. New York: Dutton (Everyman's Library), 1950.

——————. *Émile*. New York: Dutton (Everyman's Library), 1961.

Tocqueville, Alexis de. *Democracy in America*. New York: New American Library, 1956.

Tuchman, Barbara. *The Proud Tower*. New York: Bantam, 1967.

White, Theodore H. *The Making of the President 1964*. New York: New American Library, 1966.

Page   2.   "Europe was ready." Durants, *Rousseau*, p. 3.
          3.   "The indifference of children." Rousseau, *Émile*, p. 118.
          3.   Rousseau on city and country life: *Social Contract*, pp. 132 and 181.
          3.   "Men are not made." *Émile*, p. 26.
          3.   "Let us lay it down." ibid., pp. 56–58, 251, 254.
          3.   "Never punish your pupil." ibid., pp. 56–58.
          4.   "Let men learn for once." *Social Contract*, p. 139.
          4.   "The first man who." ibid., p. 207.
          4.   Antiquity of territoriality. Ardrey, p. 12.
          5.   Biringuccio's *Pirotechnia*. Quoted in Nef.
          5.   Deforestation of England. Durants, *Voltaire*, p. 49.
          6.   Coal in England, 1600–1750. Durants *loc. cit.*; Nef.
          6.   Reverberatory furnace. Harris.
          6.   Newcomen steam engine. Durants, *Louis XVI*, pp. 516–18.
          7.   Watt steam engine. Durants, *Rousseau*, pp. 671–76.
          7.   Liverpool and Manchester Railroad. Chaloner.
          7.   Railroad mileage statistics. Canfield and Anderson, pp. 234–36, 366.
          8.   Steel, banking, and oil statistics. Manchester, pp. 50–51; Becker, p. 643; Erickson, p. 132.
        8–9.   "They came from the warrens of the poor." Tuchman, p. 73.
          9.   "The red vision of revolution." Quoted in Tuchman, p. 75.
         10.   Statistics on income. Allen, p. 45; Howarth, p. 12.
         10.   Social security in 1903. Tuchman, p. 487.
         10.   Bernstein and Revisionism. Tuchman, pp. 500–3.
         10.   "Fancy living in one of those streets." Tuchman, p. 431.
         11.   "Not to content itself." Becker, p. 570.
         11.   Kaiser Wilhelm quotes. Tuchman, p. 280; Manchester, pp. 241, 244.
       11–12.   "We must strenuously combat." Bernhardi, pp. 105, 111.
         12.   "If these aims are achieved." Manchester, p. 324.
         12.   Battle of the Somme. Manchester, p. 335.
         13.   "A little man with the narrow eyes." Tuchman, p. 509.
         14.   De Tocqueville on Russia and America, p. 142.
         15.   "The Great Society rests upon abundance." White, p. 464.
         15.   Poverty statistics. Boskin; McAlpern.

CHAPTER 2. THE OIL QUESTION

Abelson, Philip H. "Organic Matter in the Earth's Crust," *Annual Review of Earth and Planetary Science*, Vol. 6, 1978. Palo Alto, Calif.: Annual Reviews, Inc., pp. 325–51.

————. "Synthetic Chemicals in South Africa," *Science*, August 17, 1979, p. 649.

————. "Synthetic Chemicals from Coal," ibid., February 2, 1980, p. 479.

"A Rail Town Booms," *Fortune*, August 25, 1980, pp. 49–52.

Atwood, Genevieve. "The Strip-Mining of Western Coal," *Scientific American*, December 1975, pp. 23–29.

Beck, Robert J. "Demand, Imports to Sink Again, But Production to Rise," *Oil and Gas Journal*, January 26, 1982, pp. 127–44.

Canby, Thomas Y. and Blair, Jonathan. "Synfuels: Fill 'er up! With What?" *Energy*, special report, *National Geographic*, February 1981, pp. 74–95.

Canfield, Leon H. and Anderson, Howard R. *The Making of Modern America*. New York: Houghton Mifflin, 1964.

Carter, Luther J. "Synfuels Crash Program Viewed as Risky," *Science*, September 7, 1979, pp. 978–79.

Cochran, Neal P. "Oil and Gas from Coal," *Scientific American*, May 1976, pp. 24–29.

David, Edward E. "Research Opportunities in Fossil Fuels," *Science*, January 6, 1978, p. 9.

Griffith, Edward D. and Clarke, Alan W. "World Coal Production," *Scientific American*, January 1979, pp. 38–47.

Hammond, Allen L., "Coal Research (III): Liquefaction Has Far to Go," ibid., September 3, 1976, pp. 873–75.

————. Metz, William D. and Maugh, Thomas H. *Energy and the Future*. Washington, D.C.: American Association for the Advancement of Science, 1973.

Hayes, Earl T. "Energy Resources Available to the United States, 1985 to 2000," *Science*, January 19, 1979, pp. 233–39.

"How the Price Is Pumped Up." *Time*, May 7, 1979, p. 79.

Hubbert, M. King. "The Energy Resources of the Earth," *Scientific American*, September 1971, pp. 60–70.

Kelly, James. "Rocky Mountain High," *Time*, December 15, 1980, pp. 28–41.

Luten, Daniel B. "The Economic Geography of Energy," *Scientific American*, September 1971, pp. 164–75.

Marshall, Eliot. "OPEC Prices Make Heavy Oil Look Profitable," *Science*, June 22, 1979, pp. 1283–87.

————. "DOE Leads Synfuels Crusade Without a Map," ibid., September 12, 1980, pp. 1208–10.

————. "Reagan's Cabinet Split on Synfuels Funding," ibid., August 14, 1981, pp. 742–43.

————. "Reagan Endorses Two More Synfuel Loans," ibid., August 21, 1981, p. 848.

Maugh, Thomas H. "Oil Shale: Prospects on the Upswing . . . Again," ibid., December 9, 1977, pp. 1023–27.

————. "Tar Sands: A New Fuel Industry Takes Shape," ibid., February 17, 1978, pp. 756–60.

————. "Work on U.S. Oil Sands Heating Up," ibid., March 14, 1980, pp. 1191–92.

Metz, William D., "Oil Shale: A Huge Resource of Low-Grade Fuel," *Science*, June 21, 1974, pp. 1271–75.

Mossop, Grant D. "Geology of the Athabasca Oil Sands," ibid., January 11, 1980, pp. 145–52.

Nulty, Peter. "The Tortuous Road to Synfuels," *Fortune*, September 8, 1980, pp. 58–64.

————. "Shale Oil Is Braced for Big Role," ibid., September 24, 1979, pp. 42–48.

O'Leary, John F. "Facing the Energy Facts," *Astronautics and Aeronautics*, February 1979, pp. 36–40.

Peirce, Neal R. "Exxon's Oil-Crisis Solution Is Monstrous—But So Is Problem," Los Angeles *Times*, September 21, 1980, Part V, p. 5.

Richards, Bill and Psihoyos, Louie. "New Energy Frontier," *Energy*, special report, *National Geographic*, February 1981, pp. 96–113.

Sampson, Anthony. *The Seven Sisters*. New York: Bantam, 1980.

Stuart, Alexander. "The Rough Road to Making Oil and Gas from Coal," *Fortune*, September 24, 1979, pp. 49–64.

"Synfuel Success." *Time*, August 20, 1979, p. 42.

"The Coal Question in England," *Science*, February 27, 1885, pp. 175–76.

"U.S., Japan, Germany Scuttle Giant Synfuel Plant; Soaring Costs Cited," Los Angeles *Times*, June 25, 1981, Part IV, p. 2.

Wade, Nicholas. "Synfuels In Haste, Repent At Leisure," *Science*, July 13, 1979, pp. 167–68.

————. "Roll with Coal," ibid., May 30, 1980, p. 1008.

————. "Oil Companies Suppress German Synfuels Formula, Says Hollywood," ibid., November 28, 1980, p. 990.

"What's Keeping Coal in the Ground," *Fortune*, August 25, 1980, p. 48.

Page   20.   "There are two things wrong with coal." Hammond, Metz and Maugh, p. 2.

          20–21.   Great Plains coal mines. Richards and Psihoyos; Atwood.

              21.   "We'll just widen the valley." Kelly.

          21–22.   Unit trains. "A Rail Town Booms."

              22.   Norfolk coal port. Luten; "What's Keeping Coal in the Ground."

              22.   Coal reserves. Hayes; Griffith and Clarke; Hubbert, 1971.

              23.   Carroll L. Wilson. Wade, May 30, 1980.

              23.   Increased use of coal. Beck; "What's Keeping Coal in the Ground."

              23.   Synfuels process. Stuart; Cochran.

              24.   German synfuels in World War II. Wade, November 28, 1980; Carter.

          24–25.   South Africa's Sasol. Abelson, August 17, 1979; Stuart; Canby and Blair; "Synfuel Success."

              25.   U.S. synfuels efforts. Nulty, 1980; Hammond, 1976; Cochran; Wade, 1979.

25. Erich Reichl. Nulty, 1980.
25. Richard Corrigan. Wade, 1979.
26. Coal gasification. Stuart; Marshall, 1980; Marshall, August 21, 1981.
26. Solvent refining. Marshall, 1980; Stuart; "U.S., Japan, Germany Scuttle Giant Synfuel Plant."
26. Sasol technology. Marshall, 1980; Nulty, 1980.
26. Synthetic chemicals. Abelson, February 2, 1980.
26–27. Synthetic gasoline. Stuart; Canby and Blair.
27–28. Athabasca tar sands. Maugh, 1978; Marshall, 1979; Mossop.
28. Syncrude project. David.
28. U.S. heavy oil sands. Maugh, 1980.
29. "For most of this century." Maugh, 1977.
29. Oil shale. Metz; Nulty, 1979.
29. Union Oil. O'Leary; Marshall, August 14, 1981.
29–30. Occidental Petroleum. Canby and Blair; Nulty, 1979.
30–31. "Each would rank among the greatest." Peirce.
31. "In 1885 . . . Mr. Hull." "The Coal Question in England."
32. Oil's wild price swings. Canfield and Anderson, p. 373; Sampson, pp. 34–35, 90–91.
32. Paul Frankel. Quoted in Sampson, pp. 90–91.
32–33. Production price of Saudi oil. "How the Price Is Pumped Up."
33. "The upper regions of the earth's crust." Abelson, 1978.

CHAPTER 3. ANTARCTICA AND MIAMI

Adler, Jerry. "The Browning of America," *Newsweek*, February 23, 1981, pp. 26–37.
Ambroggi, Robert P. "Underground Reservoirs to Control the Water Cycle," *Scientific American*, May 1977, pp. 21–27.
Ardrey, Robert. *The Hunting Hypothesis*. New York: Bantam, 1977.
Broecker, Wallace S. "Climatic Change: Are We on the Brink of a Pronounced Global Warming?" *Science*, August 8, 1975, pp. 460–63.
———. Takahashi, T.; Simpson, H. J.; Peng, T.-H., "Fate of Fossil Fuel Carbon Dioxide and the Global Carbon Budget," ibid., October 26, 1979, pp. 409–18.
Bryson, Reid A. "A Perspective on Climatic Change," ibid., May 17, 1974, pp. 753–60.
CLIMAP Project Members, "The Surface of the Ice-Age Earth," *Science*, March 19, 1976, pp. 1131–37.
Cooper, Charles F. "What Might Man-Induced Climate Change Mean?" *Foreign Affairs*, April 1978, pp. 500–19.
Crandell, Dwight R.; Mullineaux, Donal R. and Rubin, Meyer. "Mount St. Helens Volcano: Recent and Future Behavior," *Science*, February 7, 1975, pp. 438–41.
Dansgaard, W.; Johnsen, J. J.; Clausen, H. B. and Langway, C. C. "Climatic Record Revealed by the Camp Century Ice Core," in *The Late Cenozoic Glacial Ages* (Karl K. Turekian, ed.). New Haven: Yale University Press, 1971, pp. 37–56.
Emiliani, Cesare. "Quaternary Paleotemperatures and the Duration of the High-Temperature Intervals," *Science*, October 27, 1972, pp. 398–99.

————— and Shackleton, Nicholas J. "The Brunhes Epoch: Isotopic Paleotemperatures and Geochronology," ibid., February 8, 1974, pp. 511–14.

—————. "Paleoclimatological Analysis of Late Quaternary Cores from the Northeastern Gulf of Mexico," ibid., September 26, 1975, pp. 1083–88.

—————. "The Cause of the Ice Ages," *Earth and Planetary Science Letters*, Vol. 37, 1978, pp. 349–52.

Gates, W. Lawrence. "Modeling the Ice-Age Climate," *Science*, March 19, 1976, pp. 1138–44.

Hansen, J.; Johnson, D.; Lacis, A.; Lebedeff, S.; Lee, P.; Rind, D. and Russell, G. "Climate Impact of Increasing Atmospheric Carbon Dioxide," ibid., August 28, 1981, pp. 957–66.

Hughes, T. "West Antarctic Ice Streams," *Reviews of Geophysics and Space Physics*, February 1977, pp. 1–46.

Ives, Jack D.; Andrews, John T.; and Barry, Roger G. "Growth and Decay of the Laurentide Ice Sheet and Comparisons with Fenno-Scandinavia," *Naturwissenschaften*, Vol. 62, 1975, pp. 118–25.

Krishna, Raj. "The Economic Development of India," *Scientific American*, September 1977, pp. 166–78.

Ku, Teh-Lung; Kimmel, Margaret A.; Easton, William H.; and O'Neil, Thomas J. "Eustatic Sea Level 120,000 Years Ago on Oahu, Hawaii," *Science*, March 8, 1974, pp. 959–62.

Kukla, G. J. and Matthews, R. K. "When Will the Present Interglacial End?" ibid., October 13, 1972, pp. 190–91.

Madden, Roland A. and Ramanathan, V. "Detecting Climate Change due to Increasing Carbon Dioxide," ibid., August 15, 1980, pp. 763–68.

Manabe, Syukuro and Wetherald, Richard T. "The Effects of Doubling the CO_2 Concentration on the Climate of a General Circulation Model," *Journal of the Atmospheric Sciences*, January 1975, pp. 3–15.

Mercer, J. H. "West Antarctic Ice Sheet and CO_2 Greenhouse Effect; A Threat of Disaster," *Nature*, January 26, 1978, pp. 321–25.

Rawls, Wendell. "The Mississippi Short Cut," *Next*, June 1981, pp. 78–84.

Schneider, Stephen H. and Chen, Robert S. "Carbon Dioxide Warming and Coastline Flooding: Physical Factors and Climatic Impact," *Annual Review of Energy*, Vol. 5, 1980. Palo Alto, California: Annual Reviews, Inc., pp. 107–40.

Shoji, Kobe. "Drip Irrigation," *Scientific American*, November 1977, pp. 62–68.

Siegenthaler, U. and Oeschger, H. "Predicting Future Atmospheric Carbon Dioxide Levels," *Science*, January 27, 1978, pp. 388–95.

Splinter, William E. "Center-Pivot Irrigation," *Scientific American*, June 1976, pp. 90–99.

Thomas, Robert H.; Sanderson, Timothy J. O.; and Rose, Keith E. "Effect of Climatic Warming on the West Antarctic Ice Sheet," *Nature*, February 1, 1979, pp. 355–58.

Thompson, Louis M. "Weather Variability, Climatic Change, and Grain Production," *Science*, May 9, 1975, pp. 535–41.

Wade, Nicholas. "CO_2 in Climate: Gloomsday Predictions Have No Fault," ibid., November 23, 1979, pp. 912–13.

Warner, Kenneth E. "Cigarette Smoking in the 1970's: The Impact of the Antismoking Campaign on Consumption," ibid., February 13, 1981, pp. 729–31.

Page   35.   Greenhouse effect. Siegenthaler and Oeschger.
   35–36.   Calculated temperature rises. Manabe and Wetherald; Hansen et al.
     36.   "We have tried but have been unable." Wade.
     36.   Rate of rise of carbon dioxide. Broecker et al., 1979.
  37–38.   Ancient temperatures. Emiliani, 1972, 1978; Emiliani and Shackleton, 1974; Ardrey, pp. 192–93; Dansgaard; Broecker, 1975.
     38.   "When comparing the present." Kukla and Matthews.
     38.   "The repetitive nature." Crandell, Mullineaux and Rubin.
     38.   CLIMAP calculations. CLIMAP Project Members; Gates. See also *Science*, February 6, 1976, p. 455.
  39–40.   Climate in 1930–1960. Dansgaard; Broecker, 1975; Bryson.
     40.   India self-sufficient in grain. Krishna.
     40.   Iceland. Bryson; Ardrey, p. 197.
     40.   Thoroddsen's history of Iceland. In Bryson.
  40–41.   Climate in the Soviet Union. Ardrey, p. 196.
     41.   Crop yields. Thompson.
     41.   "A low-pressure region on the East Coast." Adler.
     42.   Aquifers and U.S. water usage. ibid.; Ambroggi.
  42–43.   Irrigation methods. Shoji; Splinter.
     43.   Canadian Arctic. Bryson.
  43–44.   Sea levels 124,000 years ago. Ku et al.
     44.   West Antarctic ice sheet. Mercer; Hughes; Thomas et al.
     44.   "Once this level of comparative warmth." Mercer; Hughes.
     45.   Laurentide Ice Sheet. Ives et al.
     45.   "The concomitant, accelerated rise." Emiliani, 1975.
     45.   "The flags of a half-dozen nations." Rawls.
     46.   Regions to be flooded. Schneider and Chen.
     47.   Warming of the Soviet Union. Cooper.
     47.   "The results indicate." Madden and Ramanathan.
     48.   "We are on the brink." Broecker, 1975; Wade.
  48–49.   Fine-tuning the effect. Siegenthaler and Oeschger.
     49.   U.S. cigarette consumption. Warner.
     51.   *Prometheus Bound.* Aeschylus, Episode 4.

CHAPTER 4. NUCLEAR DECISIONS

Abelson, Philip H. "World Energy in Transition," ibid., December 19, 1980, p. 1311.

"Atomic Power's Future," *Time*, April 9, 1979, p. 20.

Barnaby, Frank. "A Nuclear Engineer's Paradise," *Bulletin of the Atomic Scientists*, February 1981, pp. 55–56.

Bodansky, David. "Electricity Generation Choices for the Near Term," *Science*, February 15, 1980, pp. 721–28.

Carter, Luther J. "Radioactive Waste Policy Is in Disarray," ibid., October 19, 1979, pp. 312–14.

Chen, Francis F. "Alternate Concepts in Magnetic Fusion," *Physics Today*, May 1979, pp. 36–42.

Churchill, Winston. *The Grand Alliance.* New York: Bantam, 1962.

Clarke, John F. "The Next Step in Fusion: What It Is and How It Is Being Taken," *Science*, November 28, 1980, pp. 967–72.

Faltermeyer, Edmund. "Burying Nuclear Trash Where It Will Stay Put," *Fortune*, March 26, 1979, pp. 98–104.

———. Keeping the Peaceful Atom from Raising the Risk of War," ibid., April 9, 1979, pp. 90–96.

———. "Nuclear Power After Three Mile Island," ibid., May 7, 1979, pp. 115–22.

"Four New Tokamaks Will Each Try for a Finite Power Output," *Physics Today*, January 1978, pp. 19–20.

Furth, Harold P. "Progress Toward a Tokamak Fusion Reactor," *Scientific American*, August 1979, pp. 50–61.

Gibson, Robert. "A Nuclear Bombshell That Is Killing the Industry," *Maclean's*, October 30, 1978, pp. 47–48.

Gibbs-Smith, Charles H. *The Invention of the Airplane 1799–1909*. New York: Taplinger Press, 1965.

Grisham, L. R. "Neutral Beam Heating in the Princeton Large Torus," *Science*, March 21, 1980, pp. 1301–9.

Hohenemser, Christopher; Kasperson, Roger; and Kates, Robert. "The Distrust of Nuclear Power," ibid., April 1, 1977, pp. 25–34.

Holdren, John P. "Fusion Energy in Context: Its Fitness for the Long Term," ibid., April 14, 1978, pp. 168–80.

"International Tokamak Reactor Design," *Physics Today*, March 1980, pp. 20–22.

Jassby, D. L. and Metz, W. D. "Princeton Fusion Experiment," *Science*, October 27, 1978, pp. 370–72.

Kerr, Richard A. "Nuclear Waste Disposal: Alternatives to Solidification in Glass Proposed," ibid., April 20, 1979, pp. 289–91.

———. "Geological Disposal of Nuclear Wastes: Salt's Lead Is Challenged," ibid., May 11, 1979, pp. 603–6.

Kirk, Donald. " 'Made in Japan,' " *Saturday Review*, January 22, 1977, p. 24.

Marshall, Eliot. "Tories Prefer Nukes," *Science*, October 19, 1979, pp. 308–9.

———. "Westinghouse Feels Impact of Declining Demand," ibid., October 17, 1980, p. 295.

McDonald, Marci. "Off and Running in a Nuclear Race," *Maclean's*, October 20, 1980, pp. 30–31.

McKean, Kevin. "The Mounting Crisis in Nuclear Energy," *Discover*, May 1982, pp. 20–26.

Metz, William D. "Nuclear Fusion: The Next Big Step Will Be a Tokamak," *Science*, February 2, 1975, pp. 421–23.

———. "European Breeders (II): The Nuclear Parts Are Not the Problem," ibid., January 30, 1976, pp. 368–72.

———. "Dream Cycles." ibid., June 25, 1976, p. 1321.

———. "New York Blackout: Weak Links Tie Con Ed to Neighboring Utilities," ibid., July 29, 1977, pp. 324–25.

———. "Report of Fusion Breakthrough Proves to Be a Media Event," ibid., September 1, 1978, pp. 792–94.

"MFTF-B: A Tandem Mirror Fusion Test Facility," LLNL-TB-031, Lawrence Livermore Laboratory, 1981.

Morison, Samuel Eliot. *The Two-Ocean War*. New York: Ballantine, 1972.

Murakami, Masanori and Eubank, Harold P. "Recent Progress in Tokamak Experiments," *Physics Today*, May 1979, pp. 25–32.

Nagel, Theodore J. "Operating a Major Electric Utility Today," *Science*, September 15, 1978, pp. 985–93.

Nelkin, Dorothy and Pollak, Michael. "French and German Courts on Nuclear Power," *Bulletin of the Atomic Scientists*, May 1980, pp. 36–42.

"1980 White Paper on Atomic Energy," *Atoms in Japan*, Japan Atomic Industrial Forum, January 1981, pp. 22–23.

"Nukes: Not Nice, but Necessary," *Time*, December 22, 1980, p. 61.

Post, Richard F. "LLL Magnetic Fusion Research: The First 25 Years," *Energy and Technology Review* (Lawrence Livermore Laboratory), May 1978, pp. 1–13.

———— and Ribe, Frederick L. "Fusion Reactors as Future Energy Sources," *Science*, November 1, 1974, pp. 397–406.

"Progress of Power Plants to Date," *Atoms in Japan*, Japan Atomic Industrial Forum, January 1981, pp. 22–23.

Robertson, J. A. L. "The CANDU Reactor System: An Appropriate Technology," *Science*, February 10, 1978, pp. 657–64.

Robinson, Arthur L. "Energy Sweepstakes: Fusion Gets a Chance," ibid., October 24, 1980, pp. 415–16.

————. "Slower Magnetic Fusion Pace Set," ibid., July 16, 1982, pp. 256–57.

Shapley, Deborah. "Nuclear Weapons History: Japan's Wartime Bomb Projects Revealed," *Science*, January 13, 1978, pp. 152–57.

Stobaugh, Robert and Yergin, Daniel. *Energy Future*. New York: Random House, 1979.

"The Magnetic Fusion Energy Engineering Act of 1980," *Fusion*, January 1981, pp. 18–21.

"Uranium: Too Hot to Handle?" *Energy*, special report, *National Geographic*, February 1981, pp. 66–67.

Wade, Nicholas. "France's All-Out Nuclear Program Takes Shape," *Science*, August 22, 1980, pp. 884–89.

Walsh, John. "Nuclear Power: France Forges Ahead on Ambitious Plan Despite Critics," ibid., July 23, 1976, pp. 305–6, 340.

————. "French Nuclear Policy Only Slightly Revised," ibid., November 27, 1981, pp. 1006–7.

"Where the Atom Is Admired," *Time*, February 18, 1980, p. 77.

Will, George. "A Film About Greed," *Newsweek*, April 2, 1979, p. 96.

————. "As I Was Saying," ibid., April 16, 1979, p. 100.

Page   53.   Operations Olympic and Coronet. Morison, p. 484.
    54–55.   "The historian remembers more." Churchill, p. 360 and chapter 11.
    55.   David Howell. Quoted in Marshall, 1979.
    55.   Samuel McCracken. Quoted in Will, April 2, 1979.
    55.   George Will quote. April 16, 1979.
    55–56.   Creys-Malville. McDonald; "Where the Atom Is Admired."
    56–58.   French nuclear industry. Wade; Walsh, 1976.
    58.   Giscard d'Estaing quote. "Where the Atom Is Admired."
    58.   French nuclear politics. Wade; Walsh 1981; "Where the Atom Is Admired."
    58–59.   French nuclear law. Nelkin and Pollak.
    59–60.   British nuclear program. Metz, Jan. 30, 1976; Marshall 1979.

60.   Japan's nuclear program. Abelson 1980; "1980 White Paper"; "Progress of Power Plants"; Will, April 16, 1979; Shapley.

60.   Japan's nuclear opponents. Kirk.

61.   Soviet nuclear program. Barnaby; Metz, Jan. 30, 1976.

61.   Sweden's nuclear program. "Nukes: Not Nice, but Necessary."

61–62. Canada's nuclear program. Robertson.

62.   New York City power requirements. Metz 1977.

62.   Reactor statistics. McKean; "Uranium: Too Hot to Handle?"

62.   Risk of war from the peaceful atom. Faltermeyer, April 9, 1979.

63.   Nuclear power in the U.S. Faltermeyer, May 7, 1979; "Uranium: Too Hot to Handle?"; McKean.

63.   Amory Lovins quote. Gibson.

63n. Uranium and thorium in coal. Bodansky; Hohenemser et al; Will, April 16, 1979.

64.   "Nuclear power provides a dramatic focus." Bodansky.

64.   Electricity growth rates. Bodansky; Marshall 1980.

64.   Millstone Point 1. Faltermeyer, May 7, 1979.

65.   Indian Point 1. Hohenemser.

65.   Plant construction delays. Nagel.

65–66. Nuclear waste disposal. Faltermeyer, March 26, 1979; Kerr, April 20, 1979; Kerr, May 11, 1979; Carter.

66.   1976 California law. Stobaugh and Yergin, p. 129.

67.   French and Japanese energy imports. Abelson 1980.

68.   Lewis Strauss quote. "Atomic Power's Future."

70.   Deuterium and tritium. Holdren; Post and Ribe.

70–71. Princeton Large Torus. Furth; Murakami and Eubank; Metz, 1978; Grisham; Jassby and Metz.

71–72. Tokamak Fusion Test Reactor, other large tokamaks. Metz, Feb. 7, 1975; Furth; "Four New Tokamaks."

72.   Fusion Energy Device and Engineering Test Reactor. Robinson 1982; Clarke; "The Magnetic Fusion Act."

72.   INTOR. "International Tokamak Reactor Design."

72–73. "Assessment of fusion's full potential." "The Magnetic Fusion Act"; Clarke; Holdren.

73.   Wolf Häfele. Holdren; Post and Ribe.

73–74. Advanced fusion approaches. Furth; Chen; Post; Robinson 1980; "MFTF-B."

74.   Advanced fuel cycles. Metz, June 25, 1976; Holdren.

75.   "The prospect of adapting fusion power." Furth.

75.   Sir George Cayley. Gibbs-Smith.

CHAPTER 5. SUBARUS AND SUBWAYS

Abegglen, James C. "The Economic Growth of Japan," *Scientific American,* March 1970, pp. 31–37.

Abelson, Philip H. "Science and Engineering Education," *Science,* November 28, 1980, p. 965.

———. "East Is East and West Is West," ibid., February 6, 1981, p. 533.

Alexander, Charles. "Time to Repair and Restore," *Time*, April 27, 1981, pp. 46–49.

Alpern, David M. "Mr. Fixit for the Cities," *Newsweek*, May 4, 1981, pp. 26–30.

Boroson, Warren. "For Your Next Car, Options Unlimited," *Next*, November/December 1980, pp. 90–99.

Boyle, Patrick. "Some Americans Seeking an Automobile Built for 2," Los Angeles *Times*, February 22, 1981, Part VI, p. 1.

―――. "Ford Facing Rough Road, Seeks Detour," ibid., April 6, 1981, Part I, p. 1.

Branscomb, Lewis M. "Electronics and Computers: An Overview."

Byron, Christopher. "How Japan Does It," *Time*, March 30, 1981, pp. 54–60.

Church, George J. "The Biggest Challenge," *Time*, January 19, 1981, pp. 60–67.

Churchill, Winston. *The Grand Alliance*. New York: Bantam, 1962.

Fischler, Stanley. "The Next Time Ride the Bus—and Leave the Sagging to Us," Los Angeles *Times*, December 23, 1980, Part II, p. 11.

Goldman, John J. "Major City Transit Systems Losing Battles With Decay," ibid., March 9, 1981, Part I, p. 1.

Gray, Charles L. and von Hippel, Frank. "The Fuel Economy of Light Vehicles," *Scientific American*, May 1981, pp. 48–59.

Hammond, Allen L. "Alcohol: A Brazilian Answer to the Energy Crisis," February 11, 1977, pp. 564–66.

Holden, Constance. "Innovation: Japan Races Ahead as U. S. Falters," *Science*, November 14, 1980, pp. 751–54.

―――. "Autos: A Challenge for Industrial Policy," ibid., March 13, 1981, pp. 1141–46.

Johnston, David. "Home Stills: Alcohol Fuel at Low Cost," Los Angeles *Times*, February 6, 1981, Part I, p. 1.

Joseph, James. "Why There's a Diesel in Your Future," *Next*, September/October 1980, pp. 46–51.

Keating, Bern. "The Gasohol Gamble," ibid., September/October 1980, pp. 66–69.

McGrath, Ellie. "Rumbling Toward Ruin," *Time*, March 30, 1981, pp. 12–15.

Morrison, Philip. Book review of "Giving Up the Gun: Japan's Reversion to the Sword, 1543–1879" by Noel Perrin, *Scientific American*, July 1981, pp. 32–37.

Peirce, Neal R. "Crumbling U.S. Roads Must Be Patched With Money—Fast." Los Angeles *Times*, December 28, 1980, Part V, p. 5.

Pesta, Ben. "Is There Any Other Way to Go?" *Next*, March/April 1980, pp. 74–80.

"Proof It Works," *Time*, December 22, 1980, p. 62.

Taylor, Alexander. "Detroit's Uphill Battle," ibid., September 8, 1980, pp. 46–52.

Wade, Nicholas. "Oil Pinch Stirs Dreams of Moonshine Travel," *Science*, June 1, 1979, pp. 928–29.

―――. "Gasohol: A Choice That May Buy Grief," ibid., March 28, 1980, pp. 1450–51.

Wouk, Victor. "From Horsepower to Shanks' Mare Power," *Engineering and Science*, May–June 1980, pp. 6–12.

Page   77.   "450 horsepower in some cases." Wouk.
         78–79.   U.S. auto industry in the 1970s. Holden, 1981.
            79.   Ford Motor's 1978 decision. Boyle, April 6, 1981.
         79–80.   "In 1543 three Portuguese seamen." Morrison.
            80.   "When we sent you the beautiful products." Churchill,
                   p. 489.
         80–82.   Japanese industry. Abegglen; Byron; Holden, 1980.
            82.   "A bright heart overflowing." Time, March 30, 1981, p. 63.
         82–83.   Attorneys in Japan. Abelson, 1981.
            83.   Engineering degrees. Abelson, 1980.
            83.   Percentages of income saved. Savings and Loan Foundation,
                   Washington, D.C.
            83.   Inflation rates and deficits. Byron; Congressional Budget
                   Office data.
            84.   Tohatsu and Honda. Byron.
            84.   "We wrote off the workers." Holden, 1981.
         84–85.   Response of U.S. auto industry. Holden, 1981; Taylor.
            85.   Plymouth Volare and Dodge Aspen of 1974. Taylor.
            86.   Cars of the future. Gray and von Hippel; Boroson; Pesta;
                   Holden, 1981.
            86.   "We use plastics for beauty." Boroson.
            86.   6.5 million cars with microprocessors. Branscomb.
            87.   Demise of the V-8. Gray and von Hippel.
            87.   Diesel engines. Joseph.
         87–88.   Daihatsu Cuore. Boyle, Feb. 22, 1981.
            89.   Auto industry of the year 2000. Holden, 1981; Byron.
         89–91.   Alcohol in Brazil. "Proof It Works"; Hammond; Keating.
            90.   Gasohol in the United States. Johnston; Wade, 1979, 1980.
            90.   "More stills than service stations." Johnston.
            90.   "Real issue has been the price of corn." Wade, 1979.
            91.   Al Capone. Wade, 1979.
         91–92.   Crumbling U.S. roads and bridges. Peirce; Alexander.
            91.   "That road just tears a rig apart." Alexander.
         92–93.   Subway systems. Goldman; McGrath.
            92.   Boston door guards. McGrath.
            93.   "The picture that emerges." Goldman.
            93.   Successful streetcar and subway vehicles. Peirce.
         93–94.   Grumman Flxible buses. McGrath; Peirce.
            95.   Felix Rohatyn. Alpern.
            95.   "Transit is not the kind of place." Goldman.
         95–96.   Road taxes. Peirce.
            96.   Dallas, Texas. Alexander.

CHAPTER 6. TOWARD NEW FRONTIERS

Allen, Oliver E. The Pacific Navigators. Alexandria, Virginia: Time-Life
    Books, 1980.
Bekey, Ivan. "Big Comsats for Big Jobs at Low User Cost," Astronautics and
    Aeronautics, February 1979, pp. 42–56.
Bell, M. W. J. "Space Shuttle Vehicle Growth Options," Paper 78-1656. New
    York: American Institute of Aeronautics and Astronautics, 1978.

Chaudhari, Praveen; Giessen, Bill C.; and Turnbull, David. "Metallic Glasses," *Scientific American*, April 1980, pp. 98–117.

Clarke, Arthur C. *Profiles of the Future*. New York: Bantam, 1964.

Covault, Craig. "NASA Curtails Shuttle Flights," *Aviation Week and Space Technology*, June 21, 1982, pp. 16–18.

Dooling, Dave. "Once More—The New 'High Ground,'" *Astronautics and Aeronautics*, April 1981, pp. 4–12.

Dornberger, Walter. *V-2*. New York: Bantam, 1979.

Edelson, Burton I. "Satellite Communications: A Benefit Realized," *Science*, October 24, 1975, p. 333.

————. "Global Satellite Communications," *Scientific American*, February 1977, pp. 58–73.

Gilman, John J. "Metallic Glasses," *Science*, May 23, 1980, pp. 856–61.

Grey, Jerry. "Case for a Fifth Shuttle and More Expendable Launch Vehicles," *Astronautics and Aeronautics*, March 1981, pp. 22–26.

Jones, A. L. "Extension of Space Shuttle Capability," Paper 79-292. San Diego: American Astronautical Society, 1979.

Kerr, Richard A. "Planetary Science on the Brink Again," *Science*, December 14, 1979, pp. 1288–89.

Ley, Willy. *Rockets, Missiles, and Men in Space*. New York: New American Library, 1969.

Mansfield, Harold. *Vision*. New York: Popular Library, 1966.

"Materials Processing in Space." Washington, D.C.: National Research Council, 1978.

Mordan, G. W. "25 kW Power Module: First Step Beyond the Baseline STS," Paper 78-1693. New York: American Institute of Aeronautics and Astronautics, 1978.

Murray, Bruce C. "A New Strategy for Planetary Exploration," *Astronautics and Aeronautics*, October 1968, pp. 42–53.

————. "The Planets—So Long to All that?" ibid., July–August 1980, p. 18.

Pierce, John R. "Being Practical About Space," *Science*, February 3, 1978, p. 483.

Pritchard, Wilbur L. "The Holdup on Broadcast and Mobile Communications Satellites," *Astronautics and Aeronautics*, Setepmber 1980, pp. 40–43.

Sagan, Carl. *Other Worlds*. New York: Bantam, 1975.

Slaughter, John B. "The National Science Foundation Looks to the Future," *Science*, March 13, 1981, pp. 1131–36.

Smith, R. Jeffrey. "Uncertainties Mark Space Program of the 1980's," ibid., December 14, 1979, pp. 1284–86.

————. "Military Plans for Shuttle Stir Concern," ibid., May 1, 1980, pp. 520–21.

Stein, G. Harry. *The Third Industrial Revolution*. New York: Putnam, 1975.

"The Great Capsule Hunt." *Time*, April 27, 1959, p. 16–17.

"The Last Words of Pioneer V," ibid., July 18, 1960, p. 52.

Waldrop, M. Mitchell. "Space Science in the Year of the Shuttle," *Science*, April 17, 1981, pp. 316–18.

Walsh, John. "NASA and ESRO. A European Payload for the Space Shuttle," ibid., November 9, 1973, pp. 562–63.

Wuenscher, Hans F. "Manufacturing in Space," *Astronautics and Aeronautics*, September 1972, pp. 42–54.

Page 99. American Rocket Society. Ley, pp. 587–88.
     100. "We wanted to have done once and for all." Dornberger, pp. 21–22.
     101. Pioneer V. "The Last Words of Pioneer V."
     101. Discoverer II. "The Great Capsule Hunt"; Dooling.
     102. Arthur C. Clarke predictions. Clarke, p. xiv.
   104–5. Communications satellites. Edelson, 1975, 1977; Pierce.
     105. AIAA projections of satellite traffic. Grey.
     106. Direct broadcast and air and sea links. Pritchard.
   106–7. Electronic mail. Bekey.
     107. "A satellite of 54,000 pounds." Bekey.
     108. "In contrast, I advocate." Murray, 1968.
     108. Diane Ackerman poem. Sagan, p. 63.
     109. Planetary program prospects. Murray, 1980; Kerr; Smith, 1979; Waldrop.
     110. Exploration of Australia. Allen, chapter 2.
  110–11. Space manufacturing prospects. Stein; Wuenscher; "Materials Processing in Space."
     112. "When gravity has an adverse effect." "Materials Processing in Space."
  112–13. Metallic glasses. Chaudhari et al.; Gilman; Slaughter.
     113. "Atlas-Centaur which could hoist Intelsat V." Edelson 1977.
     113. Spacelab. Walsh.
     113. 25-kilowatt solar panel. Mordan.
     114. Shuttle improvements. Bell; Jones.
     114. Solar Electric Propulsion Stage. Waldrop.
     115. Shuttle as a military spacecraft. Dooling; Smith, 1981.
     115. "It is clear that space science." Smith, 1979.
     115. B-52 bomber. Mansfield, chapters 14, 15.
     115. Titan III costs. Martin Marietta Corp. data, 1971.

Chapter 7. The Electron and the Individual

Abelson, Philip H. "America's Vanishing Lead in Electronics," *Science*, December 5, 1980, p. 1079.
Albus, James S. and Evans, John M. "Robot Systems," *Scientific American*, February 1976, pp. 76–86.
Ardrey, Robert. *The Social Contract*. New York: Dell, 1970.
———. *The Hunting Hypothesis*. New York: Bantam, 1977.
Barrow, Harry. "Artificial Intelligence After 25 Years." Seminar, California Institute of Technology, June 2, 1981.
Berliner, Hans. "Computer Backgammon," *Scientific American*, June 1980, pp. 64–72.
Brennan, Peter J. "Advanced Technology Center: Santa Clara Valley, California," *Scientific American*, March 1981; advertisement following p. 120.
Costello, Dennis and Rappaport, Paul. "The Technological and Economic Development of Photovoltaics," *Annual Review of Energy*, Volume 5, 1980. Palo Alto, California: Annual Reviews, Inc., pp. 335–56.
Davis, Ruth M. "Evolution of Computers and Computing," *Science*, March 18, 1977, pp. 1096–1102.

————. "Computers and Electronics for Individual Services," *Science*, February 12, 1982, pp. 852–55.

Friedrich, Otto. "The Robot Revolution," *Time*, December 8, 1980, pp. 72–83.

Hammond, Allen L. "Photovoltaics: The Semiconductor Revolution Comes to Solar," *Science*, July 29, 1977, pp. 445–47.

Hardin, Garrett. "Will Xerox Kill Gutenberg?" ibid., December 2, 1977, p. 883.

Holden, Constance. "Maryland Lures Chip-Makers East," ibid., May 2, 1980, p. 479.

Holmes, Nigel. "Who's Doing Best?" *Time*, January 19, 1981, p. 65.

Jahnke, Eugene and Emde, Fritz. *Tables of Functions*. New York: Dover, 1945.

Jones, Trevor O. "Some Recent and Future Automotive Electronic Developments," *Science*, March 18, 1977, pp. 1156–60.

Kay, Alan C. "Microelectronics and the Personal Computer," *Scientific American*, September 1977, pp. 230–44.

Kelly, Henry. "Photovoltaic Power Systems: A Tour Through the Alternatives," *Science*, February 10, 1978, pp. 634–43.

Levinson, Stephen E. and Liberman, Mark Y. "Speech Recognition by Computer," *Scientific American*, April 1981, pp. 64–76.

Linvill, John C. and Hogan, C. Lester. "Intellectual and Economic Fuel for the Electronics Revolution," *Science*, March 18, 1977, pp. 1107–13.

Luff, Peter P. "The Electronic Telephone," *Scientific American*, March 1978, pp. 58–64.

Mayo, John S. "The Role of Microelectronics in Communication," ibid., September 1977, pp. 192–209.

Meindl, James D. "Microelectronic Circuit Elements," ibid., September 1977, pp. 70–81.

"Micromainframe," ibid., May 1981, pp. 92–96.

"National Photovoltaics Program," U.S. Department of Energy. Pasadena, California: Jet Propulsion Laboratory, September 1980.

Noyce, Robert N. "Microelectronics," *Scientific American*, September 1977, pp. 62–69.

Oldham, William G. "The Fabrication of Microelectronic Circuits," ibid., September 1977, pp. 110–28.

Reed, Fred. "The Robots Are Coming, The Robots Are Coming," *Next*, May/June 1980, pp. 30–38.

Robinson, Arthur L. "Automotive Electronics: Computerized Engine Control," *Science*, October 22, 1976, pp. 414–15.

————. "More People Are Talking to Computers as Speech Recognition Enters the Real World," ibid., February 16, 1979, pp. 634–38.

————. "Speech Is Another Microelectronics Conquest," ibid., February 16, 1979, p. 635.

————. "Communicating with Computers By Voice," ibid., February 23, 1979, pp. 734–36.

————. "Tournament Competition Fuels Computer Chess," ibid., June 29, 1979, pp. 1396–98.

————. "Giant Corporations from Tiny Chips Grow," ibid., May 2, 1980, pp. 480–84.

————. "Perilous Times for U.S. Microcircuit Makers," ibid., May 9, 1980, pp. 582–86.

————. "Do the Japanese Make Better IC's?" ibid., May 9, 1980, p. 585.

————. "New Ways to Make Microcircuits Smaller," ibid., May 30, 1980, pp. 1019–22.

————. "Computer Chess: Belle Sweeps the Board," ibid., October 17, 1980, pp. 293–94.

————. "Micromainframe Is Newest Computer on a Chip," ibid., May 1, 1981, pp. 527–31.

Sperling, George. "Bandwidth Requirements for Video Transmission of American Sign Language and Finger Spelling," ibid., November 14, 1980, pp. 797–99.

Spinrad, R. J. "Office Automation," ibid., February 12, 1982, pp. 808–13.

Stobaugh, Robert and Yergin, Daniel. *Energy Future.* New York: Random House, 1979.

Taylor, Alexander L. "Striking It Rich," *Time,* February 15, 1982, pp. 36–41.

"The New Age," *Time,* April 29, 1957, pp. 84–90.

Thurow, Lester C. "Why Japan Owns the Robot Field," Los Angeles *Times,* April 28, 1981, Part IV, p. 3.

Toong, Hoo-Min D. "Microprocessors," *Scientific American,* September 1977, pp. 146–61.

Tuchman, Barbara. *The Proud Tower.* New York: Bantam, 1967.

"Very Large Scale Integration: Designing 'Street Maps' of North America," *Engineering and Science,* October 1980, pp. 7–9.

Whitted, Turner. "Some Recent Advances in Computer Graphics," *Science,* February 12, 1982, pp. 767–74.

Page 119–21.   Silicon Valley. Brennan; Robinson, May 2, 1980; Taylor.
      121.   Palo Alto housing prices. Holden.
      121.   "Silicon Valley is similar." Holden.
      121–22.   "The house was like none ever." "The New Age."
      122.   Picturephone. Sperling.
      122–23.   Photovoltaics. Stobaugh and Yergin pp. 258–62; Costello and Rappaport; Hammond; Kelly.
      123.   "As Robert Noyce has noted." Noyce.
      123.   Photovoltaic houses. "National Photovoltaics Program."
      124–25.   "Creative turning of necessities into virtues." Meindl; Oldham.
      125.   Moore's law. Robinson, May 2, 1980; Noyce.
      125.   1981 Hewlett-Packard chip. Robinson, May 1, 1981; "Micromainframe."
      125.   Economics of microelectronics advances. Noyce.
      126.   Comparison of ENIAC and F8 microcomputer. Linvill and Hogan.
      126.   Complexity of microcircuitry. Robinson, May 30, 1980; "Very Large Scale Integration."
      126.   "The Table of Powers was calculated." Jahnke and Emde.
      127–28.   Electronic telephone. Mayo; Luff.
      128–29.   Automotive electronics. Jones; Robinson, October 22, 1976.
      129–30.   Home computers. Davis, 1982.
      130.   Color graphics. Kay; Whitted.
      132.   Electronic office. Spinrad.

132.   "Divisions of companies do not willingly disappear." Toong.
133.   Books as information-storage media. Hardin.
134–35.   Hearst press and Spanish-American War. Tuchman, pp. 167–74.
136–37.   Robots. Friedrich; Reed; Albus and Evans.
137.   Sheepshearing robots. Reed.
137.   "If we can bring in a robot." Friedrich, p. 78.
137.   Clyde the Claw. ibid, p. 77.
137–38.   Census of robots. ibid., p. 75.
138.   Japan's robot leasing company. Thurow.
138.   Osaka's Matsushita's TV plant. Abelson.
139.   British and Japanese unemployment rates. Holmes.
139.   Japan's computer memory chips. Abelson; Robinson, May 9, 1980.
139.   Robots as blind grabbers. Reed; Albus and Evans.
139–40.   Shakey. Albus and Evans; Barrow.
140.   Heini Hediger's monkeys. Ardrey, *Hunting Hypothesis*, pp. 78–79.
140–41.   Computer understanding of speech. Robinson, Feb. 16 and 23, 1979; Levinson and Liberman.
142.   Computer chess. Robison, June 29, 1979; October 17, 1980.
142–43.   Computer backgammon. Berliner.
143.   "In a computer the basic element of decision-making." Davis.
143.   "You are a monkey." Ardrey, *Social Contract*, p. 130.

CHAPTER 8. A CODE FOR GENES

Abelson, Philip H. "Recombinant DNA," August 19, 1977, p. 721.
Ayala, Francisco J. "The Mechanisms of Evolution." *Scientific American*, September 1978, pp. 56–69.
Bailey, James E. "Biotechnology for Fun and Profit," *Engineering and Science*, June 1981, pp. 13–17.
Berg, Paul; Baltimore, David; Boyer, Herbert W.; Cohen, Stanley N.; Davis, Ronald W.; Hogness, David S.; Nathans, Daniel; Roblin, Richard; Watson, James D.; Weissman, Sherman; and Zinder, Norton D. "Potential Biohazards of Recombinant DNA Molecules," *Science*, July 26, 1974, p. 303.
Burke, Derek C. "The Status of Interferon," *Scientific American*, April 1977, pp. 42–50.
Cairns, John. "The Cancer Problem," ibid., November 1975, pp. 64–78.
Chargaff, Erwin. "A Slap at the Bishops of Asilomar," *Science*, October 10, 1975, p. 135.
———. "On the Dangers of Genetic Meddling," ibid., June 4, 1976, pp. 938–40.
Cohen, Stanley N. "The Manipulation of Genes," *Scientific American*, July 1975, pp. 24–33.
———. "Recombinant DNA: Fact and Fiction," *Science*, February 18, 1977, pp. 654–57.
Cohn, Victor. "Gene Transferred Between Species," Los Angeles *Times*, September 8, 1981, Part I, p. 4.

"Commemorating a Heroic Act." *Time*, September 14, 1981, p. 101.

Cotzias, George C.; Miller, Samuel T.; Tang, Lily C.; Papavasiliou, Paul S.; and Wang, Ying Yao. "Levodopa, Fertility, and Longevity," *Science*, April 29, 1977, pp. 549–51.

Davis, Bernard D. "Genetic Engineering: How Great the Danger?" ibid., October 25, 1974, p. 309.

Folkman, Judah. "The Vascularization of Tumors," *Scientific American*, May 1976, pp. 58–73.

Golden, Frederick. "Shaping Life in the Lab," *Time*, March 9, 1981, pp. 50–59.

Grobstein, Clifford. "The Recombinant-DNA Debate," *Scientific American*, July 1977, pp. 22–33.

Guillemin, Roger and Burgus, Roger. "The Hormones of the Hypothalamus," ibid., November 1972, pp. 24–33.

Hayflick, Leonard. "Human Cells and Aging," ibid., March 1968, pp. 32–37.

———. "The Cell Biology of Human Aging," ibid., January 1980, pp. 58–65.

Holden, Constance. "Identical Twins Reared Apart," *Science*, March 21, 1980, pp. 1323–28.

———. "Twins Reunited," *Science 80*, November 1980, pp. 54–59.

———. "More About Cloned Mice," *Science*, April 10, 1981, p. 145.

Israel, Mark A.; Chan, Hardy W.; Rowe, Wallace P.; and Martin, Malcolm A. "Molecular Cloning of Polyoma Virus DNA in *Escherichia coli*: Plasmid Vector System," ibid., March 2, 1979, pp. 883–87.

———. "Molecular Cloning of Polyoma Virus DNA in *Escherichia coli*: Lambda Phage Vector System," ibid., March 2, 1979, pp. 887–92.

Kolata, Gina Bari. "The 1980 Nobel Prize in Chemistry," ibid., November 21, 1980, pp. 887–89.

——— and Wade, Nicholas. "Human Gene Treatment Stirs New Debate," ibid., October 24, 1980, p. 407.

Lehman, I. R. "DNA Ligase: Structure, Mechanism, and Function," ibid., November 29, 1974, pp. 790–97.

Marshall, Eliot. "Environmental Groups Lose Friends in Effort to Control DNA Research," ibid., December 22, 1978, pp. 1265–69.

———. "Gene Splicers Simulate a 'Disaster,' Find No Risk," ibid., March 23, 1979, p. 1223.

Marx, Jean L. "Restriction Enzymes: New Tools for Studying DNA," ibid., May 4, 1973, pp. 482–85.

———. "Aging Research (I): Cellular Theories of Senescence," ibid., December 20, 1974, pp. 1105–7.

———. "Gene Transfer Given a New Twist," ibid., April 25, 1980, pp. 386–87.

———. "Gene Transfer Moves Ahead," ibid., December 12, 1980, pp. 1334–36.

———. "Three Mice 'Cloned' in Switzerland," ibid., January 23, 1981, pp. 375–76.

———. "Gene Control Puzzle Begins to Yield," ibid., May 8, 1981, pp. 653–55.

———. "Globin Gene Transferred," ibid., September 25, 1981, p. 1488.

Maugh, Thomas H. "Unlike Money, Diesel Fuel Grows on Trees," ibid., October 26, 1979, p. 436.

———. "First Course for Genetic Engineering Technicians," ibid., March 13, 1981, p. 1142.

———. "Angiogenesis Inhibitors Link Many Diseases," ibid., June 19, 1981, pp. 1374–75.

Melville, Herman. *Moby Dick*. New York: Random House, 1950.

"More Magic from Gene Splicing," *Time*, June 29, 1981, p. 54.

"Put a Microbe in Your Tank," *Newsweek*, February 16, 1981, p. 88.

Singer, Maxine and Soll, Dieter. "Guidelines for DNA Hybrid Molecules," *Science*, September 21, 1973, p. 1114.

Smith, R. Jeffrey. "NIH Plan Relaxes Recombinant DNA Rules," ibid., September 25, 1981, p. 1482.

Thomas, Lewis. *The Lives of a Cell*. New York: Bantam, 1975.

———. "Hubris in Science?" *Science*, June 30, 1978, pp. 1459–62.

Toufexis, Anastasia. "The Big IF in Cancer," *Time*, March 31, 1980, pp. 60–66.

Wade, Nicholas. "Genetic Manipulation: Temporary Embargo Proposed on Research," *Science*, July 26, 1974, pp. 332–34.

———. "Genetics: Conference Sets Strict Controls to Replace Moratorium," ibid., March 14, 1975, pp. 931–35.

———. "Major Relaxation in DNA Rules," ibid., September 21, 1979, p. 1238.

———. "Waiting for the Oil Bug," ibid., November 30, 1979, p. 1053.

———. "Gene Splicing Company Wows Wall Street," ibid., October 31, 1980, pp. 506–7.

———. "How to Keep Your Shirt—If You Put It in Genes," ibid., April 3, 1981, p. 26.

Page 145–46. Introduction to DNA. Grobstein; Golden.
147. Genetic code. Ayala.
147. TRF. Guillemin and Burgus.
147–48. E. coli. Grobstein.
148–49. Recombinant DNA techniques. Grobstein; Cohen, 1975.
148. Restriction enzymes. Marx, 1973.
148. DNA ligase. Lehman.
148–49. Work of Boyer and Cohen. Grobstein; Cohen, 1975; Golden.
149–50. Paul Berg. Kolata.
150. Gordon Research Conference. Singer and Soll.
150. Berg's DNA-research moratorium. Berg et al.; Wade, 1974.
150. Asilomar conference. Wade, 1975.
151. Erwin Chargaff. Chargaff, 1975, 1976.
151. Stanley Cohen. Cohen, 1977.
151–52. Lewis Thomas. Thomas, 1978.
152. American Society of Microbiology. Abelson.
152. Cancer genes in E. coli. Israel et al.; Marshall, 1979.
153. Opposition from environmental groups. Marshall, 1978.
152–53. Relaxation of DNA research guidelines. Wade, Sept. 21, 1979; Smith.
153. Training of genetic engineers. Maugh, March 13, 1981.
153–54. Genentech on Wall Street. Wade, 1980; Golden.
153. Foot-and-mouth disease. "More Magic from Gene Splicing."

153–54.   "How to Keep Your Shirt." Wade, 1981.
   154.   Interferon. Burke; Toufexis.
154–55.   Cancer. Cairns.
   155.   TAF and AI. Folkman; Maugh, June 19, 1981.
   156.   Three stages of therapy. Thomas, *Lives of a Cell*, pp. 35–42.
156–57.   "Ring-a-ring o' roses." "Commemorating a Heroic Act."
157–58.   Hayflick limit. Hayflick, 1968, 1980; Marx, 1974.
   158.   Vitamin E. Marx, 1974.
   158.   L-dopa. Cotzias et al.
158–59.   Petrobugs. "Put a Microbe in Your Tank."
   159.   "*Alga ayatollahphobera.*" Wade, November 30, 1979.
   159.   Penicillin mutation programs. Bailey.
159–60.   Macy's *History of Nantucket*. Melville, p. xxviii.
   160.   Diesel tree. Maugh, 1979.
160–61.   Genentech yeast cells and interferon. Golden.
   161.   "Today it is easier." Marx, May 8, 1981.
   161.   Injecting genes into nuclei. Marx, December 19, 1980.
   161.   Treating thalassemia. Marx, April 25, 1980; Kolata and Wade.
   162.   Richard Axel quote. Marx, December 19, 1980.
   162.   Genes introduced into mice. Marx, September 25, 1981; Cohn.
   163.   Polygenic traits. Davis.
164–65.   Cloning. Marx, January 23, 1981; Holden, 1981.
   165.   "Perhaps four parts in a thousand." Hayflick, 1980.
   166.   Identical twins reared apart. Holden, 1980.

Chapter 9. War and Deterrence

Becker, Carl L. *Modern History*. Morristown, N.J.: Silver Burdett, 1958.
Church, George J. "Arming for the '80's," *Time*, July 27, 1981, pp. 6–21.
Durant, Will. *The Age of Faith*. New York: Simon & Schuster, 1950.
"Four to Go," *Time*, December 11, 1950, pp. 18–19.
Garwin, Richard. "Antisubmarine Warfare and National Security," *Scientific American*, July 1972, pp. 14–25.
Howarth, David A. *The Dreadnoughts*. Alexandria, Va.: Time-Life Books, 1979.
Leviero, Anthony. New York *Times*, December 1, 1950, pp. 1, 3.
Meselson, Matthew S. "Chemical and Biological Weapons," *Scientific American*, July 1970, pp. 15–25.
——— and Robinson, Julian Perry. "Chemical Warfare and Chemical Disarmament," ibid., April 1980, pp. 38–47.
Meyer, Lawrence. *Israel Now: Portrait of a Troubled Land*. New York: Delacorte, 1982.
Mostert, Noel. *Supership*. New York: Penguin, 1975.
Tsipis, Kosta. "Cruise Missiles," *Scientific American*, February 1977, pp. 20–29.
Tuchman, Barbara. *The Guns of August.* New York: Dell, 1963.
———. *The Proud Tower*. New York: Bantam, 1967.
Walker, Paul F. "Precision-Guided Weapons," *Scientific American*, August 1981, pp. 36–45.

Westoff, Charles F. "The Populations of the Developed Countries," ibid., September 1974, pp. 108–20.
Wit, Joel S. "Advances in Antisubmarine Warfare," ibid., February 1981, pp. 31–41.

Page 169–70. Countdown to 1914. Becker, pp. 652–58; Tuchman, *Guns of August*, pp. 63, 69, 91.
171–72. President Truman's press conference. Leviero; "Four to Go."
176. Crusader states. Durant, chapter 23.
176–77. Israeli demographics. Westoff; Los Angeles *Times*, August 20, 1981, Part I, p. 2.
178. "I am just now not reading but devouring." Tuchman, *Proud Tower*, p. 153.
178–79. British naval buildup. Howarth.
179. "As it grew in strength and efficiency." Tuchman, *Guns of August*, p. 368.
181. Poison gas. Meselson; Meselson and Robinson.
181. "Outlawed by the general opinion." Meselson.
181–83. Antitank warfare. Walker.
182. Cost of Soviet tanks. Church.
183. "On a heavily armed front." Walker.
183. "No other ships have been so universally important." Mostert, p. 24.
184–85. Geography of an antisub war. Wit.
185–87. Methods of antisub warfare. Garwin; Wit.
186. Tomahawk cruise missile. Tsipis.
187. "They expect me to tell them." Wit.

CHAPTER 10. POPULATIONS AND PEOPLE

Anderson, Harry. "The Crisis in Social Security," *Newsweek*, June 1, 1981, pp. 25–27.
Barr, Terry N. "The World Food Situation and Global Grain Prospects," *Science*, December 4, 1981, pp. 1087–95.
Boyd, Robert. "World Dynamics: A Note," ibid., August 11, 1972, pp. 516–19.
Church, George J. "A Debt-Threatened Dream," *Time*, May 24, 1982, pp. 16–27.
Ding, Chen. "The Economic Development of China," *Scientific American*, September 1980, pp. 152–65.
Durant, Will. *Caesar and Christ.* New York: Simon & Schuster, 1944.
——— and Durant, Ariel. *The Age of Reason Begins.* New York: Simon & Schuster, 1961.
———. *The Age of Napoleon.* New York: Simon & Schuster, 1975.
FAO Production Yearbook, 1980. Rome: Food and Agriculture Organization, United Nations, 1981.
"Fertility of American Women: June 1980 (Advance Report)." Current Population Reports, Series P-20, No. 364. U.S. Census Bureau, August 1981.
Forrester, Jay W. *World Dynamics.* Cambridge, Massachusetts: Wright-Allen Press, 1971.
Gillette, Robert. "The Limits to Growth: Hard Sell for a Computer View of Doomsday," *Science*, March 10, 1972, pp. 1088–92.

Graunt, John. "Foundations of Vital Statistics," in *The World of Mathematics*. (James R. Newman, ed.) New York: Simon & Schuster, 1956, pp. 1420–35.

Hardin, Garrett. "The Survival of Nations and Civilization," *Science*, June 25, 1971, p. 1297.

Hueckel, Glenn. "A Historical Approach to Future Economic Growth," ibid., March 14, 1975, pp. 410–14.

Krishna, Raj. "The Economic Development of India," *Scientific American*, September 1980, pp. 166–78.

Langway, Lynn. "At Long Last Motherhood," *Newsweek*, March 16, 1981, pp. 86–86D.

Lewis, Oscar. "The Possessions of the Poor," *Scientific American*, October 1969, pp. 114–24.

Mauldin, W. Parker. "Population Trends and Prospects," *Science*, July 4, 1980, pp. 148–57.

McCullough, Daniel. *The Path Between the Seas*. New York: Simon & Schuster, 1977.

McWhirter, Norris. *Guinness Book of World Records*. New York: Bantam, 1981.

——— and McWhirter, Ross. *Guinness Book of World Records*. New York: Bantam, 1973.

Meadows, Donella; Meadows, Dennis; Randers, Jorgen; and Behrens, William. *The Limits to Growth*. New York: New American Library, 1972.

Monthly Vital Statistics Report. Washington, D.C.: U.S. Census Bureau, 1982.

Morrow, Lance. "The Endless Rediscovery of the Wheel," *Time*, December 15, 1980, pp. 97–98.

Paddock, Paul and William. *Famine—1975!* Boston: Little, Brown & Co., 1967.

Parks, Michael. "Motorbikes: China Faces a Crossroads," Los Angeles *Times*, March 17, 1981, Part I, p. 1.

Playbill, July 1981. New York: American Theater Press, Inc.

Plucknett, D. L. and Smith, N. J. H. "Agricultural Research and Third World Food Production," *Science*, July 16, 1982, pp. 215–20.

"Population Profile of the United States: 1980." Current Population Reports, Series P-20, No. 363. U.S. Census Bureau, June 1981.

Revelle, Roger. "Food and Population," *Scientific American*, September 1974, pp. 160–70.

Simon, Julian L. "Resources, Population, Environment: An Oversupply of False Bad News," *Science*, June 27, 1980, pp. 1431–37.

Starr, Chauncey and Rudman, Richard. "Parameters of Technological Growth," ibid., October 26, 1973, pp. 358–64.

Statistical Abstract of the United States 1981. Washington, D.C.: U.S. Census Bureau, 1982.

Wade, Nicholas. "Limits to Growth: Texas Conference Finds None, but Didn't Look Too Hard," *Science*, November 7, 1975, pp. 540–41.

Wallechinsky, David; Wallace, Irving and Amy. *The Book of Lists*. New York: Bantam, 1978.

Wattenberg, Ben J. *The Real America*. New York: Putnam, 1976.

Westoff, Charles F. "The Populations of the Developed Countries," *Scientific American*, September 1974, pp. 108–20.

Wolf, Charles. "Third World—Myths and Realities," *Los Angeles Times*, January 27, 1981, Part II, p. 7.

Page

191. Birth Statistics. *Monthly Vital Statistics Report; Statistical Abstract*, p. 58.

191. "There is a profound baby hunger." Langway.

192. Parson Weems. Hardin.

192. Historical birth statistics. *Statistical Abstract*, p. 58; Wattenberg, pp. 152–57.

193. Fertility statistics. ibid.; "Population Profile."

193. Women's expectations for births. *Statistical Abstract*, p. 62; Wattenberg, p. 161; "Fertility of American Women."

194. "Suppose [she] doesn't get married." Wattenberg, pp. 165–66.

194–95. Population proportions, median ages. *Statistical Abstract*, pp. 25–27.

195. College enrollments. Wattenberg, pp. 75, 80; *Statistical Abstract*, p. 159.

195. Giovanni Papani. Morrow.

195–96. "Just now the baby boomers." ibid.

196. Broadway. *Playbill*, pp. 19, 43; McWhirter, 1973, p. 211; 1981, p. 232; Wallechinsky et al., p. 187.

196. "The child grows up." Wattenberg, p. 178.

197–98. Social Security. Church; Anderson.

199–201. World population statistics. Mauldin.

199–201. Year of the Fiery Horse. Westoff.

201. China. Mauldin; Ding Chen.

201–2. India. Paddock; Mauldin; Krishna; Plucknett and Smith.

202. "A diet based on 4,000 to 5,000 kilocalories." Revelle.

202–3. Per capita food production. Simon; Barr; *FAO Production Yearbook*.

203. African food production. Revelle; Barr; *FAO Production Yearbook*.

203. Sahel drought. Simon.

204. Cyprian's *Ad Demetrium*. Durant, *Caesar and Christ*, p. 665.

204. Thomas Malthus. Durants, *Age of Napoleon*, pp. 400–3.

204. H. G. Wells. McCullough, pp. 498–99.

204–5. Limits to Growth. Forrester; Meadows et al.

205. Allen Kneese. Gillette.

205. Robert Boyd. Boyd.

205–6. "Technology's exponential growth." Starr and Rudman.

206. "The history of technological advance." Hueckel.

206. Forrester's retreat. Wade.

206–7. Quotes from Wattenberg, pp. 13–22; Mailer in *Times*, May 26, 1968.

207. Responses to Forrester's work. Hueckel.

208. Conditions in Merrie England. Durants, *Age of Reason Begins*, pp. 47–48, 184–85.

208. "My first Observation is." Graunt.

209. Activities of government leaders. Wolf.

209. D. Gale Johnson. Simon.
210. Chinese roads and railroads. Ding.
211. Third World village scene. Lewis; Parks; author's personal observations of Nevis, British West Indies.

CHAPTER 11. LIVING IN THE FUTURE

"Annual Housing Survey: 1980." Current Housing Reports, Series H-150–80, Part A. Washington, D.C.: U.S. Census Bureau, February 1982.

Demarest, Michael. "He Digs Downtown," *Time*, August 24, 1981, pp. 42–53.

Durant, Will. *The Age of Faith*. New York: Simon & Schuster, 1950.

———. *The Renaissance*. New York: Simon & Schuster, 1953.

———. *The Reformation*. New York: Simon & Schuster, 1957.

——— and Durant, Ariel. *The Age of Voltaire*. New York: Simon & Schuster, 1965.

———. *The Age of Napoleon*. New York: Simon & Schuster, 1975.

Gelman, David. "The Games Teen-Agers Play," *Newsweek*, September 1, 1980, pp. 48–53.

Ginsberg, Eli. "The Professionalization of the U.S. Labor Force," *Scientific American*, March 1979, pp. 48–53.

Green Larry. "Derelict Chicago Buildings Recycled Into Class Homes," Los Angeles *Times*, March 8, 1981, Part I, p. 1.

Grose, Frances. *1811 Dictionary of the Vulgar Tongue*. Chicago: Follett, 1971.

Hauser, Philip M. "The Census of 1980." *Scientific American*, November 1981, pp. 53–61.

Klerman, Lorraine V. and Jekel, James F. "Teenage Pregnancies," *Science*, March 31, 1978, p. 1390.

McAlpern, David. "How the Poor Will Be Hurt," *Newsweek*, March 23, 1981, pp. 23–24.

"Money Income and Poverty Status of Families and Persons in the United States: 1980." Current Population Reports (Consumer Income), Series P-60, No. 127. Washington, D.C.: U.S. Census Bureau, August 1981.

Nelson, Bryce. "Minorities' Growth Rate Up in State," Los Angeles *Times*, March 27, 1981, Part II, p. 1.

"Out-of-Wedlock Births Up 50% in Last Decade in U.S.," ibid., October 26, 1981, Part I, p. 5.

Sanderson, Jim. "Speak Up, Ladies; Men Do Listen," ibid., December 28, 1980.

Sowell, Thomas. "The First Step in Fighting Poverty Is to Eliminate the Misconceptions," ibid., August 23, 1981, Part V, p. 3.

Statistical Abstract of the United States 1981. Washington, D.C.: U.S. Census Bureau, 1982.

Sternlieb, George and Hughes, James W. "The Changing Demography of the Central City," *Scientific American*, August 1980, pp. 48–53.

Wattenberg, Ben J. *The Real America*. New York: Putnam, 1976.

Page   215. Baltimore. Demarest.
            216. Refurbishing of old buildings. Demarest; Sternlieb and Hughes; Green.

219. Orange County and Los Angeles growth statistics. Nelson.
219–20. Urban and suburban growth statistics. Hauser; Wattenberg, p. 106.
220. "Sucked the blood out of the central cities." Demarest.
220–21. "Suburban Sprawl is a pejorative phrase." Wattenberg, p. 175.
221. Fraction of houses occupied by owners. "Annual Housing Survey," p. 1.
221. Median number of rooms. ibid., p. 2.
221. Median number of persons per home. ibid., p. 5.
221. "New houses have been getting larger." *Statistical Abstract*, p. 758.
221. Central air conditioning. ibid., p. 758; "Annual Housing Survey," p. 4.
221. Units lacking plumbing facilities. "Annual Housing Survey," p. 2.
221. Median income. "Money Income," p. 1; *Statistical Abstract*, p. 436.
221. Per capita income. *Statistical Abstract*, pp. 423, 436.
221. Women in the work force. "Money Income," p. 5.
221–22. Size of U.S. households. *Statistical Abstract*, p. 42.
222. Education statistics. ibid., p. 142; Ginsberg.
222. Young woman's letter. Sanderson ("Liberated Male" column, Part V).
223–24. U.S. occupations. *Statistical Abstract*, pp. 401–3.
224. Couples living together. ibid., p. 41.
224. Family households. ibid., p. 42.
224. Mary Leakey and Paul Abell. The evidence was a set of fossilized footprints formed in volcanic ash by three hominids: a male, female, and child.
224–25. "Boys reached the age of work at twelve." Durant, *Age of Faith*, pp. 821–24.
225. Jacques de Vitry. ibid., p. 927.
225. "There must have been considerable." Durant, *Renaissance*, pp. 575–79.
226. Pious lands of northern Europe. Durant, *Reformation*, pp. 760–61.
226. France of Voltaire. Durants, *Age of Voltaire*, p. 290.
226. Montesquieu. ibid., pp. 290–91.
226. "Balum rancum." Grose.
226. Legal age for marriage. Durants, *Age of Napoleon*, p. 364.
229. Teenage sex. Gelman.
229. Births to girls under 14. *Statistical Abstract*, pp. 59, 65.
229–30. Teenage pregnancies. Klerman and Jekel; *Statistical Abstract*, pp. 65, 66.
232. "Two societies separate and unequal." Wattenberg, p. 20.
232. Educational levels. *Statistical Abstract*, p. 142.
232–33. Employment categories. ibid., p. 400.
233. Poverty statistics. ibid., p. 447; "Money Income," p. 27.
233. Unemployment insurance in 1959. Wattenberg, p. xvi.
233. Contribution of benefit programs to reducing poverty. McAl-

pern; *1980 Statistical Abstract*, p. 465 as adjusted by p. 463.

233–34. Income statistics. *Statistical Abstract*, p. 436; "Money Income," p. 2.

233. James Baldwin. Wattenberg, p. 20.

233. Families headed by women. ibid., p. 137; "Money Income," pp. 8–9.

233. Median ages. *Statistical Abstract*, p. 25.

233–34. "People's incomes increase as they get older." Sowell, *Statistical Abstract*, p. 439.

234. Households headed by black women. "Money Income," pp. 3, 9.

234. Illegitimacy rates. *Statistical Abstract*, p. 65; "Out-of-Wedlock Births."

234. Families whose householder is a full-time worker. "Money Income," pp. 11–12.

234. All husband-wife families or husband and wife both work. ibid., pp. 8–9.

234. Householder under 35. *Statistical Abstract*, p. 442.

234–35. Husband-wife families outside the South. ibid., p. 443.

235. "As early as 1971." Wattenberg, p. 128.

235. West Indian blacks. Sowell.

CHAPTER 12. BEYOND THE PRESENT ERA

Abelson, Philip H. "Education, Science, and Technology in China," *Science*, February 9, 1979, pp. 505–9.

Adler, Jerry. "The Browning of America," *Newsweek*, February 23, 1981, pp. 26–37.

Amalrik, Andrei. *Will the Soviet Union Survive Until 1984?* New York: Harper & Row, 1970.

Carey, William D. "The Chinese Scene," *Science*, February 9, 1979, pp. 509–12.

d'Encausse, Hélène Carrière. *Decline of an Empire*. New York: Newsweek Books, 1979.

Durant, Will. *The Age of Faith*. New York: Simon & Schuster, 1950.

Hackett, General Sir John. *The Third World War: August 1985*. New York: Berkley, 1980.

Häfele, Wolf. "A Global and Long-Range Picture of Energy Developments," *Science*, July 4, 1980, pp. 174–82.

Jefferson, Pat and Benson, Johan. "Washington Scene," *Astronautics and Aeronautics*, September 1981, pp. 10–11.

Kantrowitz, Arthur. "The Ming Navy and the U.S. Space Program," ibid., September 1981, pp. 44–46.

Levin, Michael. "How to Tell Bad from Worse," *Newsweek*, July 20, 1981, p. 7.

Parks, Michael. "Motorbikes: China Faces a Crossroads," *Los Angeles Times*, March 17, 1981, Part I, p. 1.

————. "China Shifting Many Military Plants to Civilian Production," ibid., August 27, 1981, Part I-B, p. 1.

Pfaff, William. "Finlandization of Poland Might Solve Soviet Security Problem," ibid., June 23, 1981, Part II, p. 7.

Pillsbury, Arthur F. "The Salinity of Rivers," *Scientific American*, July 1981, pp. 54–65.

Tuchman, Barbara. *The Guns of August.* New York: Dell, 1963.

———. *The Zimmermann Telegram.* New York: Ballantine, 1979.

Page   241.   Kaiser Wilhelm's meddling. Tuchman, *Guns of August,* chapter 10; *Zimmermann Telegram,* chapters 5, 6.

          241.   Predictions of Soviet Union breakup. Amalrik; Hackett.

      241–42.   Soviet Union as an empire. d'Encausse.

          242.   Breakup and reformation of Soviet empire. ibid., chapter 1.

          242.   "Poor Mexico." Tuchman, *Zimmermann Telegram,* p. 65.

          243.   Finland. Pfaff.

      243–44.   Totalitarianism and authoritarianism. Levin.

          245.   Motorbikes in China. Park, March 27, 1981.

          245.   Chinese consumer goods. Park, August 27, 1981.

      245–46.   Chinese computer. Abelson.

          246.   "We stayed for English." Carey.

      246–48.   Energy projections. Häfele.

          249.   George Keyworth. Jefferson and Benson.

          249.   James E. Smith. Adler.

      249–51.   Western water project proposals. Pillsbury.

          253.   Bacon's *New Atlantis.* Kantrowitz.

          253.   Abelard, Exeter, Durant. Durant, p. 1082.

INDEX

Italicized numbers are figure numbers. Figures 1–16 follow page 56. Figures 17–32 follow page 104. Figures 33–48 follow page 176. Figures 49–64 follow page 224.